Construction Health and Safety in Developing Countries

The global construction sector is infamous for high levels of injuries, accidents and fatalities, and poor health and well-being of its workforce. While this record appears in both developed and developing countries, the situation is worse in developing countries, where major spending on infrastructure development is expected. There is an urgent need to improve construction health and safety (H&S) in developing countries. The improvement calls for the development of context-specific solutions underpinned by research into challenges and related solutions.

This edited volume advances the current understanding of construction H&S in developing countries by revealing context-specific issues and challenges that have hitherto not been well explored in the literature, and applying emergent H&S management approaches and practices in developing countries. Coverage includes countries from the regions of sub-Saharan Africa, Latin America, Asia and Europe. This book, which is the first compendium of research into construction H&S issues in developing countries, adds considerable insight into the field and presents innovative solutions to help address poor H&S in construction in developing nations. It is a must read for all construction professionals, researchers and practitioners interested in construction and occupational H&S, safety management, engineering management and development studies.

Patrick Manu is a Senior Lecturer in Project Management at The University of Manchester, United Kingdom. He has held academic posts in the United Kingdom and Ghana. Amongst his research interests are occupational health and safety, digital construction, sustainable construction, and procurement. He has over 70 publications including articles in leading multi-disciplinary and construction journals. His research, which has been published in several outlets, has been funded by several organisations including the Engineering and Physical Sciences Research Council.

Fidelis Emuze is a Professor and head of the Department of Built Environment, and the Unit for Lean Construction and Sustainability at the Central University of Technology, Free State, South Africa. He is a member of the Association of Researchers in Construction Management and the Lean Construction Institute.

He is the International Coordinator of CIB TG59 (People in Construction), and in 2017, he co-organised and hosted the Joint CIB W099 and TG59 International Safety, Health, and People in Construction Conference.

Tarcisio Abreu Saurin is an Associate Professor at the Industrial Engineering Department of the Universidade Federal do Rio Grande do Sul (Brazil). He has a BS in civil engineering, MS in construction management and PhD in industrial engineering. He was a visiting scholar at the University of Salford (United Kingdom) and at Macquairie University, at the Australian Institute of Health Innovation. His main research interests include the modelling and management of complex socio-technical systems, resilience engineering, safety management, lean production, and performance measurement.

Bonaventura H. W. Hadikusumo is a Professor at the School of Engineering and Technology, Asian Institute of Technology, Thailand. Safety management, system dynamics, project visualization, intelligent agent system, knowledge management tools, IT applications like building information modelling (BIM) and simulation in construction site improvement, and web-based project management are some of his key expertise areas and interests.

Spon Research

publishes a stream of advanced books for built environment researchers and professionals from one of the world's leading publishers. The ISSN for the Spon Research programme is ISSN 1940-7653 and the ISSN for the Spon Research E-book programme is ISSN 1940-8005.

Trust in Construction Projects
A. Cerić

New Forms of Procurement
PPP and Relational Contracting in the 21st Century
M. Jefferies and S. Rowlinson

Target Cost Contracting Strategy in Construction
Principles, Practices and Case Studies
D. W. M. Chan and J. H. L. Chan

Valuing People in Construction
F. Emuze and J. Smallwood

Funding and Financing Transport Infrastructure
A. Roumboutsos, H. Voordijk and A. Pantelias

Sustainable Design and Construction in Africa
A System Dynamics Approach
Oluwaseun Dosumu and Clinton Aigbavboa

Making Sense of Innovation in the Built Environment
Natalya Sergeeva

The Connectivity of Innovation in the Construction Industry
Edited by Malena Ingemansson Havenvid, Åse Linné, Lena E. Bygballe and Chris Harty

Construction Health and Safety in Developing Countries

Edited by Patrick Manu,
Fidelis Emuze, Tarcisio Abreu Saurin and
Bonaventura H. W. Hadikusumo

Routledge
Taylor & Francis Group

LONDON AND NEW YORK

First published 2020 by Routledge

2 Park Square, Milton Park, Abingdon, Oxon OX14 4RN

605 Third Avenue, New York, NY 10017

Routledge is an imprint of the Taylor & Francis Group, an informa business

First issued in paperback 2021

Copyright © 2020 selection and editorial matter, Patrick Manu, Fidelis Emuze, Tarcisio Abreu Saurin and Bonaventura H. W. Hadikusumo; individual chapters, the contributors

The right of Patrick Manu, Fidelis Emuze, Tarcisio Abreu Saurin and Bonaventura H. W. Hadikusumo to be identified as the authors of the editorial material, and of the authors for their individual chapters, has been asserted in accordance with sections 77 and 78 of the Copyright, Designs and Patents Act 1988.

All rights reserved. No part of this book may be reprinted or reproduced or utilised in any form or by any electronic, mechanical, or other means, now known or hereafter invented, including photocopying and recording, or in any information storage or retrieval system, without permission in writing from the publishers.

Notice:
Product or corporate names may be trademarks or registered trademarks, and are used only for identification and explanation without intent to infringe.

Publisher's Note

The publisher has gone to great lengths to ensure the quality of this reprint but points out that some imperfections in the original copies may be apparent.

British Library Cataloguing-in-Publication Data
A catalogue record for this book is available from the British Library

Library of Congress Cataloging-in-Publication Data
Names: Manu, Patrick, editor.
Title: Construction health and safety in developing countries / [edited by] Patrick Manu [and 3 others].
Description: First edition. | Abingdon, Oxon; New York, NY: Routledge, 2019. | Includes bibliographical references.
Identifiers: LCCN 2019019114 | ISBN 9781138317079 (hardback) | ISBN 9780429455377 (ebook)
Subjects: LCSH: Building–Developing countries–Safety measures. | Construction workers–Health and hygiene–Developing countries.
Classification: LCC TH443 .C64 2019 | DDC 690/.22091724–dc23
LC record available at https://lccn.loc.gov/2019019114

ISBN: 978-1-138-31707-9 (hbk)
ISBN: 978-1-03-217741-0 (pbk)
DOI: 10.1201/9780429455377

Typeset in Goudy
by Deanta Global Publishing Services, Chennai, India

Contents

List of figures	x
List of tables	xii
List of contributors	xiv
Foreword	xxi

1 An introduction to construction health and safety in
   developing countries   1
   PATRICK MANU, FIDELIS EMUZE, TARCISIO ABREU SAURIN AND
   BONAVENTURA H. W. HADIKUSUMO

PART I
Occupational health and safety legislation   13

2 Implications and opportunities in a complex construction health
   and safety regulatory environment   15
   NNEDINMA UMEOKAFOR, KONSTANTINOS EVANGELINOS AND ABIMBOLA WINDAPO

3 Contractors' compliance with occupational health and safety
   legislation in South Africa: The benefits of self-regulation   29
   ABIMBOLA WINDAPO, OLUWOLE OLATUNJI AND NNEDINMA UMEOKAFOR

4 A narrative review of occupational safety and health legislation
   in Pakistan   43
   ABDUL QAYOOM AND BONAVENTURA H. W. HADIKUSUMO

5 Safety law, system and regulation influencing the construction
   sector in India   59
   SRI DATTA DUDDUKURU AND BONAVENTURA H. W. HADIKUSUMO

PART II
Occupational health and safety management — 83

6 The strategies to enhance safety performance in the construction industry of Pakistan — 85
HAFIZ ZAHOOR AND ALBERT P. C. CHAN

7 Health and safety performance in the Ugandan construction industry — 103
MOSES OKWEL, HENRY ALINAITWE AND DENIS KALUMBA

8 Towards the development of an integrated safety, health and environmental management capability maturity model (SHEM-CMM) for uptake by construction companies in Ghana — 116
MILLICENT ASAH-KISSIEDU, PATRICK MANU, COLIN BOOTH AND ABDUL-MAJEED MAHAMADU

9 Ad hoc and post hoc analysis of contractors' safety risks during procurement in Nigeria — 128
OLUWOLE OLATUNJI, ABIMBOLA WINDAPO AND NNEDINMA UMEOKAFOR

10 Integrating health and safety into labour-only procurement system: Opportunities, barriers and strategies — 140
NNEDINMA UMEOKAFOR, ABIMBOLA WINDAPO AND OLUWOLE OLATUNJI

11 Tertiary built environment construction health and safety education in South Africa — 152
JOHN SMALLWOOD

PART III
Behavioural and cultural issues in safety management — 167

12 Safety communication and suggestion scheme in construction — 169
VICTOR N. OKORIE AND FIDELIS EMUZE

13 Making sense of safety in systemic and cultural contexts — 179
ANDREA Y. JIA, STEVE ROWLINSON, MARTIN LOOSEMORE, MENGNAN XU, BAIZHAN LI, MARINA CICCARELLI AND HEAP-YIH CHONG

14 Managing construction safety in developing countries: From the viewpoint of Chinese international contractors — 191
RAN GAO AND ALBERT P. C. CHAN

15 Learning from failures (LFF): A multi-level conceptual model for changing safety culture in the Nigerian construction industry 205
CLARA MAN CHEUNG, AKILU YUNUSA-KALTUNGO, OBUKS EJOHWOMU AND RITA PEIHUA ZHANG

16 Overview of safety behaviour and safety culture in the Malaysian construction industry 218
MAZLINA ZAIRA MOHAMMAD AND BONAVENTURA H. W. HADIKUSUMO

PART IV
Construction workers' well-being 235

17 Determinants of risky sexual behaviour by South African construction workers 237
PAUL A. BOWEN AND PETER J. EDWARDS

18 Construction migrants from 'developing' countries in 'developed' hosts 258
DAVID OSWALD

19 Occupational accidents predicting early retirement of construction workers in South Africa 270
JUSTUS N. AGUMBA AND INNOCENT MUSONDA

PART V
Technologies for safety management 285

20 Design of flexible horizontal lifeline systems 287
MARCELO FABIANO COSTELLA, VALÉRIA CRISTINA DA MOTTA AND LETÍCIA NONNENMACHER

21 Development of RFID automation system to improve safety on construction sites in Brazil 301
VICTOR HUGO MAZON DE OLIVEIRA AND SHEYLA MARA BAPTISTA SERRA

Index 317

Figures

2.1	Regulatory spectrum	17
2.2	Summary of the realities of the structure on CH&S regulation	21
3.1	Conceptual framework of the study	33
3.2	Self-regulation analytical framework	34
3.3	Level of self-regulation of the respondents	36
3.4	Accident frequency rates	37
3.5	Scatter plot diagram between level of self-regulation and accident frequency rate	38
4.1	PDCA cycle to improve OSH legislation	44
4.2	Literature sources	47
5.1	Overview of the BOCW Act	64
5.2	Overview of the research methodology	68
6.1	Percentage of employment vis-à-vis injuries in major industries of Pakistan	86
6.2	Research framework of the study	88
6.3	Research model and research hypotheses	90
6.4	Twenty-four items SC scale	94
6.5	Hypothesized model showing the relationship between SC and SP	95
6.6	Finalized SP measurement model	97
12.1	Example of an issue resolution procedure using a suggestion box as a tool	175
12.2	Characteristics of a positive safety culture	176
13.1	Sensemaking in a developing society	184
13.2	Sensemaking in a developed society	184
15.1	Comparison of workplace fatalities for selected countries	206
15.2	Evolving stages of health, safety and environment	208
15.3	Comparing the performance of NICN (Nigeria) and OSHRC (USA) in terms of finalised cases between 1993 and 2015	212
15.4	An LFF conceptual model	214
16.1	Total occupational accidents for all sectors in Malaysia for 2017	219
16.2	A depiction of accident causation	221

16.3	Summary of information gathered	222
17.1	Graphical overview of the conceptual model	241
17.2	The structural model	249
20.1	Anchoring system	291
20.2	Horizontal lifeline diagram without energy absorber	292
20.3	Forces calculated for the system	295
20.4	Free-fall zone (ZLQ) diagram	297
21.1	Investigative structure using the DSR for the application of RFID technology in this study	305
21.2	Software design: construction control by RFID	308
21.3	Operational sequence for the attainment of results on the CPE as reports	310
21.4	Operational sequence for graphical results	311
21.5	Comparative chart by employee group and measurement date	312

Tables

2.1	Summary of the implications of the structure and actors in CH&S regulation	24
2.2	Summary of opportunities in the structure, actors and their roles in CH&S regulation	24
3.1	Pearson correlation statistics	37
4.1	Pakistan labour force statistics	48
4.2	Sector-wise injury statistics	49
4.3	Labour legislation in Pakistan	51
5.1	Classification of companies based on number of employees	66
5.2	Definition of scale for the factors	67
5.3	Profile of the company and respondents interviewed	69
5.4	Awareness, implementation and compliance with safety regulations in the AP BOCW Rules 1999 among small companies	70
5.5	Awareness, implementation and compliance with safety regulations in the BOCW Central Rules 1998 among medium companies	71
5.6	Awareness, implementation and compliance with safety regulations in the BOCW Central Rules 1998 among large companies	72
5.7	Profile of the respondents interviewed in the Labor Department at central and state levels	73
5.8	Comparison of central and state government respondents' responses regarding the factors considered in the study	74
6.1	Research methods and analytical techniques to achieve five objectives of the study	92
7.1	Ranking of the status of H&S programmes according to coefficient of variation	108
7.2	Distribution of tradesmen, nature and causes of recorded accidents (2011) and reported accidents (2006–2010) at construction sites	109
7.3	Regression analysis of the effects of H&S programmes on H&S performance	111

9.1	CIDA's benchmark of OHS performance of construction contractors	133
9.2	Assessment instrument for construction contractors on project safety	135
10.1	Characteristics of LoPS	145
10.2	Merits of LoPS	146
10.3	Demerits of LoPS	147
11.1	Revised proposed aspects for inclusion in construction H&S modules of tertiary built environment programmes	157
11.2	Respondents to Study 1	159
11.3	Importance of the inclusion of construction H&S in the tertiary education programmes of nine built environment disciplines	160
11.4	Degree of support for the inclusion of aspects in tertiary built environment programmes' construction H&S modules	162
14.1	CICs' value of turnover fulfilled for contracted projects in particular continents and regions in 2014	194
14.2	Basic information of selected projects	195
14.3	Summary of tactical management problems	199
14.4	Summary of identified three problem levels	201
15.1	Comparison of the Nigerian safety regulatory agency to other countries	211
16.1	NPD, PD and D for occupational accidents in Malaysia for 2017	220
16.2	Case study observation information at three construction sites	225
17.1	Demographic and scale items ($n = 512$)	243
17.2	Regression models of demographic characteristics, behavioural and cognitive factors, and risky sexual behaviour	246
19.1	Pension status of occupation disabilities in construction	277
19.2	Cost of occupational accidents in construction	278
19.3	Early retirement from permanent disability	280
21.1	RFID equipment used in the tests	306

Contributors

Justus N. Agumba is a Senior Lecturer and programme coordinator for the masters programme in the Department of Building Sciences at the Tshwane University of Technology. He has worked in both the construction industry and academia in Kenya and South Africa. His research interests are in occupational health and safety, construction education, small and medium construction enterprises, construction management, construction project management, and sustainability. He has contributed to scholarly articles in local and international accredited journals and conferences.

Henry Alinaitwe is a Professor of Civil Engineering and Principal of the College of Engineering, Design, Art and Technology at Makerere University, Uganda. He has a PhD and Licentiate degree in engineering, an MSc in construction management, a master of engineering studies in civil engineering, and a bachelor of science in engineering (civil option). His research interest is in the areas of performance and productivity in construction, safety and health in construction, lean construction, construction materials and structural health and integrity. His research, which has been published in several outlets, has been funded by several organisations including Sida and Makerere University.

Millicent Asah-Kissiedu is a PhD student in the Department of Architecture and the Built Environment Department at the University of the West of England, United Kingdom. Her areas of research interest are safety, health and environmental management; safety maturity and process improvement models in construction; and construction management.

Colin Booth is the Associate Head for Research and Scholarship for the Department of Architecture and the Built Environment, which he has been since joining the University of the West of England, United Kingdom, in early 2012. He has also been a director of the Construction and Property Research Centre, and a director of the Centre for Floods, Communities and Resilience. His areas of expertise include flooding and water resources management, sustainability in the built environment, environmental management in construction, climate change mitigation and adaptation strategies, and urban pollution. He holds the distinguished titles of Visiting Professor at the University of Ibadan, Federal University of Rio de Janeiro and Vilniaus Kolegija.

Contributors xv

Paul A. Bowen is an Emeritus Professor within the Department of Construction Economics and Management at the University of Cape Town, held a B2 rating with the National Research Foundation of South Africa (NRF), and is a member of the South African Academy of Science. His research interests embrace HIV/AIDS in the South African construction industry; and workplace stress experienced by construction professionals. Research outputs include a book, and some 200 peer-reviewed journal and conference papers. He is currently Visiting Professor at RMIT University in Melbourne, Australia.

Albert P. C. Chan is the Chair Professor of Construction Engineering and Management, and Head of Department of Building and Real Estate at the Hong Kong Polytechnic University, Hong Kong. His research and teaching interests include project management and project success, construction procurement and relational contracting, construction management and economics, construction health and safety, and construction industry development. He has produced over 700 research outputs in refereed journal papers, international refereed conference papers, consultancy reports and other articles.

Clara Man Cheung is a Lecturer in Project Management at the School of Mechanical, Aerospace and Civil Engineering at the University of Manchester, United Kingdom. She has a mix of business and engineering educational background. She holds a bachelor's degree in business administration from Hong Kong University of Science and Technology, and an MSc and PhD in civil engineering from the University of Maryland, United States. Her research interest focuses on occupational health and safety. Her recent publications include applying positive psychology and organisational theories to study construction safety management and well-being. Her research has been funded and supported by public and private organisations.

Heap-Yih Chong is a Senior Lecturer (Teaching and Research) in the Discipline of Construction Management at Curtin University, Australia. His research areas are multidisciplinary in nature that are revolving around project management, ICT and sustainability studies.

Marina Ciccarelli is Associate Professor at Curtin University, Australia. She is an occupational therapist with advanced qualifications in ergonomics and occupational safety and health. With 15-plus years' experience consulting to a range of industries, including construction, in Australia and the United States, her focus is on effective injury management and participatory ergonomics to improve workplace safety and productivity.

Marcelo Fabiano Costella is a Civil Engineer and holds a PhD in production engineering. He is a Lecturer of Unochapecó's (Brazil) master's and doctorate in Technology and Management of Innovation, Faculdade Meridional's master's program in civil engineering and specializes in workplace safety engineering. He has also been involved in the construction of more than 50,000 m² of multi-story buildings.

xvi *Contributors*

Sri Datta Duddukuru is a graduate of the master's degree programme in Construction Engineering and Infrastructure Management (CEIM), Asian Institute of Technology, Thailand; he also holds a bachelor's degree from the Department of Civil Engineering from Jawaharlal Nehru Technological University, Kakinada, India. His research interests are related with construction safety and project management.

Peter J. Edwards is currently an Adjunct Professor at RMIT University in Melbourne and has authored and co-authored more than 150 peer-reviewed journal and conference papers, and 2 books and 5 book chapters in fields such as workplace health and safety, project management, risk management and sustainable development. He holds a Master of Science degree from the University of Natal and his PhD was awarded by the University of Cape Town.

Obuks Ejohwomu is Programme Director of MSc Commercial Project Management at the School of Mechanical, Aerospace and Civil Engineering (MACE), University of Manchester, United Kingdom. He holds a BEng degree in civil engineering, MSc in systems engineering, and PhD in construction engineering and management. Prior to joining MACE he worked as an engineer and an academic in various roles in the United Kingdom and in Nigeria. He has received research funding from Carnegie, Association of Project Management and MACE. He is also a grant reviewer for Newton Funds, Carnegie Africa Diaspora Fellowship programme and National Research Foundation.

Konstantinos Evangelinos is an established academic and researcher on corporate social responsibility and environmental management with more than a hundred research publications and extensive research and consultancy experience. He is also the editor of *Progress in Industrial Ecology: An International Journal*.

Ran Gao is a Lecturer at the School of Management Science and Engineering, Central University of Finance and Economics, Beijing, China. Her main research areas are construction health and safety, project management, and international construction.

Andrea Y. Jia is a Senior Researcher at the University of Hong Kong. She has an interdisciplinary background in architecture and construction project management, and has worked in Australia, Hong Kong and China. Her research is in the areas of safety, health and well-being, professionalism, climate change adaptation, productivity, and innovation in construction, with a theoretical focus on the institutional logics perspective.

Denis Kalumba is an Associate Professor in Geotechnical Engineering in the Department of Civil Engineering at the University of Cape Town, South Africa. He holds an MSc degree in geotechnical engineering from UCT. As a British Commonwealth Scholar, he completed his PhD in 2006 at the University of Newcastle upon Tyne, United Kingdom (UK). His research was on remediation of metal-contaminated soils using electro-kinetics.

After which, he took up a research fellow position at the same university. The research project he worked on, involving dewatering of liquid wastes using electro-osmosis, was funded by UK's Engineering and Physical Sciences Research Council (EPSRC).

Baizhan Li is a Dean and Professor at the School of Urban Construction and Environmental Engineering at Chongqing University, China. His research is focused on the areas of environmental science and thermal comfort in the built environment.

Martin Loosemore is a Professor of Construction Management at the University of New South Wales, Sydney, Australia. He is a Fellow of the Chartered Institute of Building and has published many books and articles in the areas of social enterprise and procurement, innovation, risk management, corporate social responsibility and corporate strategy.

Abdul-Majeed Mahamadu is a Senior Lecturer in Project Management at the University of the West of England, United Kingdom. He has previously worked in the construction industry as a project manager and quantity surveyor on road and highway infrastructure projects. He lectures in the areas of building information modelling (BIM), construction project management and quantity surveying. His research interests include application of BIM-enabled technologies for construction process and design optimisation, BIM capability, competence for supply chain integration, and occupational health and safety and integrated procurement.

Mazlina Zaira Mohammad is a Senior Lecturer in the Department of Construction Business and Project Management (CBPM) and a Coordinator of Safety, Health and Environment Committee in the Faculty of Civil Engineering, Universiti Teknologi MARA (UiTM), Malaysia. She received her PhD in Construction, Engineering and Infrastructure Management (CEIM) from Asian Institute of Technology (AIT), Thailand. She holds a master's degree in Business Administration and a first degree in Civil Engineering from Universiti Teknologi MARA (UiTM), Malaysia. Her research interests are construction safety management, safety culture, construction waste management and total quality management.

Valéria Cristina da Motta is a Civil Engineer through Unochapecó, Brazil. Motta is a Technical Manager at an earthworks company, and designer and manager of works for an engineering company in Santa Catarina, Brazil.

Innocent Musonda is an Associate Professor at the University of Johannesburg, South Africa. He holds a PhD in engineering management, and qualifications in construction management and civil engineering. He is a registered civil engineer, a professional construction manager and a member of the Chartered Institute of Building (CIOB). He has worked in Botswana, South Africa and Zambia. He is founder and director of the Centre for Applied Research and Innovation in the Built Environment (CARINBE), University of Johannesburg. He has served as a chairperson of an international conference series and he is also a roaming scholar at the University of Toronto.

Letícia Nonnenmacher is a Civil Engineer at Unochapecó, Brazil, with a short extension at the Universitat Rovira i Virgili, Spain, and pursuing a master's degree in technology and innovation management. She designs and executes works in an engineering company and manages a construction company located in the heartland of Rio Grande do Sul.

Abdul Qayoom is a Lecturer in the Department of Civil Engineering, Mehran University of Engineering and Technology, Pakistan. Currently, pursuing doctoral studies in the field of Construction Engineering & Infrastructure Management, School of Engineering and Technology, Asian Institute of Technology, Thailand, with major focus on safety culture. Besides that, he holds professional certifications to manage health and safety like NEBOSH IGC, IOSH Managing Safely, HABC Level II fire safety and workplace safety. Managing health and safety, developing safety management systems and safety audits are his key interest areas.

Victor N. Okorie is a Senior Lecturer and Assistant Dean, Faculty of Environment Sciences, University of Benin, Benin, Ugbowo Campus, Edo State, Nigeria. He received his first degree in quantity surveying from the Enugu State University of Science and Technology, Enugu State, Nigeria in 1995; a masters' degree in construction management from University of Lagos also in Nigeria; and PhD in construction management from the Nelson Mandela University, Port Elizabeth, South Africa, in 2014. His research interest is in construction health and safety (H&S) with focus on culture, behaviour and leadership.

Moses Okwel is a consulting civil engineer with Fichtner. He was formally a graduate student of civil engineering at Makerere University, Uganda, where he carried out research on construction safety and health. He has participated in the design and construction of a number of infrastructure projects including water treatment, water supply, roads and others.

Oluwole Alfred Olatunji is a Senior Lecturer and Course Lead in the School of Design and the Built Environment at Curtin University, Australia. He has published over 100 scholarly works on issues relating to construction management.

Victor Hugo Mazon de Oliveira is a civil engineer and student of the graduate program in structures and civil construction of the Federal University of São Carlos (UFSCar), Brazil. He researches construction management, working on the following topics: health and safety at construction sites; and the logistics and analytic hierarchy process (AHP).

David Oswald is a safety scientist, with a background in structural and fire safety engineering, and has written award-winning work on safety within the construction industry. He teaches undergraduate and postgraduate students within both the Construction Management and Occupational Health & Safety programs at RMIT University, Australia.

Steve Rowlinson is Chair Professor of Construction Project Management and Programme Director of the MSc Integrated Project Delivery programme at the Department of Real Estate and Construction, the University of Hong Kong. He is the co-coordinator of CIB Working Commission W092 Procurement Systems and has a long-standing track record in industry-engaged research in occupational health and safety in construction.

Sheyla Mara Baptista Serra is an Associate Professor at the Civil Engineering Department of the Federal University of São Carlos (UFSCar), Brazil, since 1994. She researches construction management, working on the following topics: planning and production control, subcontractor management, construction site design, lean construction, sustainable construction, building information modelling (BIM), and safety and health at work.

John Smallwood is a Professor of Construction Management in the Department of Construction Management, Nelson Mandela University, South Africa, and the Principal at Construction Research Education and Training Enterprises (CREATE). In both capacities he specialises in, among others, but mostly construction health and safety, ergonomics, health and well-being, occupational health, primary health promotion, quality management, and risk management.

Nnedinma Umeokafor is a Lecturer at the University College of Estate Management, Reading, United Kingdom. Among his research interests is occupational safety and health, an area where he has published extensively.

Abimbola Windapo is an Associate Professor at the Department of Construction Economics and Management, University of Cape Town, with more than 30 years of experience in practice, teaching and research. She is a C2 Rated researcher with the National Research Foundation (NRF) and a Professional Construction Project Manager and Mentor registered with the South African Council for the Project and Construction Management Professions (SACPCMP) and Registered with the Council of Registered Builders of Nigeria (CORBON). Her research is interdisciplinary and focusses on construction industry development, business and project management from a practice and performance perspective. She is also the Editor, Journal of Construction Business and Management (JCBM).

Mengnan Xu was a PhD candidate in thermal comfort at the School Urban Construction and Environmental Engineering, Chongqing University, China. After graduating in 2014, he is now working in the finance sector.

Akilu Yunusa-Kaltungo is a Lecturer in Plant Reliability and Maintenance Engineering at the School of Mechanical, Aerospace and Civil Engineering

(MACE) at the University of Manchester, United Kingdom. Prior to his lectureship, he managed various industrial positions including health and safety manager, maintenance manager, training and development manager, plant operations champion, reliability engineer, mechanical engineer and project core team lead for the world's leading building materials manufacturing corporation (LafargeHolcim Cement PLC). He is the author of a book and more than 30 peer-reviewed technical articles in highly reputable publication outlets.

Hafiz Zahoor is currently an Assistant Professor at the Department of Construction Engineering and Management, National University of Sciences and Technology, Risalpur Campus, Khyber Pakhtunkhwa, Pakistan. He completed his PhD from The Hong Kong Polytechnic University in December 2016. His research interests include construction safety management, enhancing project productivity using building information modelling (BIM), earned value management and risk assessment in public–private partnership infrastructure projects.

Rita Peihua Zhang is a Senior Lecturer in Construction Management in the School of Property, Construction and Project Management (PCPM) at RMIT University, Australia. She conducts research in the areas of construction work health and safety (WHS), safety governance and social responsibility. She actively engages with industry for research development and undertakes research of practical implications. Her research has been funded by construction organizations and government agencies. She has also been disseminating her research findings to a global audience through actively publishing in top-ranked international journals and conference proceedings.

Foreword

Blood is the same colour everywhere in the world and our bodies and minds are damaged and hurt, irrespective of the colour of our skin, nationality, religion or world view. Construction and the production and development of the built environment play an essential role in every society across the world and with this comes a responsibility to care for the people involved at all stages of the process. Sadly, across the globe, construction is one of the worst-performing sectors for both safety and occupational health, whatever the stage of societal development in the particular region. In addition, whilst the fundamentals of construction technology are consistent worldwide, the processes, procedures, customs and practices vary widely among regions, in particular when related to the cost of labour.

Much of the research has been focused on more developed countries and regions with relatively high gross domestic product. This has led to some cases of research snobbery, with very worthy investigations in developing countries being discounted, as they do not reflect the current trends in other parts of the world. Furthermore, there has also been a trend for the inappropriate adoption in less developed countries of certain techniques and approaches that seem to work in more developed countries.

The International Council for Research and Innovation (CIB) has a working commission on Health and Safety in Construction (W099) that has developed the theory by Pybus (1996) seeking to demonstrate the differing challenges faced by different regions of the world (Finneran et al., 2013). Pybus describes three stages in the evolution of a culture of safety – traditional, transitional and innovative, with each stage building on rather than replacing the previous stage. The traditional stage is founded on rules and enforcement, and the transitional stage focuses on procedures and programmes. Therefore, before trying to implement health, safety and well-being innovations, these fundamental building blocks need to be in place. There is a significant risk for developing countries attempting to apply innovative approaches that may have some relevance in more developed countries, without having secure foundations in place.

The challenges and opportunities of managing the construction process, including health and safety aspects, are different when the location is in the developing world rather than those regions that are considered as more developed.

I am very keen to support this publication and hope that it makes an important contribution to improving the lives of construction workers across the world.

Professor Alistair Gibb
Professor of Construction Engineering and Management at Loughborough University, UK
Past Coordinator of CIB Working Commission on Health and Safety in Construction (W099)

References

Pybus, R. (1996) *Safety Management: Strategy and Practice*, Butterworth-Heinemann.

Finneran, A.M. & Gibb, A.G.F. (2013) Safety and Health in Construction Research Road Map, CIB Publication 376, CIB General Secretariat, International Council for Research and Innovation, The Netherlands.

1 An introduction to construction health and safety in developing countries

Patrick Manu, Fidelis Emuze, Tarcisio Abreu Saurin and Bonaventura H. W. Hadikusumo

Summary

In many countries, the construction industry has an unenviable reputation in terms of the occupational safety and health of the workforce. Occupational injuries and illnesses are highly prevalent in the construction industry compared to other industries. The situation in developing countries is worse than in developed countries. However, the bulk of construction occupational safety and health (OSH) literature has focused on issues in developed countries. Seeking to address this imbalance, this edited collection, which is the first of its kind, focuses on construction OSH in developing countries. This introductory chapter highlights the disparity in OSH performance in developed and developing countries, and juxtaposes this against the limited body of research on construction OSH in developing countries. The chapter presents some benefits, challenges and opportunities regarding construction OSH in developing countries and then closes with a summary of the chapters in the book.

Introduction

In many countries, the construction sector is notorious for high levels of occupational accidents, injuries and illnesses (see Health and Safety Executive [HSE], 2018a; Bureau of Labour Statistics, 2018). It not uncommon to hear of tragic incidents in the construction sector that result in physical harm, death or illness to workers and members of the general public. While this state of affairs lingers in both developed countries (described here as high-income economies) and developing countries (described here as low- to middle-income economies) (World Bank, 2019), the occurrence of occupational injuries and illnesses is direr in developing countries (Hämäläinen et al., 2006; Takala et al., 2014). With global construction estimated to grow by an unprecedented 85% to about US$15.5 trillion by 2030 (Global Construction Perspectives and Oxford Economics, 2015), it is worth noting that the current status of occupational safety and health (OSH) in construction in developing countries could get worse with growing investments in the construction sector to address infrastructure needs. There is thus an urgent need to improve OSH in construction in developing countries. When implementing action for improvement, however, cognisance should be taken of the fact that the construction industries of developing countries are at different phases of OSH maturity compared with their developed counterparts (Finneran and Gibb, 2013). Therefore, while some of the established OSH management practices in developed countries may work in the context of developing countries with little or no modifications, others may not. This then also calls for the development of context-specific solutions underpinned by research to deal with OSH challenges in developing countries. Against this backdrop, this book, which is the first edited book to focus on construction OSH issues in the context of developing countries, includes chapters that (1) reveal context-specific challenges that have hitherto not been well explored in the existing literature; (2) show the application of OSH management practices in known and novel ways to help improve the OSH situation; and (3) advance innovative practices and thinking to improve construction OSH in developing countries. Coverage includes countries from regions in sub-Saharan Africa, Latin America, Asia and Europe.

This introductory chapter highlights the status of OSH in the construction sector while drawing attention to the poor performance in developing countries. This is further juxtaposed against the relatively limited body of research on construction OSH in developing countries, thereby providing a sound case for a dedicated book collection on the subject. Subsequently, we present some benefits, challenges and opportunities regarding construction OSH in developing countries and then close with a summary of the chapters in the book.

The status of construction occupational safety and health performance

In both developed and developing countries where the construction industry makes major contributions to the gross domestic product (GDP), the negative impact of OSH performance on the working population is evident. For example, in the United Kingdom (UK), the industry recorded the second highest number of occupational fatal injuries (i.e. 38) in 2017–2018, while in Malaysia the industry recorded the highest number of fatalities in 2017 (i.e. 111) (Department of Occupational Health and Safety, 2018; HSE, 2018b). In terms of the occupational health of workers, in the UK, the prevalence rate of self-reported illnesses in construction is higher than the average rate in all industries (HSE, 2018a). Generally, these statistics reflect the ubiquity of OSH hazards (e.g. dust, noise, manual handling, working in confined spaces and working at heights) in the construction industry.

For the workers and their associates, especially in construction in developing countries, involvement in the industry has produced unimaginable loss through occupational diseases and injuries (Smallwood and Emuze, 2016). As rightly illustrated by Takala et al. (2014), OSH incidence rate is associated with the socio-economic development of a country, such that less developed countries record higher prevalence of OSH incidence than more developed countries.

Furthermore, the project-based nature of the construction industry is a significant contextual issue in relation to OSH. For instance, the way in which work is organised determines how OSH challenges could be tackled on site because of dispersed project control systems, temporary worksites, multiplicity of trades and cultures, high level of subcontracting and casual employment (Lingard and Rowlinson, 2005; Emuze and Sherratt, 2018; Emuze and Mollo, 2018). Additional barriers to OSH improvement include traditional separation of design and construction through the excessive use of the design–bid–build delivery method, competitive tendering that leads to the awarding of contracts to the lowest bidders at the expense of standard services and deliverables, the mushrooming of small contractors with dismal OSH records, reliance on subcontracting for the physical execution of work on sites and the lack of good communication as well as cooperation among project actors (Manu et al., 2013; Neale, 2013). Another significant barrier to OSH improvement is the overemphasis on project and business performance parameters in construction at the expense of the health, safety and well-being of people in the industry.

While the general poor status of OSH in construction has instigated a plethora of construction OSH studies over the past three to four decades (e.g. papers in CIB W099 conference proceedings, OSH-themed papers in the Association of Researchers in Construction Management Conference proceedings, and many articles in OSH and construction journals), construction OSH studies have largely focussed on issues pertaining to developed countries. To illustrate, a review of design for OSH studies in construction (published in journals from 1992 to 2016) showed that an overwhelming majority of the studies (i.e. over 90%) focussed on developed countries (Manu et al., 2018, 2019). This state of affairs has resulted in significant gaps in the knowledge of construction OSH issues in developing countries, which this book makes an important contribution towards addressing.

Improving construction OSH in developing countries: Benefits, challenges and opportunities

Improved OSH performance benefits all stakeholders involved in the construction industry (Hughes and Ferrett, 2008; Ikpe, 2009; HSE, 2017; Gibb et al., 2018). There is no dispute that workers are the primary beneficiaries, given that their physical integrity and long-term labour capability are at stake. From a business perspective, construction companies and owners also reap the fruits of better OSH, since OSH incidents impose high costs and affect productivity and quality (Wanberg et al., 2013; Love et al., 2015; HSE, 2017; Gibb et al., 2018). In addition, safer construction sites are less likely to be shut down by government inspectors. Last, government and society as a whole are positively affected by better OSH outcomes in many ways, namely health care systems will be less burdened due to fewer OSH incidents; social security systems will be less burdened due to lower compensations and early retirements; large-scale public infrastructure and housing projects may be less expensive in the long run; and the construction industry will be more attractive to young people.

However, obtaining these benefits is not without challenges, which tend to be greater in developing countries. Examples of these challenges include weak institutions for setting policy and enforcement; illiteracy; lack of or inadequate coverage of OSH in the education of construction professionals and workers; lack of or inadequate OSH training for construction professionals and workers; low awareness of OSH issues among industry stakeholders; lack of government leadership; and lack of client leadership (Baram, 2009; Awwad et al., 2016; Umeokafor et al., 2018; Manu et al., 2018, 2019). Furthermore, in developing countries the construction industry is traditionally seen by governments as a potential employer of high numbers of unskilled workers, whose qualifications are often compatible mostly with artisanal (and often intrinsically unsafe) construction technologies. As such, some governments discourage the use of modern technologies by over-taxing these, in order to pursue short-term goals of lowering unemployment rates.

On the other hand, developing countries present some unique opportunities to be explored and leveraged in order to address poor construction OSH. To some extent, these opportunities may be highly contextual and derive from the local culture, such as the involvement of community leaders and religious institutions that could help to raise the general awareness of OSH importance, given the important role religious leaders play in some developing countries (Umeokafor et al., 2018). Similarly, collaboration between academic and industry institutions offers the opportunity of win-win relationships, which at the same time may contribute to solving practical OSH industrial problems while educating students and professionals. Furthermore, developing countries may benefit from tested technologies and established regulations in developed countries, adapting them to their local context without starting from scratch. Lastly, given the higher OSH incidence in developing countries, it is possible that short-term positive results can be quickly obtained in some developing countries, setting a basis and motivation for sustainable results.

Overview of chapters

This book comprises 21 chapters grouped into 5 thematic parts as follows:

Chapter 1, 'An introduction to construction health and safety in developing countries': This introductory chapter underscores the need to pay attention to construction OSH in developing countries while highlighting the benefits, challenges and opportunities regarding addressing poor OSH performance in construction in developing countries. In so doing, the chapter sets the stage for the remaining chapters of the book.

Part I: Occupational health and safety legislation

Chapter 2, 'Implications and opportunities in a complex construction health and safety regulatory environment': In this chapter, Umeokafor and colleagues argue

An introduction 5

that the Nigerian construction industry is self-regulated, contrary to previous reports of the industry being unregulated. Through the use of semi-structured interviews with industry stakeholders, they evidence and explain the mechanism of self-regulation in the Nigerian construction industry while explaining the role played by various regulatory actors. Umeokafor and colleagues offer both positive and negative consequences of self-regulation of the industry. They argue that despite the poor status of construction OSH in Nigeria, self-regulation has not failed, but rather it is evolving. Nonetheless they acknowledge that the regulatory regime needs improvement and suggest some practical recommendations to achieve this.

Chapter 3, 'Contractors' compliance with occupational health and safety legislation in South Africa: The benefits of self-regulation': In this chapter, Windapo and colleagues explain how contractors are meeting legislation and also self-regulating their activities over and above compliance requirements in South Africa. The chapter presents the levels of self-regulation and compliance to OSH laws with corresponding influence on the extent and severity of construction accidents on sites. It emerged that there is a major linear negative relationship between the level of contractor self-regulation and mishaps on site. In other words, the chapter reinforces the notion that OSH performs better on a site when project actors go beyond the remit of legislation in preventing harm to people in construction.

Chapter 4, 'A narrative review of occupational safety and health legislation in Pakistan': In this chapter, Qayoom and Hadikusumo study the OSH legislative framework in Pakistan related to the construction industry. They adopt a narrative review of research articles, annual reports and OSH legislation in developing countries to find the shortfalls of OSH legislation in Pakistan. They reveal inadequate legislative structure, outdated and fragmented laws, and lack of database management. Based on the findings, recommendations are suggested which include legislative reforms, capacity-building training, upgrading the reporting system and involvement of trade unions.

Chapter 5, 'Safety law, system and regulation influencing the construction sector in India': In this chapter, Duddukuru and Hadikusumo study safety laws and regulations in India. Case studies are adopted to investigate construction firms' level of awareness, implementation and compliance with safety regulations. From the cases investigated, Duddukuru and Hadikusumo reveal that most of the small firms are still not aware, not implementing and not complying with safety regulations, whereas medium and large firms are fully aware of all the safety regulations but do not fully implement and comply with all the safety regulations. The lack of awareness of safety laws and regulations are caused by some factors which are also discussed in the chapter.

Part II: Occupational health and safety management

Chapter 6, 'The strategies to enhance safety performance in the construction industry of Pakistan': Against the backdrop of a high prevalence of accidents on building projects in Pakistan, this chapter presents research that aims to develop strategies for enhancing safety performance on building projects in Pakistan. Through a multi-pronged research approach involving a questionnaire survey, a two-round Delphi technique and semi-structured interviews, Zahoor and Chan determined contributory factors of accident causation and significant safety climate factors that affect safety compliance, safety participation and number of accidents/injuries. They recommend several safety improvement measures that are aligned to the contributory factors of accident causation and the safety climate factors. The measures include actions to be taken by an industry professional institution, government agencies, clients and contractors, therefore emphasising the need for joint efforts by multiple stakeholders in order to address poor safety performance in Pakistan's construction industry.

Chapter 7, 'Health and safety performance in the Ugandan construction industry': In this chapter, Okwel and colleagues contend that in Uganda, there has been poor management of construction OSH on sites. The poor state of OSH management produces increased frequencies of incidents, which beget loss in various forms. More telling is the reactive way of addressing hazards that are transformed to risks in the industry. The reactive manner of OSH management relegates preventative measures to the background at the expense of productivity and other performance measures. The case study of Ugandan construction amplifies the need to eradicate unsafe practices on sites. A similar observation was made by the Department of Labour in South Africa almost a decade ago. People are increasingly exposed to unsafe construction site practices in sub-Saharan Africa. While countries such as South Africa have legislation comparable to that of developed countries, the gaps in implementation and compliance are causing havoc in the industry. In other words, legislation is a good tool to combat unsafe site practices, but workers and management must look beyond it to engender optimum OSH outcomes.

Chapter 8, 'Towards the development of an integrated safety, health and environmental management capability maturity model (SHEM-CMM) for uptake by construction companies in Ghana': Although legislation is a tool for OSH improvement, managers of the business and project aspects of construction must also deploy other tools that are cost effective in developing countries where access to capital is a major economic constraint. In this chapter, Asah-Kissiedu and colleagues reinforce the notion that the high cost of adoption of separate management systems (e.g. OSH and environmental management systems) remains a major improvement hurdle. They go on to propose the development of an integrated safety, health and environmental management tool that could potentially improve OSH and environmental management with less cost and simplicity. Such a tool is desirable in developing countries where access to business capital is limited for contractors that also rely on a low-educated workforce. The proposed tool for contractors in Ghana could serve the same purpose in other developing countries.

An introduction 7

Chapter 9, 'Ad hoc and post hoc analysis of contractors' safety risks during procurement in Nigeria': As opposed to a positive OSH culture that engenders the achievement of project deliverables, a negative safety culture is a challenge in many developing countries. The culture in Nigeria and other developing countries constitute a mockery to the notion of respect for people. In this chapter, Olatunji and colleagues underline this reality with the comment that while Nigerian laws compel contractors to mindfully address OSH on work sites, the media constantly report collapse of structures, site accidents and fatalities. The chapter presents an assessment of pre-award OSH assessment processes and standards concerning contractors. The authors note a major gap between the pre-award assessment outcome and OSH practices on site after the contract is awarded. The chapter then proposes a new framework for assessing contractors' OSH capability. The proposed framework is performance oriented as it could measure capabilities at the pre-qualification stage of bid evaluation quantitatively.

Chapter 10, 'Integrating health and safety into labour-only procurement system: Opportunities, barriers and strategies': Labour-only procurement (LoP) is a form of contracting between a client and contractor whereby the client provides materials and the contractor provides labour to execute the construction work. This contracting approach is common in developing countries, including those in sub-Saharan Africa, yet its impact on construction OSH management in this region is less known. In this chapter, Umeokafor and colleagues, through an exploration of literature, describe the characteristics of LoP, its merits and demerits. They also suggest ways by which OSH can be integrated into LoP, while recognising opportunities and barriers to integration.

Chapter 11, 'Health and safety in tertiary, built environment, construction education in South Africa': The notion of 'enlightenment' refers to the ability of scholars to use reason to address issues in society. Reason helps in the application of context-specific knowledge. It therefore implies that preventing OSH incidents with education that positively directs the thoughts and actions of relevant persons is a must in the construction industry. In this chapter, Smallwood presents the status of OSH in higher education curricula to engender prevention strategies in the minds of practitioners and researchers. For the umpteenth time, the chapter shows that tertiary built-environment education is inadequate in terms of construction OSH content in relation to teaching and learning. As a result, the gaps in tertiary built-environment education are not helping graduates to contribute to construction OSH improvement in South Africa.

Part III: Behavioural and cultural issues in safety management

Chapter 12, 'Safety communication and suggestion scheme in construction': Safety communication, being the sharing of information between management personnel and the workforce regarding workplace safety, is an important aspect of effective safety management. In line with this, safety suggestion schemes offer a proactive communication mechanism that allows workers to contribute to decision-making regarding the management of safety by an organisation. In this chapter, Okorie and Emuze therefore propose a safety communication and suggestion scheme to improve safety communication on construction projects in developing countries, where there is likely to be high illiteracy among workers as well as a lack of confidence among rural and unskilled workers to be vocal about workplace safety issues.

Chapter 13, 'Making sense of safety in systemic and cultural contexts': Safety can be interpreted differently by relevant actors (e.g. workers and employers) in different cultural contexts, and this has implications for how safety is effectively managed. In this chapter, Jia and colleagues use a comparative study between a developed and a developing society to explore the constructed meaning of safety. They find that in the developing society, safety could mean workers' self-interest and employers' benevolence, while in the developed society, both employers and workers interpret safety as compliance to rules. These insights suggest that in a developing country, rules that develop bottom-up could tackle safety issues better than top-down coercive rules.

Chapter 14, 'Managing construction safety in developing countries: From the viewpoint of Chinese international contractors': In this chapter, Gao and Chan present a study on Chinese contractors' viewpoints related to managing construction safety in developing countries. By using six construction projects as case studies, they identify three levels of problems of managing construction safety in overseas projects. The three levels are environmental and societal level, organisational level, and personal level. They recommend that Chinese international contractors must be ready to encounter and strive to resolve these problems, and conduct effective safety management without neglecting local culture.

Chapter 15, 'Learning from Failures (LFF): A multi-level conceptual model for changing safety culture in the Nigerian construction industry': Against the backdrop of a high record of occupational fatalities in Africa, this chapter takes the view that it is important to understand how different manifestations of safety culture influence safety performance in the construction industry. Using Nigeria as a case study, Cheung and colleagues explain the different manifestations of safety culture in Nigeria's construction industry (namely, at the national level, industry level and organisational level) and propose a multi-level conceptual model for enhancing safety performance.

Chapter 16, 'Overview of safety behaviour and safety culture in the Malaysian construction industry': In this chapter, an overview of safety behaviour and safety culture of the construction industry in Malaysia is presented by Mohammad and Hadikusumo. The issue of safety behaviour in Malaysia is complicated because of many foreign workers from countries such as Indonesia, Bangladesh, Nepal and the Philippines. Three construction projects were used in a case study to explore safety behaviour and safety culture. In one case, the authors observed a positive safety culture, which can be identified from top management commitment, provision of personal protective equipment (PPE), adoption of rewards and penalties related to safety performance, and zero accidents. The findings could help promote positive safety behaviour and safety culture in construction projects.

Part IV: Construction workers' well-being

Chapter 17, 'Determinants of risky sexual behaviour by South African construction workers': The high rate of HIV infection in certain African countries is not a news item today. However, when risky sexual behaviour proliferates, it feeds the pipeline of infection with significant harm to the well-being of the general workforce. In this chapter, Bowen and Edwards present an explanatory framework for the risky sexual behaviour of construction workers and the factors that influence it in South Africa. The theoretical model presented in the chapter indicates certain risky sexual behaviour patterns that were analysed with regression analysis and structural equation modelling. It is notable that age, gender, acquaintance with an HIV+ person, drug usage and level of AIDS-related knowledge were direct determinants of risky sexual behaviour. The authors also observe that higher literacy levels, ethnicity and acquaintance with an HIV+ person could predict better AIDS-related knowledge.

Chapter 18, 'Construction migrants from "developing" countries in "developed" hosts': Migration from developing countries to developed countries has resulted in the growth of migrant workers from developing countries in the construction industries of several developed countries. While this situation is beneficial for both the migrants and their host industries (e.g. migrants earning relatively better wages and host industries/countries addressing skills shortage), it poses challenges to managing the OSH of migrant workers. In this chapter, Oswald draws upon an ethnographic study in the United Kingdom (as a developed host environment) regarding migrant workers from lower-income countries in Eastern Europe to unpack the exploitation and OSH risks to which migrant workers are exposed.

Chapter 19, 'Occupational accidents predicting early retirement of construction workers in South Africa': In this chapter, Agumba and Musonda identify, based on the analysis of about 90,000 accidents at South African construction sites, the types of occupational accidents most likely to lead to construction workers' early retirement. Although being struck by an object was the most frequent accident type, electrocutions were more likely to result in permanent disability and early retirement. Other major trends were identified, such as motor vehicle accidents being the major cause of fatalities and electrocutions being the most expensive accident type. Overall, the findings of this study contribute to the definition of the safety priorities to be tackled in the South African construction industry.

Part V: Technologies for safety management

Chapter 20, 'Design of flexible horizontal lifeline systems': In this chapter, Costella and colleagues describe a protocol for the design of flexible horizontal lifeline systems on construction sites. This protocol was developed based on a diversity of information sources, including standards and best practices from industry. As such, the protocol is valuable for practitioners and researchers alike, supporting the prevention of falls from height, which is one of the leading causes of fatalities in construction worldwide.

Chapter 21, 'Development of RFID automation system to improve safety on construction sites in Brazil': In this chapter, Oliveira and Serra report on the development and testing of a monitoring and control system for site workers and collective protection equipment (CPE) using RFID technology. The application to CPE is relevant given that the location of this equipment needs to be changed on a frequent basis due to the dynamism of the construction stages. However, these changes do not always occur as planned, and therefore the real-time tracking of the CPE location may support the early detection of undesired changes. As for workers, the RFID technology allows for the real-time identification of access to unsafe areas, followed by an aural warning. The system described in the chapter was tested both in laboratory and at real construction sites, which indicated their potential benefits.

References

Awwad, R., El Souki, O. and Jabbour, M. (2016), 'Construction safety practices and challenges in a Middle Eastern developing country', *Safety Science*, 83, pp. 1–11.

Baram, M. (2009), 'Globalization and workplace hazards in developing nations', *Safety Science*, 47(6), pp. 756–766.

Bureau of Labor Statistics (2018), Census of fatal occupational injuries charts, 1992–2017 (final data). Available online at www.bls.gov/iif/oshcfoi1.htm (accessed 8 January 2019).

Department of Occupational Health and Safety (2018), Occupational accidents statistics by sector 2017. Available online at www.dosh.gov.my/index.php/en/archive-statistics/2017/2003-occupational-accidents-statistics-by-sector-2017 (accessed 5 April 2019).

Emuze, F. A. and Mollo, L. G. (2018), 'Disrespecting people with working conditions in South Africa', Institution of Civil Engineers: Management, Procurement and Law, Paper 180029, pp. 1–6.

Emuze, F. A. and Sherratt, F. (2018), 'Making zero harm work for the construction industry', In: Emuze, F. A. and Smallwood, J. J. (Eds.) *Valuing People in Construction*, Oxon: Routledge, pp. 226–241.

Finneran, A. and Gibb, A. (2013), W099 - Safety and health in construction research roadmap – Report for consultation, CIB Publication 376. Rotterdam: CIB General Secretariat.

Gibb, A., Drake, C. and Jones. W. (2018), Costs of occupational ill-health in construction. Institution of Civil Engineers.

Global Construction Perspectives and Oxford Economics (2015), Global construction 2030. Available online at www.globalconstruction2025.com/ (accessed 30 July 2015).

Hämäläinen, P., Takala, J. and Saarela, K. L. (2006), 'Global estimates of occupational accidents', *Safety Science*, 44(2), pp. 137–156.

HSE (2017), Health and safety at work – Summary statistics for Great Britain 2016. HSE.

HSE (2018a), Construction statistics in Great Britain, 2018. HSE. Available online at www.hse.gov.uk/statistics/industry/construction.pdf (accessed 8 January 2019).

HSE (2018b), RIDHIST – Reported fatal and non-fatal injuries in Great Britain from 1974. Health and Safety Executive. Available online at www.hse.gov.uk/statistics/tables/index.htm (accessed 5 April 2019).
Hughes, P. and Ferrett, E. (2008), *Introduction to Health and Safety in Construction*, 3rd edn. Amsterdam: Butterworth-Heinemann.
Ikpe, E. (2009), Development of cost benefit analysis model of accident prevention on construction projects. PhD thesis, University of Wolverhampton.
Lingard, H. and Rowlinson, S. (2005), *Occupational Health and Safety in Construction Project Management*. London: Routledge.
Love, P. E. D., Teo, P., Carey, B., Sing, C.-P., and Ackermann, F. (2015), 'The symbiotic nature of safety and quality in construction: Incidents and rework non-conformances', *Safety Science*, 79, pp. 55–62.
Manu, P., Ankrah, N., Proverbs, D. and Suresh, S. (2013), 'Mitigating the health and safety influence of subcontracting in construction: The approach of main contractors', *International Journal of Project Management*, 31(7), pp. 1017–1026.
Manu, P., Poghosyan, A., Agyei, G., Mahamadu, A. M. and Dziekonski, K. (2018), 'Design for safety in construction in sub-Saharan Africa: A study of architects in Ghana', *International Journal of Construction Management*. doi:10.1080/15623599.2018.1541704.
Manu, P., Poghosyan, A., Mshelia, I. M., Iwo, S. T., Mahamadu, A. M. and Dziekonski, K. (2019), 'Design for occupational safety and health of workers in construction in developing countries: A study of architects in Nigeria', *International Journal of Occupational Safety and Ergonomics*, 25(1), pp. 99–109.
Neale, R. (2013), 'Ten factors to improve occupational safety and health in construction projects', *African Newsletter on Occupational Health and Safety*, 23, pp. 52–54.
Smallwood, J. J. and Emuze, F. (2016), 'Factors that promote zero fatalities, injuries and disease in construction', In: Chan, P. W. and Neilson, C. J. (Eds.) *Proceedings of the 32nd Annual ARCOM Conference*, 5–7 September 2016, Manchester, Association of Researchers in Construction Management, pp. 587–596.
Takala, J., Hämäläinen, P., Saarela, K. L., Yun, L. Y., Manickam, K., Jin, T. W., Heng, P., Tjong, C., Kheng, L. G, Lim, S. and Lin, G. S. (2014), 'Global estimates of the burden of injury and illness at work in 2012', *Journal of Occupational and Environmental Hygiene*, 11(5), pp. 326–337.
Umeokafor, N., Windapo, A. O. and Manu, P. (2018), 'Country context-based opportunities for improving health and safety', In: Saurin, T. A., Costa, D. B., Behm, M. and Emuze, F. (Ed.) *Proceedings of Joint CIB W099 and TG59 International Safety, Health, and People in Construction Conference*, 1–3 August 2018, Salvador, Marketing Aumentado, pp. 177–186.
Wanberg, J., Harper, C., Hallowell, M. R. and Rajendran, S. (2013), 'Relationship between construction safety and quality performance', *Journal of Construction Engineering and Management*, 139(10), p. 04013003.
World Bank (2019), World Bank analytical classification. Available online at http://databank.worldbank.org/data/download/site-content/OGHIST.xls (accessed 6 February 2019).

Part I
Occupational health and safety legislation

2 Implications and opportunities in a complex construction health and safety regulatory environment

Nnedinma Umeokafor, Konstantinos Evangelinos and Abimbola Windapo

Summary

One of the ways of improving construction health and safety (CH&S) remains its regulation, the foundation and catalyst of other improvement measures. However, in Nigeria, it is poorly understood. Consequently, this chapter reports an examination of how it is regulated (including the roles of various parties), and the implications and opportunities therein for improving CH&S. Adopting semi-structured interviews and thematic analysis, the study shows that the Nigerian construction industry is self-regulated in various ways (including by other industries such as the oil and gas sector), contesting earlier held views that the industry is unregulated. However, the regulatory structure is distributed and fragmented. There is also evidence that the nature of the legislation adopted from other countries and the local environmental laws that cover some construction activities, to name but a few, present the industry with multiple actors, bias in regulation and control, and a complicated regulatory environment, which are counterproductive to the entire regulatory regime. However, the findings also include opportunities in the regulatory systems such as risk ownership, suggesting the likelihood of an effective and accepted regulatory system where the regulated take ownership of the risk they create. Country context-based CH&S legislation and a centralised regulatory system are recommended.

Introduction

Construction health and safety (CH&S) has significantly developed over the years resulting in various ways of improving it, for example, designing out hazards at the early stages of construction and involving the client in health and safety (H&S) statutorily (Health and Safety Executive [HSE] 2015) and applying technology to construction safety. H&S legislation and their regulation are one of the essential foundations or pillars of all the H&S improvement measures (cf. Finneran and Gibb 2013; Umeokafor 2017). They are regularly updated to ensure currency, enabling H&S improvement measures to function effectively.

However, the case is different in most developing countries. According to Abubakar (2016), many of the developing countries still operate a distributed occupational safety and health (DOSH) regulatory framework as against the (so-called) developed ones that have moved to a consolidated occupational safety and health regulatory framework. The

> DOSH regulatory framework leverages on multiple and less coherent legal provisions dispersed in various related laws … (while) the consolidated occupational safety and health regulatory framework refers to a relatively harmonised regulatory and enforcement framework which comes with mandate expansion, enrichment of regulations, increased regulator powers, more budgetary allocations as well as enhanced executive and financial independence.
>
> (Abubakar 2016: 61–62)

These, according to Abubakar (2016), provide a platform for a regulatory agency to function properly. By implication, DOSH is the recipe for the lack of an adequate comprehensive H&S policy and fragmented H&S regulatory system in developing countries such as Nigeria and Ghana.

Authors criticise and evidence the poor H&S regulatory framework and practices in many developing countries (Abubakar 2016; Annan et al. 2015; Dabup 2012; Idoro 2011; Umeokafor 2017; Windapo et al. 2018). For example, in Nigeria, the DOSH regulatory framework with multiple actors is evident. Indeed, the Nuclear Safety and Radiation Protection Act of 1995 (functional in 2001) established the Nigerian Nuclear Regulatory Authority (NNRA) that oversees nuclear safety and radiation protection. The Factories Act CAP F1 L.F.N 2000 covers the Factories in Nigeria with the responsibility of enforcement on the Federal Ministry of Labour and Employment Inspectorate Division (FMLEID) (Federal Republic of Nigeria 2004a). According to Dabup (2012), 'some laws relative to H&S are embedded within the country's environmental laws' (Dabup 2012: 37). The Engineers (Registration etc.) Act, CAP E11 of 2004 empowers the Council for Regulation of Engineering to regulate and control engineering practices and training, and monitor construction processes including building structure safety and building collapse investigation (Federal Republic of Nigeria 2004b).

The Factories Act is criticised for its discontent, obsolescence and lack of subsequent secondary legislation (Abubakar 2016; Idoro 2011; Umeokafor 2017). Section 87 of the Factories Act omits construction sites and activities in its definition of premises. As a result, authors such as Idoro (2011) argue that the industry is unregulated. However, drawing on the definitions of self-regulation in literature, and the evidence and assertions in studies such as Idoro (2011) and Omeife and Windapo (2013), Umeokafor (2016, 2017) demonstrates that the industry is self-regulated in various ways and not unregulated. Self-regulation is the practice of adopting and administering standards and legislation by an organisation or industry to control its activities with little or no external involvement (Castro 2011; Gunningham 2011). While this contributes to the DOSH regulatory framework in that the construction industry has a different regulatory framework, it offers optimism to H&S improvement, a unique opportunity in the current complex regulatory environment.

Health and safety regulation in Nigeria   17

The preceding arguments show that there are many actors in the CH&S space; therefore, their interests and structure of regulation are likely to vary with opportunities and implications for H&S. CH&S regulatory issues in Nigeria have been addressed from the unregulated perspectives; hence, the reality in this area is poorly understood and unexamined. Consequently, reporting part of a more extensive study, the lead author's PhD research, this chapter seeks to fill this vast gap in knowledge. In doing this, the objectives of the chapter are to:

- Explore and explain how CH&S regulation works in Nigeria.
- Identify and explain the roles of the actors in CH&S regulation in Nigeria.
- Identify and explain the implications and opportunities presented by the structure and actors in CH&S regulation in Nigeria.

In the current study, regulation is conceptualised as the action or process of controlled '(something, especially a business activity) by means of rules and laws' (Oxford Dictionary 2018). The laws here are the primary legislation, for example, the Health and Safety at Work Act 1974 and secondary legislation Construction Design and Management (CDM) Regulations 2015 of the United Kingdom.

Conceptual foundation

Regulation including self-regulation

Self-regulation is based on flexible regulation, a contrast of the command-and-control system, which is fixed (see Figure 2.1). It is designed to involve the regulated in the regulatory process. The definition of self-regulation takes various forms and varies from industry to industry and country to country (Castro 2011; Gunningham 2011; Windapo et al. 2018). Hence, there is no satisfactory definition, and the distinction can take many forms (Gunningham and Rees 1997).

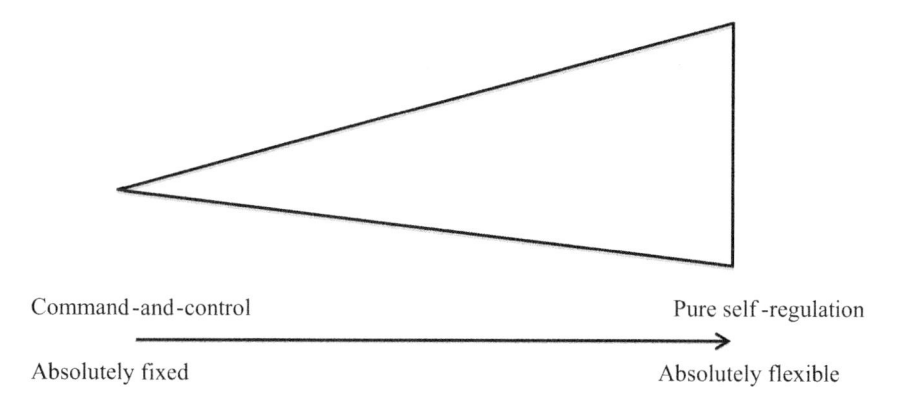

Figure 2.1 Regulatory spectrum.

Self-regulation is a spectrum (Gunningham and Rees 1997; Windapo et al. 2018). On one end of the spectrum is the command-and-control system and pure self-regulation is at the other end (see Figure 2.1).

The command-and-control system is traditionally based on the government authority solely and directly regulating an industry or organisation by setting the standards and controlling them with sanctions and prosecutions (cf. Gunningham and Rees 1997). While in pure self-regulation, there is no external involvement, and the standards or laws are self-imposed and administered by the practice community (Bartle and Vass 2005). Castro (2011) and Gunningham and Rees (1997) posit that this can be at the organisational or group industry level through self-policing, professional code of ethics and industry best practices. Various other distinctions of self-regulation fall between both ends of the spectrum (see Figure 2.1). For example, co-regulation—the regulatory process is jointly administered by the industry and government; mandated full self-regulation—the setting of rules and enforcement is by the firms or industry but overseen and approved by the government (Gunningham 2011); mandated partial self-regulation—rulemaking or enforcement is by the industry or firm but not both (Gunningham 2011); and industry self-regulation—the industry agrees on standards and its administration (Gunningham and Rees 1997).

The performance of self-regulation

Despite it critics (Bartle and Vass 2005), self-regulation has recorded success in the improvement of H&S of which the British construction industry is one (HSE 2018). For example, while the HSE's (2018) provisional data shows that the British construction industry recorded the highest level of fatal injury across other industries in 2017–2018 with 38 fatal injuries (an increase from the 30 fatal injury record of 2016–2017), this is still below the 2015–2016 fatal injury record of 47. Ironically, Britain's H&S record is the second best in the EU (HSE 2018). Conversely, in New Zealand, Walls and Dryson (2002) have reported the underperformance or failure of H&S self-regulatory programmes in the manufacturing industry.

CH&S regulation in Nigeria

Drawing on background established so far, it is indicative that CH&S in Nigeria is self-regulated in many ways. For example, the self-contribution of contractors regarding adopting and administering H&S legislation from other countries (Idoro 2011) can be pure self-regulation. Famuyiwa (2011) shows the contribution of local authorities in Lagos State regarding the regulation of H&S in the state; this can be mandated self-regulation. Pure self-regulation can also be seen in the voluntary adoption of the National Building Code of 2006 where some states adopt minimum standards for the building industry in Nigeria including H&S despite having no legal backing (Omeife and Windapo 2013). The standards set by the oil and gas sector that contractors adhere to point to industry self-regulation. Despite this, there is little known about these regulatory processes in Nigeria. Considering the poor state of CH&S in developing countries, this is very important.

Methodology

Drawing on Saunders et al. (2009) who note that the research questions or problems inform the research paradigm and approach, contractors and key informants were interviewed face-to-face and telephonically. This is because the research questions are how questions with exploratory purposes (Saunders et al. 2009), the ontological position is that there are multiple truths, and the epistemological position is that adequate knowledge is subjective and a social phenomenon (Saunders et al. 2009). According to Merrian (1998 in Zohrabi 2013), interviews excel in how-and-what explanations. Above all, the complexity of H&S in Nigeria and the project-oriented nature of construction projects make a case for adopting the methods (Umeokafor 2017).

Following a review of relevant literature and informal discussions with key informants in Nigeria, the semi-structured interview guide was developed. The questions include the profile of the respondents and their current organisations, and how they self-regulate regarding CH&S. The interview guide was refined in line with the interview refinement framework of four phases by Castillo-Montoya (2016). It involved 'ensuring that interview questions align with research questions; constructing an inquiry-based conversation; receiving feedback on the interview protocol; piloting the interview protocol'. Before piloting the interview protocol on five contractors and four key informants, the draft was pretested on eight people including academics with significant experience in construction and H&S. The interview protocol was revised afterwards for the primary study. Then in the main phase of the data collection and analysis, 53 participants were interviewed; in the framework of recommendations evaluation (for workability or practicality) phase, 11 (including some of the 53) participated. However, due to the scope and objectives of this chapter, only the main phase is reported, but a few relevant recommendations are outlined here. The framework (including methodology) is detailed in Umeokafor (2017).

Mixed purposeful sampling (stratified purposeful and snowballing sampling strategies) was adopted where the strata were developed based on the size of the contractors—large-sized contractors (LCs), medium-sized contractors (MCs), small-sized contractors (SCs) and micro-sized contractors (MiCs)—and each stratum was homogeneous. LCs have over 199 employees, MCs have 30–199 employees, SCs have 10–29 employees and MiCs have fewer than 10 employees (Kheni 2008). The interviews lasted for 65–117 minutes; translated verbatim, it was analysed thematically in line with Braun and Clarke's (2006) six-phase thematic analysis process. This was conducted with NVivo for Mac, and involved exploring the manifest and latent meanings through a combination of deductive

20 Umeokafor, Evangelinos and Windapo

and inductive approaches (Bradley at al. 2007). The analytical groups were key informants (that have the direct or indirect association with the industry in terms of health, safety and the environment) and 'contractors that constantly self-regulate' (CSRCs), those that 'do not self-regulate' (non-SRCs) and those that 'do not constantly self-regulate' (NCSRCs). In improving the trustworthiness of the study, the following steps were taken: triangulation of persons, analytical triangulation and methods triangulation, among many, peer debriefing.

Results

Description of sample

In this phase, of the 53 participants, 46 interviews were used because of saturation during the data analysis. Thirty-one were contractors (eight LCs, nine MCs, nine SCs and five MiCs) and fifteen were key informants. The scope of the operations include the oil and gas sector, civil engineering, contracting and consultancy, renovation, and building. All analytical groups were indigenous contractors except CRSCs, where seven were both foreign and Nigerian-owned firms and three wholly indigenous firms. The analytical groups are as follows: 10 CSRCs (7 LCs, 2 MCs and 1 SC); 14 NCSRCs (1 LC, 6 MCs, 5 SCs, 2 MiCs); 7 non-SRCs (1 MC, 3 SCs and 3 MiCs). All six geopolitical zones in Nigeria were covered. The participants include H&S managers, project managers, civil engineers, architects and owners/managers. CSRCs have a lot in common, namely, geographic location and scope of operation.

Concerning the remaining 15 participants, there were key informants from 13 organisations that had an association with the construction industry, for example, a higher education institution, insurance firm, law firms, regulatory bodies, H&S consultancy firms, public institutions, trade associations and client organisations.

The experiences of many of the respondents are not limited to their current organisations and scope of operations, as many have worked or currently worked in other organisations. For example, the academic also worked as a quantity surveyor alongside the academic role.

The reality of CH&S regulation in Nigeria

In the study, there is evidence that the Nigerian construction industry is self-regulated in various ways, including pure self-regulation and client-led self-regulation (see Figure 2.2). Figure 2.2 shows that there are many actors and regulatory instruments including local code/standards and international legislation. In Figure 2.2, all in bold or brackets in 'Key actors' and 'Outcomes' denote the types of approaches to self-regulation with which the corresponding texts align or relate. For example, in 'Key actors', contractors are the key actors in pure self-regulation; in 'Outcomes' pure self-regulation and part of industry self-regulation are voluntary.

Regulatory instruments	Key actors	Types of self-regulation	Outcomes
International standards and legislation; Local code/standards; Internally developed standards	Contractors (**PSR**); Banking, oil and gas, and telecommunication sectors (**ISR**); **M/ESR:** LSSC; COREN; CORBON; The Fire service; LSBCA; MWI; NESREA; MOE; Local authorities; MC; **CLSR:** International & corporate organisations, government; Social actors: Health and safety crusader-led (**H&SCSR**); Communities (**CSR**)	Pure; Industry; Mandatory/enforced; Client-led; H&S crusader-led; Community-led	Voluntary (**PSR** and **ISR**); International/local statute based (**ISR & M/ESR**); Non-state actor based actor (**CLSR, H&SC, CSR**)

Legend: MOE: Ministry of Environment; MWI: Ministry of Works and Infrastructure; LSBCA: Lagos State Building Control Agency; COREN: Council for Regulation of Engineering in Nigeria; CORBON: Council of Registered Builders of Nigeria; LSSC: Lagos State Safety Commission; NESREA: National Environmental Standards and Regulations Enforcement Agency. PSR: pure self-regulation; M/ESR: mandatory/enforced self-regulation; CSR: community-led self-regulation; CLSR: client-led self-regulation; ISR: industry self-regulation.

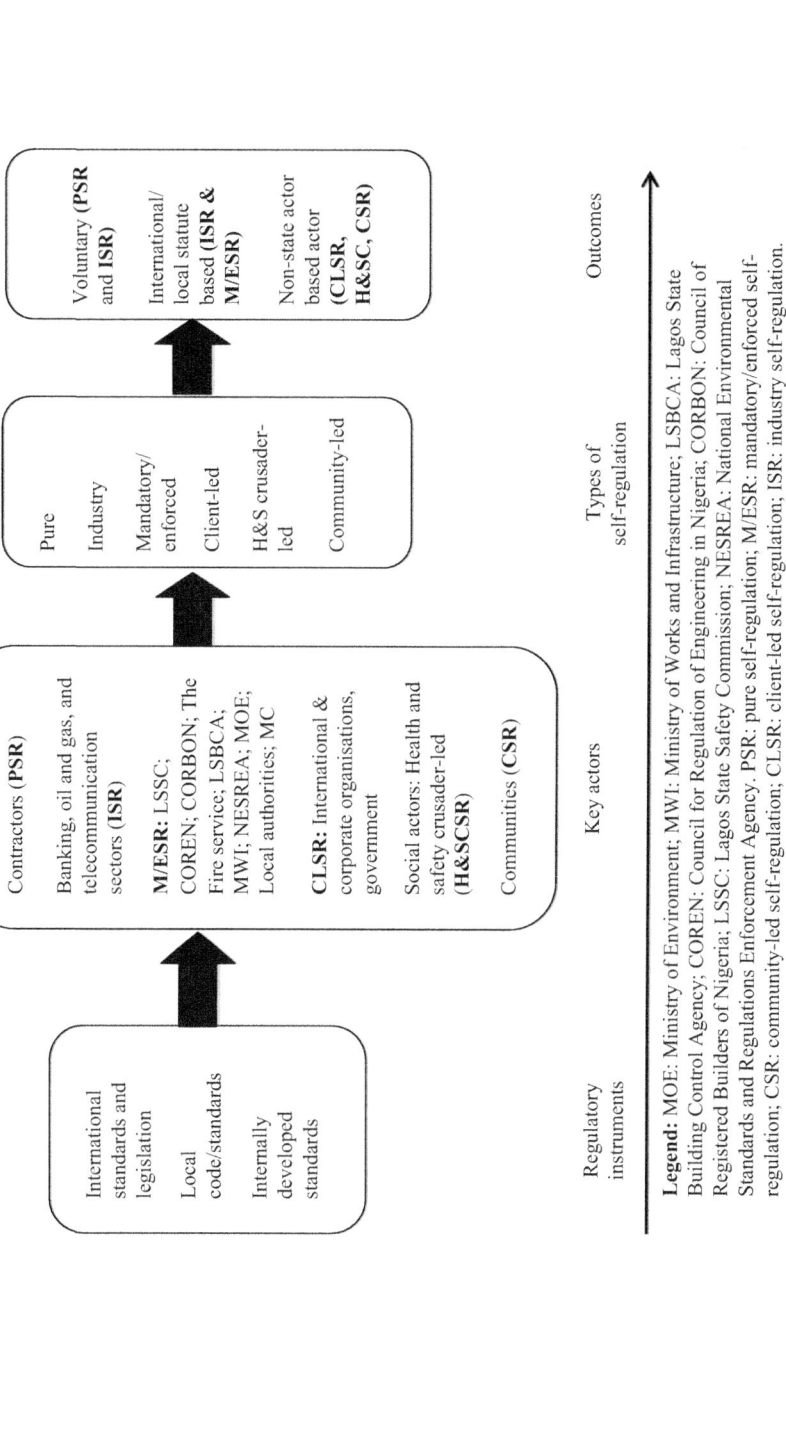

Figure 2.2 Summary of the realities of the structure on CH&S regulation.

How CH&S regulation works in Nigeria: The reality

Pure self-regulation

There is evidence that contractors adopt and administer H&S standards, programmes and legislation without external intervention. These H&S legislation and standards are not limited to fire safety standards, personal protective equipment requirements and CDM Regulations. However, some NCSRCs report what can best be described as 'cherry-picking' in that they only adopt regulatory instruments that are convenient, clear and affordable to them. Also, NCSRCs also note the incomplete implementation of the adopted regulatory instruments.

Mandatory/enforced self-regulation

There are three variations here.

The state encouraged compliance/co-regulation: The evidence here that meets the definition of co-regulation is that according to some CSRCs, NCSRCs and key informants, in Lagos State, the Lagos State Safety Commission Law 2011 CAP 6 oversees all socio-economic activities. This law empowers the Lagos State Safety Commission to enforce it in conjunction with other regulatory authorities such as Lagos State Building Control Agency and Ministry of Works and Infrastructure. Further evidence here relates to the organisations with overlapping responsibilities in the construction industry, for example, the Ministry of Environment, National Environmental Standards and Regulations Enforcement Agency, and local authorities in Lagos State.

Foreign or parent company policy-oriented H&S self-regulation: According to many respondents, many LCs have foreign or parent company policies on H&S that apply to all the parties hence have to adhere to them as the minimum standard in any country including where there are no local H&S legislation.

Chain effect: The contractors that have to comply with foreign or parent company policies get the MCs and SCs to self-regulate as a condition to work on their sites, according to some NCSRCs and few key informants. A few CSRCs expand this—they work 'hand in hand' with the sub-contractors regarding H&S including ensuring that the policies of both parties are interwoven.

Industry self-regulation

Almost all the respondents suggested, demonstrated, mentioned or agreed that industries, such as telecommunication, oil and gas, and banking, set and administer standards for contractors that work for them. There are indications that the regulatory instruments and methodology are homogeneous norms, a fundamental difference with client-led self-regulation where this is not applicable.

Client-led self-regulation

Here, the client sets standards, policies, programmes and then administer them and even sanction many offenders, according to many NCSRCs. The collaboration between the client and contractors also occurs here. However, the level of implementation and thoroughness varies, according to the contractors. There was no evidence of this in MCs, only in domestic contracting firms.

H&S crusader-led self-regulation

A few NCSRCs and key informants claim that social actors enforce standards, policies and laws through ways not limited to consultation, inspection, and naming and shaming organisations with ostensible H&S. However, the extent and scope of their activities are limited.

Community-led H&S self-regulation

Almost half of the respondents, including key informants, evidence that communities are actively involved in H&S, negotiating or insisting that contractors adopt H&S procedures. These are mainly in the oil-producing states and little in the non-oil producing states of Nigeria. The enforcement instruments are not limited to kidnapping workers, reporting to the client, shutting down sites and using diabolic means.

The roles of the actors in CH&S regulation in Nigeria

Following on from Figure 2.2, the roles of the actors are coercive and non-coercive; however, only the contractors are regulated and, to some extent, also regulators. For example, LCs regulate the activities of subcontractors but are also regulated. The MCs are also regulated in that they may have to comply with the standards set by other external parties where applicable. The actors with coercive roles are all the actors in CLSR, CSR, M/ESR and ISR, while those with non-coercive roles are in H&SCSR.

The implications of the structure and actors in CH&S regulation for H&S

Table 2.1 shows the summary of the implications of the current CH&S regulatory regime in Nigeria. Negative implications are emphasised more than the positive implications. The implications also extend to the attitudes of the regulated. The counterproductive regulatory effect is revealing, while the multiple actors in regulation, lack of adequate H&S laws and a low threat of state regulation are expected.

The structure, actors and their roles in CH&S regulation for H&S: The opportunities

Table 2.2 presents the opportunities that the current H&S regulatory regime presents. In the table, there are inter-theme interrelationships. While the flexibility opportunity is expected because it is characteristic of self-regulation, 'appreciation of the contexts' of Nigeria, 'innovation' and 'risk ownership' are revealing, encouraging and attractive.

Table 2.1 Summary of the implications of the structure and actors in CH&S regulation

Theme	Evidence
Biased regulatory and control system	Biased regulatory and control system at the organisational level.
	Bias in regulation and control at state actor level.
Counterproductive regulatory effect	Inconsistent and complex regulatory regimes.
	Non-state actors have little power.
	Counterproductive attitudes: the context determines the level or whether contractors self-regulate, H&S self-regulation is unorganized.
Excessive external intervention	Excessive external involvement and actors.
	Multiple parties in the regulatory process.
Ineffective regulatory regime	Low threat of state regulation.
	State actors: inability to effectively regulate.
	Lack of adequate H&S laws.
	Poor regulation of H&S.
Coercive shared responsibility	The threat of regulation from state and non-state actors.

Table 2.2 Summary of opportunities in the structure, actors and their roles in CH&S regulation

Theme	Evidence
Risk ownership	The risk creators (contractors) control them in their own ways.
	One state, Lagos, takes ownership of the risks and works towards addressing them.
	Communities and H&S crusaders take part in the regulatory process.
Innovation	Alternatives to the traditional 'wait' on the government approach.
	Opportunities to see the impact of other drivers of H&S other than regulatory drivers.
	Contractors can choose to go above standards including being innovative.
Appreciation of context	Appreciation of the contribution of contextual matters, enablers and actors, e.g. communities and culture.
	Opportunities for CH&S regulation in the Nigeria context towards establishing workable strategies.
Flexibility	The regulated are flexible in the regulation of the H&S, which can result in innovation.
	Some MCs and SCs choose to go above standards, which also give them leverage during the procurement process.

Discussion

The demographics of the respondents show that the contractors that continuously engage in H&S are not only the MCs but also indigenous contractors who have some things in common, for example, the scope of operation and geographic location. Consequently, it is naïve to ascribe, explain or assume that adequate H&S compliance is solely based on size or ownership of the contractors, the common stance in literature.

The findings of the study validate the inference in Umeokafor (2016, 2017) that the Nigerian construction industry is self-regulated and not unregulated as authors such as Idoro (2011) argue. Some of the approaches to CH&S self-regulation in Nigeria are consistent with the literature in the 'conceptual foundation' section while few are novel. For example, pure self-regulation is consistent with Bartle and Vass (2005) and mandatory self-regulation is consistent with Gunningham (2011), but community-led self-regulation and H&S crusader-led self-regulation were revealed.

Concerning the implications of the structure and actors in CH&S regulation, while self-regulation to improve H&S is a positive step, the outcomes in the study are both positive and negative (see Table 2.1). A possible explanation for the undesired outcomes is the complex regulatory systems where there are various parties, H&S legislation, methodologies, interests and values (Abubakar 2016).

By implication, it is tempting to conclude that given the poor CH&S record in Nigeria, self-regulation has failed. Conversely, for many reasons, it can be argued that it is developing. For example, drawing on the stages of evolution of safety culture (Pybus 1996), Nigeria is still at the early stage—traditional phase—where attention is on 'rules, discipline and enforcement', 'individual controls' and 'emphasis on the acute effect of injury risk'. The measures here are reactive and countries like Nigeria are considered to be among the lagging regions, while others such as Britain (in the leading region) are in the innovative phase of the safety culture evolution where H&S is integrated into business, and elimination or minimisation of risk with technologies are the priority, having passed through the traditional phase and transitional phase (Finneran and Gibb 2013). Second, HSE (2018) shows that countries such as Britain with centralised H&S regulatory systems (which Abubakar 2016 views as more effective than what obtains in Nigeria) and self-regulated continue to record fatalities, accidents and injuries, suggesting that regulation should not be the sole determinant for H&S. Therefore, for a scientific judgement, using CH&S self-regulation performance indicators may be a better way to test the performance of self-regulation in Nigeria.

Nevertheless, the H&S regulatory system needs significant improvement to align with currency; the opportunities in Table 2.2 offer some optimism for this. The responsible steps of 'risk ownership' and 'innovation' suggest that if the regulatory environment is improved, the involvement and acceptance of the regulated is likely to be high, a recipe for improved regulation. Given the unfavourable regulatory environment for H&S, the self-regulatory efforts of the actors in Figure 2.2 are commendable. Most important, the current regulatory system shows the importance of factoring in the contexts of countries in H&S improvement without which the improvement in developing countries would be farfetched (Kheni 2008).

Implications for practice and research

The implications of the study for developing countries include the need for adequate consideration of their contexts in H&S, appreciation of any non-governmental effort to improve H&S, but most important, providing a conducive regulatory environment for H&S. Furthermore, the need for a regulatory system that significantly involves the regulated and third parties such as industry, where they can take ownership of the risks they create, is emphasised in the study.

Based on the data analysis, the following are recommended:

- Adequate contextualised CH&S legislation that will be mainly goal-based but also (arguably) prescriptive or with detailed guidance.
- A homogeneous regulatory system for H&S with a central regulator for CH&S who will monitor and enforce the laws, advise, educate and motivate the self-regulated. However, the private complementary roles of industries (such as oil and gas) and H&S crusaders are needed. Details, including empirical evidence, are in Umeokafor (2017).
- Further research can empirically test or explore the performance of the self-regulatory approaches to show which is more effective and why.

Conclusions

This chapter reports part of a more extensive study—the lead author's PhD research—aimed at understanding the current realities of CH&S regulation in Nigeria, a poorly understood and unexamined area. The literature indicative of self-regulatory approaches to CH&S are empirically validated, and H&SCSR and CSR are revealed. While this is meritorious, the regulatory system is fragmented with many actors from the public and private sectors including communities. The implications of this include a biased regulatory and control system, counterproductive regulatory effects and excessive external intervention. However, the opportunities include the appreciation of the Nigerian context and the ownership of the risks shown by the contractors and industries where the contractors work, for example, the oil and gas sector. In conclusion, the CH&S regulatory system in Nigeria is emerging and needs significant improvement. The implications of the study for developing countries include the need for appreciating their contexts in H&S.

References

Abubakar, U. (2016) 'Structural and implementation issues around the new Nigerian Labour, Safety, Health and Welfare Bill (2012): Lessons from UK, USA, Australia and China', *Transactions of the VŠB – Technical University of Ostrava, Safety Engineering Series*, Vol. 11(1), pp. 61–71.

Annan, J., Addai, E. K. and Tulashie, S. K. (2015) 'A call for action to improve occupational health and safety in Ghana and a critical look at the existing legal requirement and legislation', *Safety and Health and Work*, Vol. 6(2), pp. 146–150.

Bartle, I. and Vass, P. (2005) 'Self-regulation and the regulatory state — A survey of policy and practice'. Centre for the Study of Regulated Industries, University of Bath. Research Report 17.

Bradley, E. H., Curry, L. A. and Devers, K. J. (2007) 'Qualitative data analysis for health services research: Developing taxonomy, themes, and theory', *Health Service Research*, Vol. 42(4), pp. 1758–1772.

Braun, V. and Clarke, V. (2006) 'Using thematic analysis in psychology', *Qualitative Research in Psychology*, Vol. 3(2), pp. 77–101.

Castillo-Montoya, M. (2016) 'Preparing for interview research: The interview protocol refinement framework', *The Qualitative Report*, Vol. 21(5), pp. 811–831.

Castro, D. (2011) 'Benefits and limitations of industry self-regulation for online behavioral advertising', Washington, DC, The Information Technology & Innovation Foundation. Retrieved on 10-07-2018 from www.itif.org/files/2011-self-regulation-online-behavioral-advertising.pdf.

Dabup, N. L. (2012) 'Health, safety and environmental implications in Nigeria's oil and gas industry,' Philosophiae doctor thesis, Nelson Mandela Metropolitan University, South Africa. Retrieved on 12-02-17 from www.masterbuilders.co.za/resources/docs/OHS-PDFs/Final%20HSE%20Implications%20040613%20in%20Nigerias%20Oil%20and%20Gas%20Industry%20PhD%20Constr%20Man%20Thesis.pdf.

Famuyiwa, F. (2011) 'An investigation into the legal infrastructure framework for safety and compensation policies on site accidents', In: *Proceedings CIB W099 International Health and Safety Conference*, 24–26 August 2011, Washington, DC.

Federal Republic of Nigeria (2004a) Factories Act, CAP F1, LFN 2004. Federal Government Press, Abuja, Nigeria.

Federal Republic of Nigeria (2004b) Engineers (Registration etc.) Act, CAP E11 2004. Government Press, Abuja, Nigeria.

Finneran, A. and Gibb, A. (2013) 'W099: Safety and Health in construction: Research Roadmap report for consultation'. Retrieved on 05-08-2018 from https://dspace.lboro.ac.uk/dspace-jspui/bitstream/2134/12523/3/pub%20376.pdf.

Gunningham, N. (2011) 'Investigation of industry self-regulation in workplace health and safety in New Zealand'. Gunningham & Associates Pty Ltd. Retrieved on 16-07-18 from http://regnet.anu.edu.au/sites/default/files/publications/attachments/2015-04/NG_investigation-industry-self-regulation-whss-nz_0.pdf.

Gunningham, N. and Rees, J. (1997) 'Industry self-regulation: An institutional perspective', *Law & Policy*, Vol. 19(4), pp. 363–414.

Health and Safety Executive (HSE) (2015) 'Managing health and safety in construction: Construction design and management'. Retrieved on 06-07-18 from www.hse.gov.uk/pubns/priced/l153.pdf.

Health and Safety Executive (HSE) (2018) Workplace fatal injuries in Great Britain 2018. Retrieved on 10-07-18 from http://www.hse.gov.uk/statistics/pdf/fatalinjuries.pdf.

Idoro, G. I. (2011) 'Comparing occupational health and safety (OHS) management efforts and performance of Nigerian construction contractors', *Journal of Construction in Developing Countries*, Vol. 16(2), pp. 151–173.

Kheni, N. A. (2008) 'Impact of health and safety management on safety performance of small and medium-sized construction business in Ghana', Doctoral thesis, Loughborough University, UK. Retrieved on 03-08-18 from https://dspace.lboro.ac.uk/dspace-jspui/bitstream/2134/4079/3/Kheni%27s%20PhD.pdf

Omeife, C. A. and Windapo, O. A. (2013) 'The impact of national building code on professionalism', In: *Proceedings 43rd Builders Conference/AGM*, 26–30 2013, Abuja, Nigeria, pp. 1–12.

Oxford Dictionary (2018) Oxford University Press.

Pybus, R. (1996) *Safety Management: Strategy and Practice*. Oxford: Butterworth Heinemann.

Saunders, M., Lewis, P. and Thornhill, A. (2009) *Research Methods for Business Students* (5th edition), London: Prentice-Hall.

Umeokafor, N. I. (2016) 'Approaches, drivers and motivators to health and safety self-regulation in the Nigerian construction industry: A scoping study', *Architectural Engineering and Design*, Vol. 12(6), pp. 460–475.

Umeokafor, N. I. (2017) 'Realities of construction health and safety regulation in Nigeria', Doctoral thesis, University of Greenwich, UK.

Walls, C. B. and Dryson, E. W. (2002) 'Failure after 5 years of self-regulation: A health and safety audit of New Zealand engineering companies carrying out welding', *Occupational Medicine*, Vol. 52(6), pp. 305–309.

Windapo, A. O., Umeokafor, N. I. and Olatunji, A. (2018) 'Self-regulation to occupational health and safety legislative requirements by contractors in South Africa', In: Saurin, T. A., Costa, D. B., Behm, M. and Emuze, F. (Eds.), *Proceedings Joint CIBW99 and TG59 Conference*, 1–3 August 2018, Salvador, pp. 187–196.

Zohrabi, M. (2013) 'Mixed method research: Instruments, validity, reliability and reporting findings', *Theory and Practice in Language Studies*, Vol. 3(2), pp. 254–262.

3 Contractors' compliance with occupational health and safety legislation in South Africa
The benefits of self-regulation

Abimbola Windapo, Oluwole Olatunji and Nnedinma Umeokafor

Summary

Occupational health and safety (OHS) laws are not always abreast of their times in developing countries. Thus, complying with OHS laws might be considered to be pedantic and superficial by contractors. In addition to meeting legislative requirements, evidence suggests South African contractors also self-regulate, and this further affects their health and safety performance beyond the remit of legislative guidelines. However, what do a commitment to self-regulation and the transition between self-regulation and compliance with OHS regulations entail in a typical construction company in South Africa? In this study the various levels of self-regulation and compliance to OHS legislative requirements in South Africa have been examined and how these affect the number of accidents on construction sites. The aim of this chapter is to answer the research question using a 20-item scale to develop a conceptual framework that helps to explain the relationship between contractors' commitment to a work-safety culture, self-regulation and accident frequency rates (AFRs). From the study, it was found that there is a high level of self-regulation ranging from 65% to 97%, and an average AFR of 1.02 accidents per 100,000 hours in South Africa. It also emerged that there is a significant, negative, linear relationship between the level of contractor self-regulation and AFR. It is concluded that the more contractors self-regulate, the lower their AFR. It is recommended that public and private sector clients encourage the use of voluntary self-regulation towards strengthening contracting organisations' ability to prevent accidents on construction sites.

Factors affecting adherence to occupational health and safety

The aim of self-regulation is to implement ways that help to lower the number of accidents on construction sites. According to Gunningham (2011), self-regulation provides a means whereby the government prescribes an outcome but does not outline how and what the industry must do to achieve that particular outcome. In the case of the construction industry, this outcome is to enhance health and safety, and decrease the number of accidents on site. An examination of occupational health and safety (OHS) in the South African construction industry reveals that there are two primary acts. The first is the Occupational Health and Safety Act No. 85 of 1993 (OHSA), and the second is the complementary Compensation for Occupational Injuries and Diseases Act No. 130 of 1993 (COID Act). The OHSA was implemented by the Department of Labour (DoL) and superseded the Machinery and Occupational Safety Amendment Act No. 40 of 1989, the Machinery and Occupational Safety Amendment Act No. 40 of 1989 and Act No. 97 of 1991 in an attempt to give more importance to health and safety in the construction industry.

The South African construction industry does not lack health and safety regulations and legislation. However, the number of fatalities, accidents and injuries in the industry is still on the rise (FEM, 2014). According to the Federated Employers' Mutual Assurance Company (FEM) (2014), by December 2014, there had been a total of 2,797 accidents in South Africa. The Construction Industry Development Board reports that the construction industry has the third highest frequency of fatalities per 100,000 workers (cidb, 2009). The main issues that contribute to poor OHS compliance and high accident frequency rate on site are, first, the poor mindset of contractors regarding a safety culture (Windapo, 2013), second, a lack of enforcement of safety regulations in practice (cidb, 2009) and, third, the impact of the human element on OHS (Hamid et al., 2008).

Windapo (2013) addresses the first issue relating to OHS, which is the mindset of the leaders of construction companies who tend to favour financial gain rather than a safety culture. Geminiani et al. (2008) and cidb (2009) address the second issue, which is the lack of enforcement of the health and safety regulations in practice. The OHS Inspectorate, positioned within the DoL in South Africa, is responsible for the enforcement of OHSA and, according to the cidb (2009), the department conducts site visits/blitz inspections infrequently and ineffectively. The effect of the human element is the third issue considered as affecting the poor record of health and safety among contractors in the construction industry (Hamid et al., 2008). The human factor theory, as explained by Goetsch (2009), emphasises that accidents are caused by human error because of the inappropriate response of workers, negligence, lack of experience and poor attitude towards training. To bring about a change in the mindset of construction company leaders regarding safety culture and to promote compliance to OHS in practice, there needs to be a paradigm shift to instil health and safety culture into the mentality of the management of contracting firms. In this study it is argued that self-regulation to OHSA legislative requirements provides one of the ways through which the problem of accidents on construction sites in South Africa can be addressed.

Studies conducted in the US and New Zealand by Gunningham (2011) and Scharrer (2011), respectively, demonstrate that the self-regulatory approach to OHS has resulted in a reduction in fatalities/accidents on construction sites. Windapo et al. (2018) established that there is a high level of self-regulation amongst South African construction organisations even though these companies operate under poor incentive regimes. One view about enforcement of OHS legislative requirements might be that contractors are not only complying with regulations because of their legal obligations but are in fact self-regulating. However, there is limited research into the use of self-regulation as an approach to compliance with OHS legislative requirements in South Africa and a lack of knowledge about whether self-regulation impacts on the number of accidents towards improving health and safety (H&S) performance on construction sites. Since there is a dearth of research on the impact of self-regulation by contractors on performance in South Africa, the arguments put forward by researchers, such as Gunningham (2011) and Scharrer (2011), that self-regulation of OHS results in a reduction in fatalities/accidents on construction sites remains purely theoretical in the South African context. The study reported in this chapter examines the various levels of self-regulation and compliance to OHS legislative requirements in South Africa and how these affect the number of accidents on construction sites. Specifically, the purpose of the chapter is to answer the question, What do a commitment to self-regulation and the transition between self-regulation and compliance with OHS regulations entail in a typical construction company in South Africa and how does this behaviour affect performance with regards to the number of accidents?

The concept of self-regulation, and occupational health and safety

Levels of self-regulation

'Direct' or command-and-control regulation, which is based on the state setting and enforcing standards with punitive measures, has not performed to expectation, hence the advent of self-regulation (Aalders & Wilthagen, 1997). Unlike command-and-control regulation, which is fixed, self-regulation is flexible and is a spectrum (Windapo et al., 2018) with pure self-regulation at one end and command-and-control on the other end (Sinclair, 1997). Within this spectrum are various levels or categories of self-regulation including co-regulation, mandatory self-regulation and pure or voluntary self-regulation. Self-regulation varies from country to country and industry to industry and so can be categorised into various levels. Rees (1988) argues that an entity's commitment to self-regulation can be assessed at three main points covered later. According to Gunningham (2011), in 'voluntary self-regulation', there is no external involvement in rule-making and enforcing, all are within the organisation or the industry; in the 'mandated full self-regulation', rule-making and enforcement are privatised. While both are privatised, the government formally sanctions the regulatory programme in the mandated full self-regulation but not the voluntary self-regulation. This external involvement is for monitoring the effectiveness of the regulatory system towards any possible modifications. The third level is the 'mandated partial self-regulation', where either rule-making or enforcement is privatised. According to Korosec (1990), the options here are 'public enforcement of rules written privately' and 'internal enforcement of rules written privately as mandated or moderated by government'.

Self-regulation practices

In the current study, the self-regulation practices at the organisational level are based on those identified in Levinson (1984), namely, H&S policy, H&S plan, OHS management system, H&S training and personal protective equipment (PPE). Health and safety policy is a written statement of the principles and goals representing an organisation's commitment to maintaining a safe and healthy workplace. The individuals responsible for certain actions and the details of how to achieve the aim of the policy are included. An OHS management system or OHS programme is a part of an extensive organisational management system used to establish OHS policies of an organisation and to manage OHS risks (International Standards Organisation, 2018). Most OHS management systems are based on the Plan, Do, Check, Act Model. The education and awareness aspect is the OHS training which, according to Robson et al. (2012), is the planned efforts to facilitate the learning of competences that are specific to OHS.

Overview of the OHS regulatory framework in South Africa

OHS in the South African construction industry is overseen by two legislative acts. First, the OHSA No. 85 of 1993 oversees the protection from hazards and the health and safety of persons at work and of persons other than persons at work. The second is the COID Act, which covers compensation for accidents and diseases relating to health and safety. Key secondary legislation from the OHSA Act is the Construction Regulations (CR) of 2014, introduced because of the poor H&S statistics in the construction industry. The CR recognises and allocates specific responsibilities to construction stakeholders such as clients and designers. For example, clients or project owners are required to include H&S specifications in tender documents and to ensure that the principal contractor makes the right allowance for H&S. The enforcement of the OSHA is the responsibility of the OHS Inspectorate of the DoL, but it performs below expectations (cidb, 2009). This is where the frequency and efficacy of site visits and blitz inspections by DoL are questionable.

The relationship between levels of self-regulation and OHS safety performance

Extant literature (Levinson, 1984; Gunningham & Rees, 1997; Castro, 2011; Scharrer, 2011; Gunningham, 2011) has shown that self-regulation has numerous benefits over the traditional state-imposed regulation. These benefits include speed, flexibility and lower costs. Rees (1988) established that industry self-regulation led to a significant drop in accident rates on a nuclear plant project on which about 4500 workers were employed.

The conceptual framework of the study is presented in Figure 3.1. The study is based on the concept that the aggregate levels of self-regulation adopted by a construction company on site are related to accident rates. The conceptual

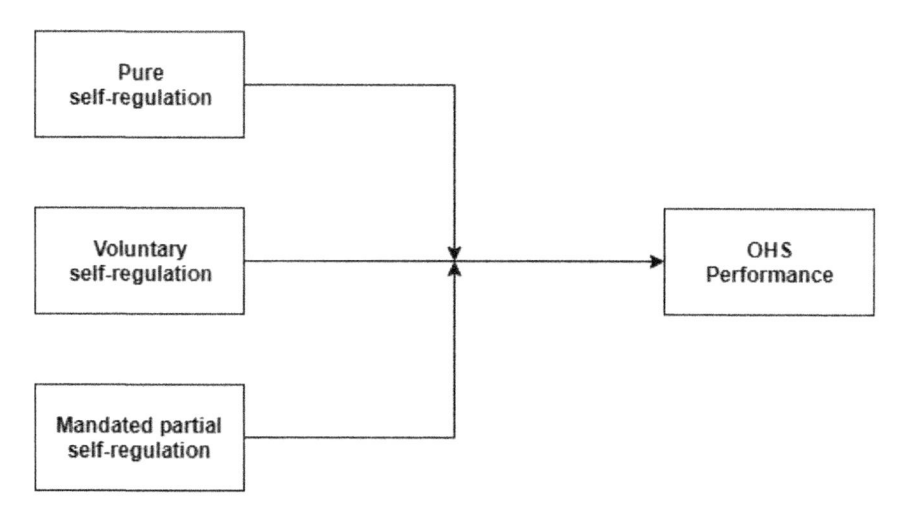

Figure 3.1 Conceptual framework of the study.

framework is based on earlier studies by Levinson (1984), Gunningham and Rees (1997), Castro (2011), Scharrer (2011) and Gunningham (2011). There is limited research in South Africa on whether self-regulation affects the number of accidents on construction sites and therefore stimulates improved H&S performance.

Figure 3.1 shows that self-regulation (comprising pure self, voluntary and mandated partial) does affect the number of accidents on construction sites.

Methodology

Since the research questions meet the criteria for the pragmatic research paradigm, a mixed-methods research approach (involving interviews and surveys) was used to collect data. Here, the research questions determined the epistemological, ontological and axiological position of the research (Saunders et al., 2009). An organisational level of regulation is a requirement that all organisations listed on the Professionals and Project Register must meet; hence, the scope of the research is limited to this.

Following an extensive literature review, the data collection instrument was developed. The 20 questions, as shown in Figure 3.2, were based on the self-regulation practices in Levinson (1984), namely, H&S policy, plan, management system, training and PPE use. While there were questions on the profile of the respondents and their organisations, the 20 questions were centred on enquiring about the organisation's commitment to creating and maintaining safe working conditions, an established health and safety plan and safety awareness initiatives. Having developed the questionnaire, experts vetted and pre-tested it objectively,

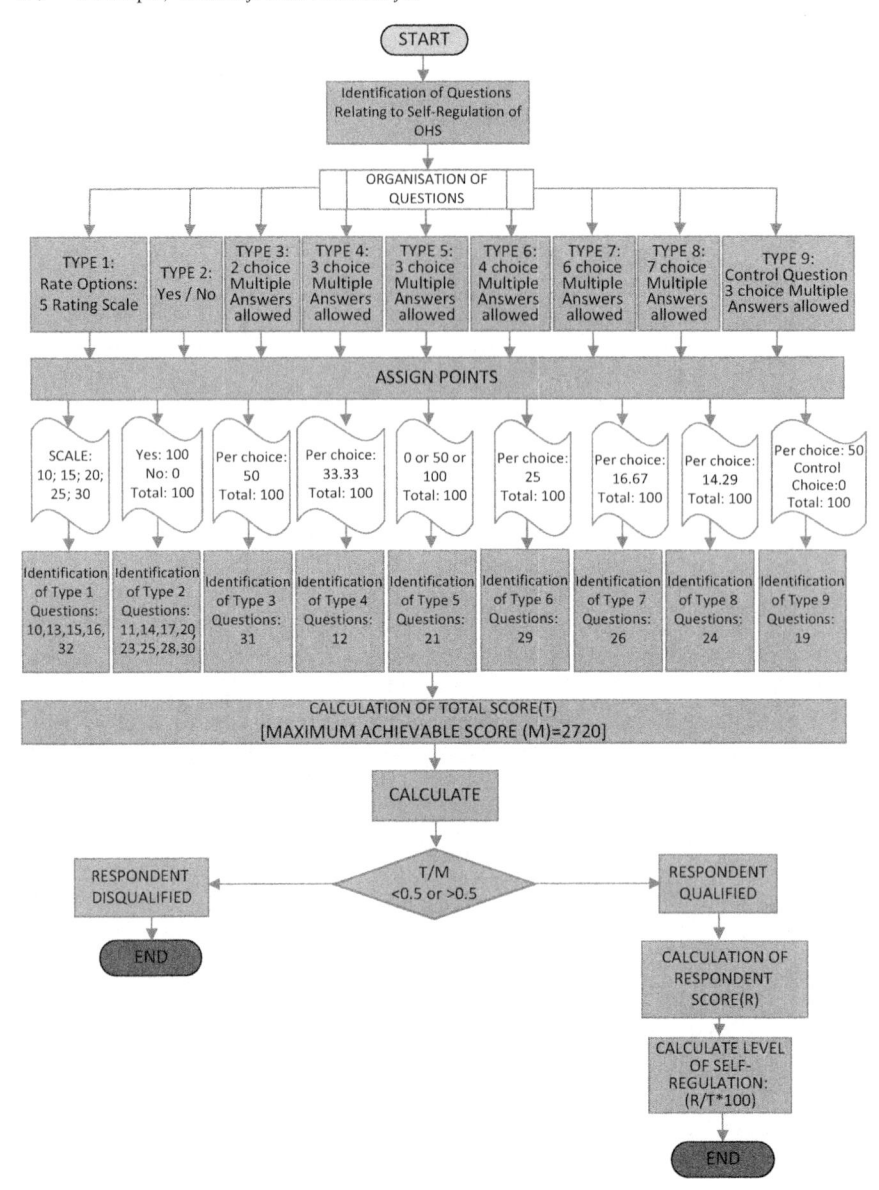

Figure 3.2 Self-regulation analytical framework.
Source: *Windapo et al. (2018)*.

paying attention to its relevance to the subject of self-regulation, amongst others, and its coverage of the entire topic for the purpose of internal validity. Further, a control question, although not related to self-regulation, was used to test the respondent's consistency in answering the questionnaire. The external validity was based on comparing the results of the study with previous research.

Using a probability sampling technique, the sample frame was obtained resulting in a sample size of 617 organisations. The population ($N = 1234$) included all organisations listed in the Professionals and Project Register 2014. For the survey, questionnaires were sent to 617 organisations to determine the level of self-regulation by construction contractors in South Africa as they strive to meet OHS legislative requirements. Three construction organisations were selected for follow-up interviews based on their willingness to participate in the sessions.

An analytical framework, detailed in Windapo et al. (2018), was developed to compute data relating to levels of self-regulation. The quantitative data had to be analysed such that a level of self-regulation for each respondent could be determined. A particular sequence of processes outlined in Windapo et al. (2018) was considered in determining levels of self-regulation by South African construction contractors towards meeting their considered OHS in-house objectives and meeting operational requirements mandated by the government (see Figure 3.2).

The accident rates used as a performance measure were obtained from the contractors surveyed. The question posed allowed the respondents to give the accident frequency rates (AFR) on an identified project per 100,000 hours. Pearson's rho correlation coefficient analysis was used in determining whether there was a significant relationship between the level of self-regulation by the South African construction companies surveyed and AFR. When Pearson's r is close to one, it indicates that there is a strong relationship between the level of self-regulation by construction companies and AFR, and a weak relationship when r is close to zero. A negative Pearson's r indicates an inverse relationship between the variables (Fellows & Liu, 2015).

Results

Having distributed questionnaires online, 59 were returned and used, giving a response rate of 9.72%. The low response rate obtained was probably a result of the sensitive nature of the information requested on H&S practice and performance. This cast doubts on the generalisability of the results to the study population. This limitation was compensated by the use of interviews. The majority of the respondents (19) worked for Grade 7 contractors listed in the General Building and Civil Engineering categories but all respondents were employed by contractors listed in Grades 4 to 9 of the cidb Register of Contractors in South Africa. The respondents were well-experienced in the construction industry and educated. For example, 78% had over 10 years of experience in the construction industry and at least 47% had a bachelor's or higher level of academic qualification. While it can be argued that this would have skewed the data collected to represent experienced persons only, it should be noted that, while the level of experience varies, the experienced respondents 'worked' their way up. By implication, they once had a lower level of experience, on which they were likely to draw. The profile of the respondents showed the potential of the respondents to provide valuable, relevant and meaningful information that was useful for this study and that the demographics of the population were represented.

Level of self-regulation

Figure 3.3 shows the graphical presentation of respondents and their corresponding level of self-regulation to OHS requirements, derived from the analysis of questions relating to self-regulation using the framework detailed in Windapo et al. (2018). The study showed a mean level of self-regulation of 80.35% and a standard deviation of 7%. There was a high level of self-regulation among the respondents, ranging from 65% to 97%. This suggests a very high level of self-regulation for the responding companies.

Accidents frequency rates (accidents per 100,000 hours)

The AFR are presented in Figure 3.4. The AFR ranged from 0 to 6.94 accidents per 100,000 hours, while an average of 1.02 accidents per 100,000 hours was obtained for the 23 respondents that provided information. Six respondents reported zero accidents, which might suggest misreporting by the management of organisations.

The relationship between the level of self-regulation and accident frequency rates on construction sites

The purpose of the study was to identify the type of relationship between the two variables – the level of self-regulation to OHS regulations and AFR. The type of relationship between the two variables was established by analysing the data presented in Figures 3.3 and 3.4, using Pearson's rho correlation analysis. A graphical presentation of the relationship is presented in Figure 3.3, while the correlation statistics are presented in Table 3.1.

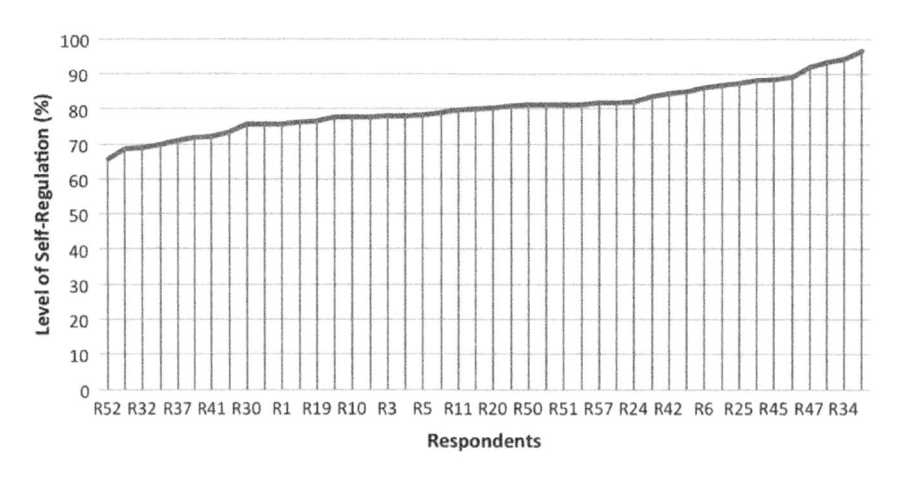

Figure 3.3 Level of self-regulation of the respondents.
Source: *Field Survey*.

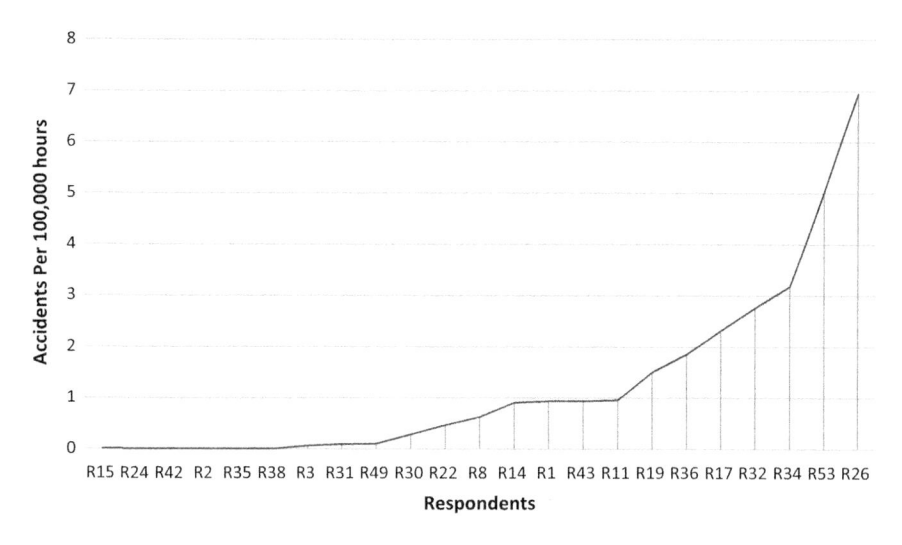

Figure 3.4 Accident frequency rates (accidents per 100,000 hours).
Source: *Field Survey*.

Table 3.1 Pearson correlation statistics (level of self-regulation and AFR on construction sites)

		Level of self-regulation
AFR	Pearson's correlation	−0.405*
	Sig. (1-tailed)	0.027
	N	23

* Correlation is significant at the 0.05 level (1-tailed).

Figure 3.5 and Table 3.1 show that there is a negative, linear relationship between the level of self-regulation and AFR. The Pearson's correlation coefficient is −0.405 (see Table 3.1) which indicates a negative correlation. It can be deduced from these findings that the higher the level of self-regulation of a contractor, the lower the AFR and that the correlation, though below average, is statistically significant at the 0.05 level (1-tailed).

Interview results

The three semi-structured interviews did not only triangulate the results of the survey but also offered new insight into the subject. The purpose of the study sought to ascertain the views of the respondents concerning the concept of self-regulation in South Africa, the level of adoption of self-regulation by construction organisations, and whether there was a relationship between the level of

[Scatter plot showing Accidents per 100,000 hours (AFR) on y-axis (−1 to 8) versus Level of Self Regulation on x-axis (0.65 to 1), with trend line equation y = −10.096x + 9.4928]

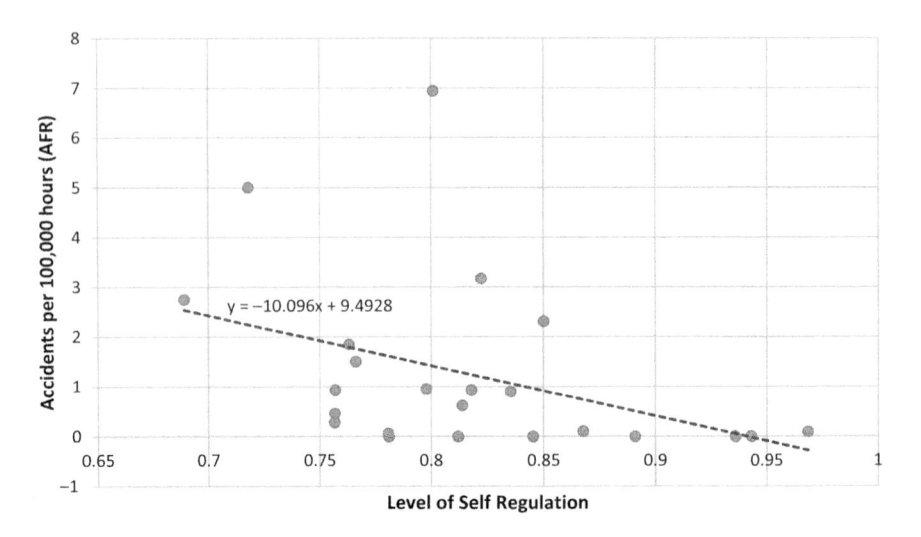

Figure 3.5 Scatter plot diagram between level of self-regulation and accident frequency rate (AFR).

Source: *Field Survey*.

self-regulation and AFR on construction projects. The interview respondents were well experienced in construction and OHS. In particular, Interviewee 1 was a professional project manager with over 20 years of experience in construction, including heading the OHS and tender management unit of a prominent contracting firm in the Western Cape. A medium-sized building construction firm in Johannesburg was represented by Interviewee 2, an H&S officer with more than 5 years of experience in the construction and building industry. Last, the third interviewee, a registered quantity surveyor and an acting chief executive officer of a construction firm that specialises in civil and mining works, had more than 18 years of experience in managing various sizes of multi-disciplinary engineering projects from inception to client handover. The implication of the various scopes of experience was that one respondent presents more than one point of experience. In other words, the points of experience are more than three. Patton (1990) describes respondents such as these as 'information-rich cases' because they provide information-rich data.

While Interviewees 1 and 2 agreed that self-regulation at the organisational level is critical for OHS improvement, Interviewee 1 was of the view that a prerequisite for organisational level self-regulation was dedicated training and a better appreciation of the purpose of OHS legislative requirements. However, Interviewee 2 stressed the obligation for H&S in that the contractors have a duty to ensure that their workers comply with H&S procedures. By implication, this interviewee suggested that H&S of workers should be non-negotiable. Interviewee 3 added an interesting perspective about the size advantage of larger contractors. The respondent noted that small contractors have limited resources compared with the large contractors who are able to establish, for example, an H&S administrative unit. Interviewees 1 and 2 held the view that the higher the level of self-regulation by construction companies, the lower the AFR would be. Interviewee 1 commented further that the level of self-regulation was supported by construction institutions such as the Master Builders Association (MBA), South African Federation of Civil Engineering Contractors (SAFCEC), Occupational Hygiene Safety and Associated Professionals (OHSAP), and South African Institute of Occupational Safety and Health (SAIOSH) who provide assistance to safety managers to ensure compliance.

When questioned about whether the level of self-regulation could be used as a panacea to accidents on construction sites, Interviewee 1 said it should be but only 'with suitable training and planning'. Likewise, Interviewee 2 contended that self-regulation would not eliminate accidents absolutely, but that it is a leading factor that drives stakeholders towards good H&S practices. Interviewee 3 recommended outsourcing H&S to specialised organisations that are more knowledgeable about health and safety as a way of reducing accidents, injuries and fatalities on construction sites.

Discussion

The cross-sectional survey and interview results were found to be similar when compared with each other. The evidence of the high level of self-regulation to OHS requirements within contracting firms in South Africa, with a mean score of 80.36% and a standard deviation of 7.10%, encourages efforts to improve OHS. The interviews indicated that higher-grade contractors self-regulate better than other forms of construction organisations. The quest to save cost by these contractors is a possible explanation for the high level of self-regulation. In particular, Windapo (2013: 78) found that 'perceived cost savings are unrelated to the degree of risk, which the regulation is trying to prevent'. OHS regulation remains the foundation on which OHS improvement strategies are based, without which the strategies are ineffective (Finneran & Gibb, 2013). Many findings of the current study were consistent with previous studies. For example, while it validates the self-regulation practices, identified by Levinson (1984), such as OHS policy, it is consistent with the findings of Umeokafor (2016). In examining the extent of various types of OHS self-regulation in construction in Nigeria, including pure, industry and enforced self-regulation, Umeokafor (2016) indicated that Nigerian contractors most often adopted pure HS self-regulation, followed by enforced self-regulation. However, the OHS self-regulatory regime in Nigeria differs from that of South Africa in many ways, for example, there are local OHS laws for the construction industry in South Africa but no local ones for the Nigerian construction industry at the time of the study by Umeokafor (2016).

It is interesting that the interviews indicated that some contractors view OHS as a duty in keeping with the findings of Umeokafor (2016), where viewing OHS as a duty is a key driver of OHS self-regulation in Nigeria. The views of the interviewee on the imperativeness of training on OHS for self-regulation were also consistent with the findings of Umeokafor (2017). Umeokafor (2017) found that the lack of awareness of OHS is a major barrier to OHS self-regulation. A finding of this study of the inability of smaller contractors to self-regulate to the same extent as the large ones was consistent with the literature, for example, Finneran and Gibb (2013). It is tempting to agree with the respondents who recommend outsourcing H&S to experts. However, this is likely to undermine efforts towards integrating OHS into the management system of organisations. There is also a risk of OHS being a stand-alone system instead of it being integrated into the management systems of organisations or projects.

The study established that there is a significant relationship between self-regulation and AFR, which was confirmed by both the quantitative and qualitative research approaches. The interview respondents were of the view that higher self-regulation by contractors will result in lower AFR on construction sites and this was confirmed by the survey results and the extant literature (Castro, 2011; Scharrer, 2011; Gunningham, 2011; Gunningham & Rees, 1997; Levinson, 1984) where it has been shown that self-regulation has numerous benefits over the traditional state-imposed regulation. Furthermore, it emerged that OHS managers oversee the self-regulation role in construction companies and take responsibility for AFR, aligning with Levinson's (1984) view that self-regulation can be achieved through internal monitoring.

Conclusions and recommendations

This research examined the level of self-regulation to OHS requirements among contractors in South Africa and whether this is related to AFR on construction sites. It emerged that the level of self-regulation to OH&S requirements by South African contractors is high. Also, it was found that the average AFR on construction sites is 1.02 accidents per 100,000 and that there is a statistically significant, negative relationship between the level of self-regulation by construction companies to OHS legislative requirements and AFR in South Africa. Based on these findings, it was concluded that construction companies in South Africa are not indifferent to H&S regulations, suggesting that the more contractors self-regulate, the lower the AFR.

Despite the high level of self-regulation found in the study, the emphasis on training and planning being imperative for adequate self-regulation, in line with the literature, the regulator of OHS and the government should increase the emphasis on training and creating awareness of OHS. The government should support and inspire contractors to self-regulate through measures such as tax incentives and giving preference to contractors that self-regulate in terms of OHS in the pre-qualification of contractors on public projects. The limited sample for the qualitative aspect of the study is acknowledged as a limitation, and thus the results should be viewed as being indicative. Also, the reliability of the accident statistics is open to debate because of confidentiality, reputation, lack of accident record enforcement and the resource-intensive nature of reporting construction accidents. Through a comparative study, further research can improve and validate the analytical framework as a tool for measuring self-regulation. Furthermore, a comparative study can also be used to investigate the extent of leading and lagging H&S performance indicators and how this happens.

References

Aalders, M. & Wilthagen, T. (1997), Moving beyond command-and control: Reflectivity in the regulation of occupational safety and health and the environment, *Law and Policy*, Vol. 19(4), pp. 415–443.

Castro, D. (2011), Benefits and limitations of industry self-regulation for online behavioral advertising. Washington, DC: The Information Technology & Innovation Foundation. [Online]. Available at: www.itif.org/files/2011-self-regulation-online-behavioral-advertising.pdf [accessed 10 July 2018].

cidb (2009), Construction health and safety in South Africa. [Online]. Available at: www.cidb.org.za/publications/Documents/Construction%20Health%20and%20Safety%20in%20South%20Africa.pdf [accessed 7 December 2018].

Fellows, R. & Liu, A. (2015), *Research Methods for Construction* (4th ed.). West Sussex: Wiley Blackwell.

FEM (2014), The Federated Employers' Mutual Assurance Company accident statistics 2014. [Online]. Available at: www.fema.co.za [accessed 17 June 2018].

Finneran, A. & Gibb, A. (2013), W099: Safety and health in construction: Research roadmap report for consultation. [Online]. Available at: https://dspace.lboro.ac.uk/dspace-jspui/bitstream/2134/12523/3/pub%20376.pdf [accessed 5 August 2018].

Geminiani, F. L., Smallwood, J. J. & van Wyk, J. J. (2008), The effectiveness of the occupational health and safety (OH&S) inspectorate in South African construction. In: Dainty, A. (Ed.), *Proceedings of the 24th Annual ARCOM Conference*, 1–3 September 2008, Cardiff, UK, Association of Researchers in Construction Management, pp. 1113–1121.

Goetsch, D. L. (2009), *Construction Safety and the OSHA Standards*. Upper Saddle River, NJ: Prentice Hall

Gunningham, N. (2011), Investigation of industry self-regulation in workplace health and safety in New Zealand, *Journal of Safety Research*, Vol. 31(13), pp. 120–143.

Gunningham, N. & Rees, J. (1997), Industry self-regulation: An institutional perspective, *Law and Policy*, Vol. 19(4), pp. 363–414.

Hamid, A. R., Majid, M. Z. & Singh, B. (2008), Causes of accidents on construction sites, *Malaysian Journal of Civil Engineering*, Vol. 20(2), pp. 242–259.

International Standards Organisation (2018), ISO 45001: Occupational safety and health. [Online]. Available at: www.iso.org/iso-45001-occupational-health-and-safety.html [accessed 7 December 2018].

Korosec, R. P. (1990), Reforming the workplace: A study of self-regulation in occupational safety, *American Political Science Review*, Vol. 84(4), pp. 1406–1407.

Levinson, A. (1984), Self-regulation of health and safety in a local authority with particular reference to safety representatives, supervisors and safety committees. PhD thesis, Aston University, UK.

Patton, M. (1990), *Qualitative Evaluation and Research Methods*. Beverley Hills, CA: Sage.

Rees, J. V. (1988), *Reforming the Workplace: A Study of Self-Regulation in Occupational Safety*. Philadelphia: University of Pennsylvania Press.

Robson, L. S., Stephenson, C. M., Schulte, P. A., Amick, B. C. I., Irvin, E. L., Eggerth, D. E. & Grubb, P. L. (2012), A systematic review of the effectiveness of occupational health and safety training, *Scandinavian Journal of Work, Environment and Health*, Vol. 38(3), pp. 193–208.

Saunders, M., Lewis, P. & Thornhill, A. (2009), *Research Methods for Business Students* (5th ed.). London: Prentice-Hall.

Scharrer, A. (2011), Command vs self-regulation in construction safety: A case study of CHASE. MSc thesis, Rensselaer Polytechnic Institute, New Mexico.

Sinclair, D. (1997), Self-regulation versus command and control? Beyond false dichotomies, *Law and Policy*, Vol. 19(4), pp. 529–559.

Umeokafor, N. I. (2016), Approaches, drivers and motivators to health and safety self-regulation in the Nigerian construction industry: A scoping study, *Architectural Engineering and Design*, Vol. 12(6), pp. 460–475.

Umeokafor, N. I. (2017), Barriers to construction health and safety self-regulation: A scoping case of Nigeria, *The Civil Engineering Dimension (Dimensi Teknik Sipil)*, Vol. 19(1), pp. 44–53.

Windapo, A. (2013), Relationship between degree of risk, cost and level of compliance to occupational health and safety regulations in construction, *Construction Economics and Building*, Vol. 13(2), pp. 67–82.

Windapo, A. O., Umeokafor, N. I. & Olatunji, A. (2018), Self-regulation to occupational health and safety legislative requirements by contractors in South Africa. In: Saurin, T. A., Costa, D. B., Behm, M. & Emuze, F. (Eds.), *Proceedings Joint CIBW99 and TG59 Conference*, 1–3 August 2018, Salvador, pp. 187–196.

4 A narrative review of occupational safety and health legislation in Pakistan

Abdul Qayoom and Bonaventura H. W. Hadikusumo

Summary

Safety is one of the key success factors besides time, cost and quality in a project. Loss of time, cost and quality causes delays, cost overrun and loss in quality, but failure to manage safety causes an immeasurable loss of human life. The other losses can be overcome but the loss of human life can never be compensated. It is the basic right of every individual to get safe working place and according to International Labour Organization (ILO) Convention C155, it is the duty of the employer to provide a reasonably practicable safe working environment. The main objective of this chapter is to review and analyse the occupational safety and health legislations in Pakistan. In order to achieve this, a narrative review of research articles, annual reports, and health and safety legislation in developing countries is carried out, and safety legislation in Pakistan is analysed to find the shortfalls. Inadequate legislative structure, outdated and fragmented laws, and lack of database management were found to be some of key findings. Based on the findings, recommendations are suggested which include legislative reforms, capacity building trainings, upgrading the reporting system and involving trade unions.

Introduction

Occupational safety and health (OSH) is one of the topics of interest for developing countries these days (Ncube and Kanda, 2018). OSH has been recognized as an integral part of sustainable development and considered as basic human right. Due to increasingly changing social, economic and political environment and advancement in technology, the OSH scope is evolving continuously and gradually. There have been incessant efforts by international labour organizations (International Labour Office, 2006) and trade unions to pinpoint the importance of OSH, but besides these untiring efforts, International Labour Organization (ILO) statistics for 2016 show that approximately 2.7 million work-related fatalities and 374 million work-related accidents occur worldwide each year (ILO, 2016). The rates are even higher in developing and underdeveloped countries (Raheem and Hinze, 2012).

Pakistan is a low-middle-income country with a population of 207.7 million. It is the sixth most populous country in the world with a growth rate of 2.4%.

Principally, agriculture is the most contributing sector besides industries, mining, oil and gas and construction (Pakistan Bureau of Statistics, 2014). There is no independent OSH law governing all the sectors of Pakistan. The main law which governs the issues of OSH and labour rights is the Factories Act of 1934. Under this act, the provincial governments have devised their own rules and regulations (Pasha, 2003). Although a signatory member of ILO, Pakistan has not yet ratified 62 conventions including the most important like C148 (working environment), C155 (occupational safety and health), C160 (labour statistics), C167 (safety and health in construction) and C187 (promotional framework for occupational safety and health at work place) (ILO NORMLEX, 2018b).

The perception of occupational risk and hazard is global but the mitigation strategies to reduce/eliminate the risk as low as reasonably practicable varies from situation to situation depending on the geographical location of the country. As far as Pakistan is considered, the main reasons of safety non-conformance include absence of separate OSH laws, lack of enforcement actions on existing laws, non-appearance of awareness regarding OSH, need for automation in industrial sector and scarcity of professional management practices (Ahmed, 2013).

The rationale behind this study is to highlight the importance of OSH legislation, create awareness on OSH legislative structure, feature the inadequacies of OSH laws in developing countries (South Asia) and present an overview of OSH laws of Pakistan.

A four-stage model for continuous improvement as proposed by Deming (Duijm et al., 2008) is applied to improve the quality of occupational safety and health legislation. Figure 4.1 shows the Plan–Do–Check–Act (PDCA) Cycle, also known as the Deming Cycle, applied to safety legislation. The planning stage includes the formulation of acts, regulations, codes of practice and guidelines as

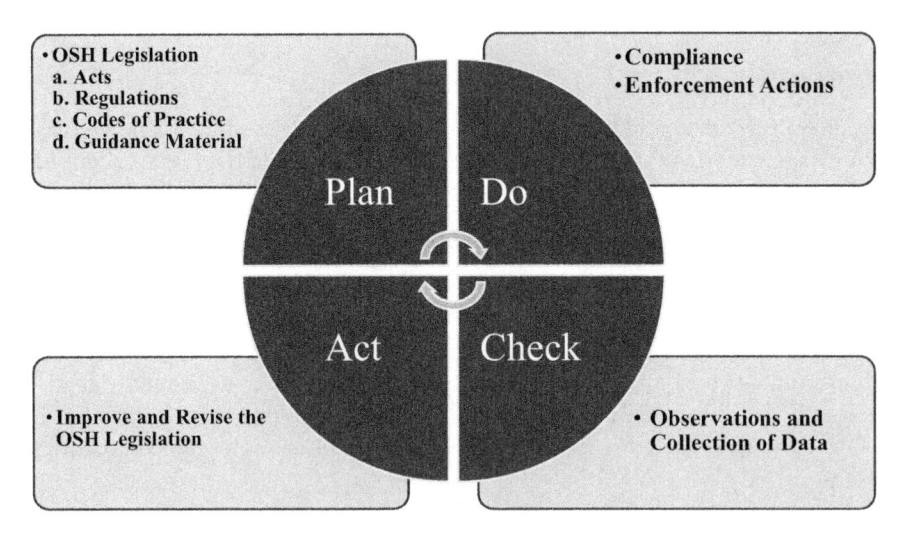

Figure 4.1 PDCA cycle to improve OSH legislation.

approved by the concerned authorities. The second stage involves the enforcement of legislation by creating awareness or through a reward/penalty system. The third stage deals with observations, collection and analysis of data. The fourth and final stage provides room for improvement and directions/recommendations for the concerned authorities to formulate new legislations.

Legislation structure (Plan)

Legislation can include national, state or provincial legislation, or conventions and recommendations of the ILO (ILO, 1996). It is the responsibility of each government to provide minimum standards of OSH at the workplace. These minimum standards can however be more protective with stronger legislation. Besides government, many trade unions have contributed to make legislation stronger and pressured the government to enforce the OSH laws (ILO, 1996). Even today, trade unions have to fight for stronger and protective OSH laws to ensure safety at the workplace. Although OSH laws provide the legal backbone to protect workers, they lose their importance if there is no enforcement action. Most employers do not take workplace regulations seriously, since many governments do not prosecute employers that violate workplace OSH regulations.

The structure and terminology used for legislation varies in different countries. However, there are some common terms used internationally. Some of these terms include act, regulation, code of practice and guidance material (ILO, 1996). Figure 4.1 shows the hierarchy of these terms.

1. *Act*: An act is basically a legal statement covering general occupational health and safety principles and responsibilities (ILO, 1996). It is approved by the government or parliament of the country. Most countries have acts regarding occupational health and safety; for example, the Health and Safety at Work Act (HASWA) and Factories Act of the United Kingdom. Acts are wholly supported by law, therefore they have great deal of power, but they are only effective if there is compliance and adequate enforcement action.
2. *Regulations*: Regulations is a second tier in the hierarchy of legal importance to comply with. After an act is passed by the parliament or government, the objectives and obligatory minimum standards to achieve a safe working environment are developed by the concerned ministry or provincial government. Regulations are more specific to industry and cannot be legally stronger than the act which they accompany (ILO, 1996).
3. *Code of practice*: After the mandatory minimum standards and objectives stated in the regulations part, the third tier in the hierarchy covers the detailed guidance to the employer and employee to achieve a safe workplace. Usually, the ministry of labour adopts and amends the codes. The International Labour Organization has also developed codes of practice that are followed as guidelines by many developing countries (ILO, 1996). Although, a code of practice has no legal power, it can be used as evidence to support regulations and acts.

4. *Guidance material*: Guidance material or guides are the elaborative notes with detailed technical information and recommendations for the employer and employee to comply with OSH laws and regulations. Like codes of practice, OSH guidelines do not have legal power, but they serve as parameters to abstain from penalties. Every country has developed guidance material for each specific industry. For countries that do not have an OSH legislative structure, the ILO has developed recommendations, codes of practice and guidance material through a tripartite system (ILO, 1996).

Enforcement (Do)

After the planning stage of the Plan–Do–Check–Act Cycle, the next step is to ensure that the employers and employees comply with the OSH regulations. The OSH regulations need legal enforcement by the government to be effective. Ideally, a satisfactory enforcement demands sufficiently trained personnel, equipped with all necessary requirements and access to the sources of information to inspect workplaces and enforce law. These personnel are known as health and safety inspectors or factory inspectors authorized by the Ministry of Labour to inspect, monitor and enforce the OSH laws in workplace. The safety inspector makes sure that employers comply with the minimum legal health and safety standards set by the competent authority. However, feeble, ineffective and outdated safety legislations limit the authority of the safety inspector resulting in no action or little action to improve the workplace conditions.

Observations and data recording (Check)

Habitually, it's human nature to learn from past experiences. The observations and data stored in our subconscious minds enable us to analyse and make decisions. Similarly, for maintaining good OSH conditions, the recording and measuring of the data is as equally important as planning. The things which cannot be measured cannot be improved. Most of developed countries have well established regulations to report accidents. For example, the UK has the Reporting of Injuries, Diseases and Dangerous Occurrences Regulations (RIDDOR) 2013 (Health and Safety Executive, 2013). Historical statistics set a benchmark for improvement. ILO Convention C160 (Labour Statistics Convention of 1985) bounds the signatory countries to collect, compile and publish the basic labour statistics including occupational injuries and as far as possible occupational diseases (ILO, 1985).

Improve and revise the OSH legislation (Act)

No matter how perfect and ideal the situation, there is always room for improvement. In modern days, the world is changing so fast due to advancement in technology and complexity of problems. The solutions of yesterday cannot solve the problems of tomorrow. Similarly, OSH regulations need to be updated and improved to address the current issues. Some countries have very strong OSH legislation, while others have outdated and weak legislation that mean employees cannot rely on legislation for their satisfactory protection. Since in many developing countries, occupational safety is of less concern, the legislations are outdated and irrelevant to current situations. For example, in Pakistan the law that governs occupational health and safety is Chapter 3 of the Factories Act 1934. Section 22 of the Factories Act 1934 addresses the welfare facilities "provision of Spittoons". Subsection 4 states "Whosoever spits in contravention of subsection (3) shall be punishable with a fine not exceeding two rupees" ('Pakistan: The Factories Act, 1934', 1997). A two rupees fine, which is equivalent to approximately 1 cent (USD), for spitting anywhere in the factory makes fun of the law.

Research methodology

A narrative review of different primary studies conducted on occupational safety and health legislation in developing countries was carried out. Figure 4.2 shows the literature sources of the study. At the beginning, previous studies with key words including safety legislation, developing countries and safety laws were compiled, and a narrative review was carried out to understand the importance of safety legislation. A total of 30 research articles on developing countries, 5 annual reports on health and safety legislations of Pakistan, and 11 labour laws were included. Additionally, the ILO database assisted in creating the country profile of Pakistan, and getting a copy of Pakistan labour force standards, and occupational safety and health legislations of Pakistan. Annual reports from the Pakistan Bureau of Statistics, and Ministry of Overseas Pakistanis and Human

Research Articles
- Occupational Health & Safety Laws
- Developing Countries
- Safety Laws in Pakistan

ILO Database
- Country Profile Asia Pacific Region
- NORMLEX: Information System on International Labour Standards
- LEGOSH: Global Database on Occupational Safety & Health Legislation

Government of Pakistan (Annual Reports)
- Ministry of Overseas Pakistanis and Human Resource Development
- Labour & Human Rights Department
- Pakistan Bureau of Statistics

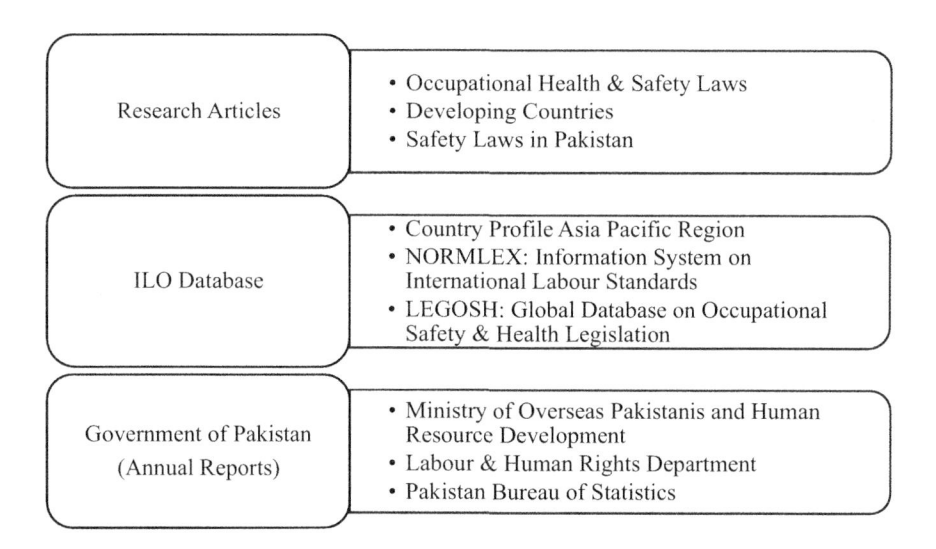

Figure 4.2 Literature sources.

Resource Development (OPHRD) supplemented the big picture of the current status of occupational safety and health in Pakistan and current applicable laws. After a comprehensive review of major legislations pertaining to OSH, a general scrutiny was carried out to check the timeliness, adequacy and applicability of laws to major sectors of the Pakistan economy.

Pakistan labour force statistics

The Islamic Republic of Pakistan is the sixth most populated country of the world with populace of approximately 200 million. The projected labour force is 61 million (30.5%) including males and females of age 10 years and older. About 57.42 million (94.13%) of the labour force is employed, while 3.62 million (5.86%) are unemployed (International Labour Standards Unit, 2015). Since Pakistan has fertile lands, most of the labour force works in the agriculture/forestry sector. Manufacturing, wholesale and trade, and construction industries are other major sources of employment. Table 4.1 shows the labour force statistics by sectoral share in employment, gross domestic product (GDP) and total injuries.

Although the agriculture sector holds the major employment share in Pakistan, contributing 20.88% to the GDP and 47.95% of the total injuries/disease (Pakistan Bureau of Statistics, 2014), there is no law related to occupational health and safety of workers. Additionally, Pakistan has not yet ratified the priority instrument C129 (Labour Inspection Agriculture Convention of 1969) (International Labour Organization, 1985). The construction industry holds 7.31% share in employment, ranking fourth among major sectors, it has relatively double (1:2.23) the share in total injuries/diseases than the agriculture sector (1:1.13). It is the unsafest sector in terms of occupational injuries/disease followed

Table 4.1 Pakistan labour force statistics

Sector	Sectoral share in employment (%)	Sectoral share in GDP (%)	Sectoral share in total injuries/diseases (%)
Agriculture/forestry and fishing	42.28	20.88	47.95
Mining and quarrying	0.16	2.92	0.27
Manufacturing	15.34	13.27	15.90
Electricity and gas	0.40	1.67	0.32
Construction	7.31	2.44	16.27
Wholesale and trade	14.64	18.26	7.07
Transport, storage, and information and communication	5.4	13.36	7.57
Financial and insurance	0.59	3.14	0.09
Others	13.89	24.06	4.55

Source: Pakistan Bureau of Statistics, 2014, Labour force statistics, available at: www.pbs.gov.pk/content/labour-force-statistics.

by mining and quarrying, and unpredictably the transport, storage, and information and communication sector (International Labour Standards Unit, 2015). There is no independent OSH law for the construction sector and Pakistan has not ratified Convention C167 (Safety and health in Construction Convention, 1988) (ILO NORMLEX, 2018b). Article 6 of C167 states, "Measures shall be taken to ensure that there is co-operation between employers and workers, in accordance with arrangements to be defined by national laws or regulations, in order to promote safety and health at construction sites" (ILO, 1988). One of the reasons attention has not been paid to occupational health and safety in construction might be its relatively low contribution to the GDP of the country.

Accident statistics in Pakistan

According to ILO estimates of 1999, about 250 million occupational accidents and over 1 million work-related fatalities occur annually worldwide (Pasha, 2003). Hundreds of millions of employees are exposed to hazardous substances and suffer from occupational diseases. The situation is graver in developing countries like Pakistan (Zahoor et al., 2015; Ncube and Kanda, 2018) because of the lack of education, illiteracy, lack of reliable information and sources, inadequate medical facilities and absence of OSH laws and enforcement action.

Section 33-N (Notice of certain accidents) of the Factories Act 1934 of Pakistan bounds the manager of a factory to send notice of any nature of accident to the concerned authority within the prescribed time on the prescribed form. For failure to notify, the manager must face a penalty according to Section 64 (penalty for failure to give notice of the accident), which states, "A manager of a factory who fails to give notice of an accident as required under section 33-N shall be punishable with fine which may extend to five hundred rupees" ('The Pakistan: The Factories Act 1934', 1997). However, no significance to this legislation has been observed because of lack of enforcement action and obsolete laws. The provincial directorates of the Labour Welfare and Human Resource Department are responsible for executing OSH laws and maintaining the database for accident statistics. Table 4.2 shows the major sector-wise injuries reported from year 2010 to 2016.

Table 4.2 Sector-wise injury statistics

Sector	2010	2011	2012	2013	2014	2015	2016
Construction	31	19	15	0	0	12	6
Services	87	55	10	03	4	27	36
Manufacturing	87	38	563	16	59	232	70
Mines	0	09	0	0	20	24	34

Source: International Labour Standards Unit, 2015, 'Occupational safety & health: Legal framework & statistical trend analysis (2010–2015), Ministry of Overseas Pakistanis and Human Resource Development, available at: http://ophrd.gov.pk/.

But the real circumstances are different from what is shown. For certain reasons, fatal accidents or otherwise are not often reported. The accident statistics may vary due to underreporting. The ILO has estimated 7400 fatal accidents in Pakistan as against 72 reported in year 2002 (Pasha, 2003).

OSH legislation in Pakistan

The constitution of the Islamic Republic of Pakistan 1973 gives assurances of safe and caring working conditions at the workplace. Article 37(e) of the constitution states, "The State shall make provision for securing just and humane conditions of work, ensuring that children and women are not employed in vocations unsuited to their age or sex, and for maternity benefits for women in employment". Article 38(a–d) further elaborates the assurance of a just and safe working place (National Assembly of Pakistan, 2012). Occupational safety and health in the workplace in Pakistan is covered by different laws. There is no single comprehensive law that applies to all sectors. The OPHRD is instructed with "coordination of labour laws" and to "keep a watch on labour legislation from international perspective" (International Labour Standards Unit, 2015). The list of legislations covering various aspects of the occupational safety and health perspective is given in Table 4.3.

Besides these, there are various acts and ordinances that cover the occupational safety and health of workers (ILO NORMLEX, 2018). Some of them are

1. West Pakistan Hazardous Occupation Rule, 1963
2. Workmen Compensation Act, 1961
3. Provincial Employees Social Security (Occupational Diseases) Regulations, 1967
4. Employment of Children Act, 1991
5. Pakistan Public Works Department Contractor's Labour Regulation
6. Fatal Accidents Act, 1855
7. Electricity Act, 1910
8. Explosives Act, 1884
9. The Pakistan Plant Quarantine Act, 1976

All these acts are obsolete these days and were devised before the existence of Pakistan (1947) by British rule, except some that were revised and implemented during the industrial revolution in Pakistan (1960–1980). There are no further rules and regulations, codes of practice, or guidelines for the employer or employee to help them to comply with these acts.

Pakistan's Labour Policy 2002 was revised in 2010. Occupational safety and health is given high priority and there is the recommendation to establish the National Tripartite Council for Occupational Safety & Health that will guide legal, judicial and administrative actions, and develop OSH standards and review them periodically. The current health and safety laws are specific to sectors and segregated into different legislations and ordinances. On the recommendation of

Table 4.3 Labour legislation in Pakistan

Act	Section	Aspects on safety	Focus
The Factories Act 1934 (as amended to 1997) Punjab Factories Rule 1978 Sindh Factories Rule 1975 KPK Factories Rule 1975 Baluchistan Factories Rules, 1978	Chapter III	Health and safety	• Working conditions • Welfare facilities • Fire, first aid and emergency • Machine/equipment • Crane, lift, hoist, pressure plant • Ergonomics • Chemicals and dangerous gases and fumes, explosives • Incident reporting • Infectious diseases
	Chapter VI	Penalties and procedure	• Obstructing instructor • Failure to give notice of commencement • Failure to give notice of accidents • Using false certificates • Duties and responsibilities of owners, agents and managers
The Mines Act 1923	Chapter IV	Mining operation and management of mines	
	Chapter V	Provisions as to health and safety	• Welfare facilities • First aid and medical facilities • Accident reporting • Accident investigation
	Chapter VII	Regulations, rules and bye-laws	• Power of government to make rules and regulations
	Chapter VIII	Penalties and procedures	• Obstruction • Falsification of records • Accident reporting • Prosecution of owner, agent or manager

(Continued)

Table 4.3 Continued

Act	Section	Aspects on safety	Focus
Pakistan Environmental Protection Act (PEPA) 1997		Hazardous Substances Rule 2003	• Environmental impact assessment (EIA) • Packing and labelling of hazardous substances • Conditions of premises • General safety precautions • Safety precautions for workers • Validity, renewal and cancellation of license • Safety plan • Safety monitoring • Accident reporting • Waste management • Import, export and transport guidelines • List of hazardous substances
Dock Labourers Act 1934	Section 5	Power to federal government to make regulations	• Workplace conditions • Machinery/equipment safety • Emergency procedures • First aid
	Section 9	Penalties	• Person responsible • Obstructing inspector

Source: ILO NATLEX, 2018, Occupational safety and health laws in Pakistan, available at: www.ilo.org/dyn/natlex/natlex4.listResults?p_lang=en&p_country=PAK&p_classification=14.

new labour policy, it is proposed to consolidate the current laws into the following six laws (Government of Pakistan, 2010):

1. Industrial Relations Ordinance
2. Conditions of Employment Ordinance
3. Wages Ordinance
4. Human Resource Development Ordinance
5. Occupational Safety & Health Ordinance
6. Labour Welfare and Social Protection Ordinance

Inadequacies in OSH legislations and enforcement actions

Ideally, a legislation should be adequate enough to protect workers and compel employers to comply with not only minimum standards (reasonably practicable) of occupational safety and health but to prevent occupational injuries and diseases. The legislation should have provisions for an adequate number of occupational safety and health inspectors who are well trained, well equipped and well authorized. In addition, the legislation should hold strong enforcement action and solid penalties for employers who do not follow the regulations.

An ILO module on occupational health and safety 'Legislation and Enforcement' has identified more than a few limitations of health and safety legislation through tripartite discussions with trade unions of many developing countries (ILO, 1996). Some of the common limitations include the following:

- The existing legislation is obsolete and does not talk about modern economic, communal and technological challenges regarding occupational health and safety in the region. For example, the safety legislation of most of commonwealth countries like India and Pakistan is derived from the British Factories Act 1961, which does not fully address the needs of these countries today.
- According to Chapter 1, Section 2h (definition of worker) of the Factories Act, a worker is defined as any person working in a factory excluding the person solely employed to do clerical work. This definition leaves many employees unprotected because they do not fit into the narrow category of workers. Also, workers in the public sector and those who are self-employed are not usually covered by present legislation.
- The penalties ordained by the law on the employer in case of breaching the legislation or not complying with the rules and regulations are very low as compared to the damage caused by serious offences.
- There is scarce number of resources (man, material and machinery) for inspection and enforcement of law. For example, according to a report by the International Labour Standards Unit of the OPHRD, there are only 14 OSH inspectors for 249 registered factories in Pakistan. It is difficult for them to visit and inspect each factory. Most of them either visit only the biggest factories or factories with known occupational safety and health problems.

Also, most inspectors are not provided with appropriate equipment, so they just inspect the workplace virtually.
- Most of the prevailing legislations cover only occupational safety, leaving occupational health as a needless item. The reason might be that most of the health effects are chronic rather than acute.
- Although obsolete acts exist, there are no detailed regulations, codes of practice or guidelines pertaining to those acts.
- There is solemn deficiency of awareness of OSH legislation among employers and workers.
- Workers, employees and trade and labour unions are not involved in the process of standard setting in any way. The unreasonable and extraneous standards are forced upon employers and employees, which results in noncompliance.
- Many standards are ambiguous from the workers' point of view and might very readily be misinterpreted, so cannot be used to defend the law.

Shortcomings of OSH legislation of Pakistan

As mentioned earlier, while reviewing the legislation of Pakistan, it was found that there is no general OSH law that applies to every sector of Pakistan. Since Pakistan was a commonwealth country, most laws are inherited from British rule. And the history reveals that the main point of convergence was industrialization. That is the reason that most of OSH laws are bits and pieces of the Factories Act 1961. The Pakistan Factories Act 1934 is specific to workers in factories and is not applicable to factories employing fewer than 10 workers. After systematically scanning the legislations of Pakistan downloaded from NATLEX (the ILO database of national labour, social security and related human rights legislation), LEGOSH (the ILO global database on occupational safety and health legislation) and the OPHRD, the following general deficiencies were pinpointed:

- Lack of comprehensible, systematic and lucid law with general appositeness to all sectors of the economy
- Absence of regulations, codes of practice and guidelines for further elaboration of the act to help employers to understand how to comply with laws
- Obsolete and outdated laws that do not put the spotlight on current scenarios
- Absence of institutional framework and structured organogram at the national and provincial level
- Unavailability of reliable data on occupational injuries, fatal accidents and diseases
- Paucity of awareness of OSH legislation among employers and employees
- Incoherence among the industrial sector and the academic institutions
- Redefining the scope of OSH laws and covering microindustries, SMEs (small and medium enterprises) and self-employed domestic workers
- Feeble and powerless enforcement authorities, and insubstantial and inefficient enforcement mechanisms

To keep national aims in allegiance with the ideologies of international labour organization standards, the Government of Pakistan has implemented labour and workplace protection laws. Pakistan has endorsed 36 ILO conventions out of 189, including all 8 fundamental conventions enclosed in the Charter of Worker's Rights, 2 out of 4 governance conventions, and 26 out of 177 technical conventions. Thirty-one conventions are in force out of 36 ratified (ILO NORMLEX, 2018b). Perceiving the Pakistan economy and the current occupational safety and health situation of major sectors contributing to the economy, the Government of Pakistan needs to ratify certain conventions, which include:

- C129 — Labour Inspection (Agriculture) Convention, 1969
- C77 and C78 — Medical Examination of Young Persons (Industry/Non-industry Occupations) Convention, 1946
- C139 — Occupational Cancer Convention, 1974
- C141 — Rural Workers' Organizations Convention, 1975
- C150 — Labour Administration Convention, 1978
- C155 — Occupational Safety and Health Convention, 1981
- C160 — Labour Statistics Convention, 1985
- C167 — Safety and Health in Construction Convention, 1988
- C174 — Prevention of Major Industrial Accidents Convention, 1993
- C184 — Safety and Health in Agriculture Convention, 2001
- C187 — Promotional Framework for Occupational Safety and Health Convention, 2006

Conclusion

Bringing up the rear arguments and reviews, it can be concluded that safety legislation is one of the most neglected aspects in developing countries. Most developing countries do not have an adequate legislative structure, enforcement/inspecting authorities and lack database management for accident statistics. Pakistan is no different. Most labour laws of Pakistan are outdated, fragmented and disincentive. There is no organizational structure for managing occupational safety and health. Although, the Labour Policy introduced in 2002 advocated the formation of a solo OSH Tripartite Council, there has been no implementation until now. Pakistan was a signatory member of ILO since its independence (1947), but it has not yet ratified 62 conventions, some of which are in extreme need. The chapter presents an overview of the OSH legislations in Pakistan, and passable and actionable recommendations that can be adopted to achieve the expedient improvements in the legislative system.

Recommendations

Based on the findings of the study, the recommendations are broadly categorized into upgrading legislation, capacity building through training and education, reporting mechanism of incidents, and involvement of trade unions.

Upgrading legislation and enforcement actions

Although the OSH legislation does exist in Pakistan in bits and pieces as part of different acts and ordinances, most of them are antiquated and irrelevant to the current situation. There is a strong need for standalone occupational safety and health legislation addressing all sectors of the economy. As discussed earlier, the construction and agriculture sectors are the most dangerous sectors of the economy together contributing 64% of accidents/injuries and occupational diseases. There is dire demand to establish OSH laws, regulations, codes of practice and guidelines specific to these sectors. The Government of Pakistan should immediately ratify Conventions C167 and C184, and develop comprehensive guidelines to show employers how to comply with law.

Globally, the increasing rate of unemployment and growth in modern technology has allowed the expansion of several start-up businesses on a small scale. The current legislations fail to encompass these small-sized enterprises. There should be a separate law for microbusiness SMEs, and self-employed and home workers.

Any legislation, until it is advocated by professionals and technical implementing agencies, remains impotent. Monitoring, evaluating and a penalty/incentive reward is the back bone of legislative enforcement. There is a demand to create separate inspection legislation and an autonomous body (separate from the labour department) to watch over the enforcement of OSH laws.

Capacity building through training and education

As the domino theory states, unsafe acts and unsafe conditions are the main causes of accidents. Workplace conditions can be improved by implementing OSH laws, but promoting safe worker behaviour, education and training also play a vital role. In addition to technical workers, non-technical staff members such as supervisors, managers and decision makers should also be bound to take basic training related to occupational safety and health. This is only viable when there is enough legislation for support. For example, in Thailand, it is the obligation of the employer to impart occupational safety and health training to each employee in companies of 100 or more workers. Although the National Institute of Labour Administration Training (NILAT) in Karachi, The Centre for the Improvement of Working Conditions & Environment (CIWC&E) in Lahore, and a few other training centres are delivering occupational safety training, enforcement by legislation should make it obligatory for each employer to train its workers. Another possibility can be setting up semi-government training centres by involving the private sector, such as the National Institute of Occupational Safety and Health (NIOSH) in Malaysia, the Korea Occupational Safety & Health Agency (KOSHA), and Japan Industrial Safety & Health Association (JISHA).

Incident reporting system

Although there is formal paper-based incident reporting system in Pakistan and as per Section 33-N of the Factories Act 1934, a manager is responsible to give notice of certain accidents to the concerned provincial authority. Failure to give such notice can cause penalty to the manager, which may exceed up to Rs 500 PKR.

After 2000, there has been technological revolution around the globe and Pakistan also has benefitted in certain fields like software development and big data management. The use of mobile applications to report incidents can make it easier and more approachable to employers and employees. Further, there is prerequisite to change the approach from penalty to reward. The legislation or law-enforcing authorities can provide incentives on the reporting of accidents. For example, a worker reporting an accident on a site would get a reward that may be in the form of mobile phone credit top-up. This approach has been successful in Pakistan to increase the number of survey respondents in many studies.

Trade union involvement

Due to legal penalties and fear of damage to the reputation of the organization, the top management tries to hide occupational safety and health discrepancies. In that case, it is best to use labour unions and trade unions as the first line of defence against poor working conditions and reporting of incidents. Also, in developing countries there is no structure of safety and health, and OSH is not considered as career to excel in, thus there is a lack of passably trained OSH inspectors. Trade unions can fill the gap with, for example, the establishment of health and safety committees. Ideally these should be joint labour/management committees, but if the employer resists participating, then workers should set up their own committees. Another benefit of a safety committee is to break the traditional top–down approach of establishment of OSH laws. Usually, the upper house or senate approves the legislation and there is no hearsay of OSH experts/trade unions in the development of legislation; this creates the gap between design and enforcement. In that case, trade unions/safety committees can codify reasonably practicable OSH laws and pass them bottom up for the approval from the senate/parliament. This would also overcome the hindrance of legislative enforcement.

References

Ahmed, S. M. (2013) 'A strategic framework to improve construction safety practices in Pakistan', *International Conference on Safety, Construction, Engineering and Project Management (ICSCEPM 2013) "Issues, Challenges and Opportunities in Developing Countries"*, 19–21 August, pp. 9–17. Available at: http://goo.gl/r0FPHP.

Duijm, N. J., Fiévez, C., Gerbec, M., Hauptmanns, U., and Konstandinidoum, M. (2008) 'Management of health, safety and environment in process industry', *Safety Science* 46(6), pp. 908–920.

Government of Pakistan (2010) Labour Policy 2010. Available at: www.ilo.org/dyn/travail/docs/995/GovernmentofPakistanLabourPolicy2010.pdf.

Health and Safety Executive (2013) RIDDOR – Reporting of Injuries, Diseases and Dangerous Occurences Regulations 2013. Available at: www.hse.gov.uk/riddor/index.htm.

ILO (1985) Conventions and recommendations. International Labour Organization. Available at: www.ilo.org/global/standards/introduction-to-international-labour-standards/conventions-and-recommendations/lang--en/index.htm.

ILO (1988) C167 – Safety and Health in Construction Convention, 1988 (No. 167). International Labour Organization. Available at: www.ilo.org/dyn/normlex/en/f?p=NORMLEXPUB:12100:0::NO:12100:P12100_INSTRUMENT_ID:312312:NO.

ILO (1996) Your health and safety at work (series). International Labour Organization. Available at: www.ilo.org/safework/info/instr/WCMS_113080/lang--en/index.htm.

ILO (2016) Statistics and databases. International Labour Organization. Available at: www.ilo.org/global/statistics-and-databases/lang--en/index.htm.

ILO NATLEX (2018) Occupational safety and health laws in Pakistan. Available at: www.ilo.org/dyn/natlex/natlex4.listResults?p_lang=en&p_country=PAK&p_classification=14 (accessed 21 August 2018).

ILO NORMLEX (2018) Ratifications for Pakistan. International Labour Organization. Available at: www.ilo.org/dyn/normlex/en/f?p=NORMLEXPUB:11200:0::NO::P11200_INSTRUMENT_SORT,P11200_COUNTRY_ID:2,103166#Occupational_safety_and_health (accessed 20 August 2018).

International Labour Office (2006) 'Promotional framework for occupational safety and health International Labour Office Geneva', *International Labour Conference, 95th Session*, (1), pp. 1–8. Available at: www.ilo.org/public/english/standards/relm/ilc/ilc95/pdf/rep-iv-1.pdf.

International Labour Standards Unit (2015). 'Occupational safety & health: Legal framework & statistical trend analysis (2010–2015). Ministry of Overseas Pakistanis and Human Resource Development. Available at: http://ophrd.gov.pk/.

National Assembly of Pakistan (2012) The Constitution of the Islamic Republic of Pakistan. Available at: http://na.gov.pk/uploads/documents/1333523681_951.pdf.

Ncube, F. and Kanda, A. (2018) 'Current status and the future of occupational safety and health legislation in low- and middle-income countries', *Safety and Health at Work* 9(4), pp. 365–371.

Pakistan Bureau of Statistics (2014) Labour force statistics. Available at: www.pbs.gov.pk/content/labour-force-statistics.

Pasha, T. S. (2003) Country profile on occupational safety and health in Pakistan. Centre for the Improvement of Working Conditions and Environment.

Raheem, A. and Hinze, J. (2012) 'Injury/fatality data collection needs for developing countries', In: *Third International Conference on Construction in Developing Countries (ICCIDC–III) "Advancing Civil, Architectural and Construction Engineering & Management"* 4–6 July, 2012, Bangkok, Thailand, pp. 308–313.

'Pakistan: The Factories Act, 1934' (1997). Available at: www.ilo.org/dyn/natlex/docs/WEBTEXT/35384/64903/E97PAK01.htm.

Zahoor, H., Chan, A. P. C., Utama, W. P., and Gao, R. (2015) 'A research framework for investigating the relationship between safety climate and safety performance in the construction of multi-storey buildings in Pakistan', *Procedia Engineering* 18, pp. 581–589.

5 Safety law, system and regulation influencing the construction sector in India

Sri Datta Duddukuru and Bonaventura H. W. Hadikusumo

Summary

In India most construction firms ignore safety rules and regulations before start of any work, and parallelly enforcement from the government side is lacking. So, to overcome this issue a case study research was conducted to investigate firms' level of awareness, implementation and compliance with safety regulations of the Building and Other Construction Workers Act, 1996. The case study comprised three cases each of small companies, medium companies and large companies. Apart from these, interviews were conducted with government agencies to investigate the challenges faced while enforcing the safety regulations in construction projects. The results of the study revealed that most of the small firms are not registered under the BOCW Act, whereas medium and large firms are registered under this act. Most of the small firms are still not aware, not implementing or not complying with safety regulations, whereas medium and large firms were fully aware of all the safety regulations and not full implementing or fully complying with all the safety regulations. Lack of inspectors, awareness, registrations, training and online inspection system are the major challenges faced by government agencies while enforcing the safety regulations in construction projects. Further research should be carried out to find why firms in India are not implementing and complying with safety regulations. Apart from this, further research should also be carried out to find what are the challenges faced by firms when implementing and complying with safety regulations. From the government side, further research should be carried out as to why government inspectors are unable to fully enforce the safety regulations in construction projects. Apart from this, further research should also be carried out to find why government inspectors are giving low priority to safety during the inspections.

Introduction

One of the most hazardous industries in world is the construction industry in which construction workers are more prone to hazards (Dorji & Hadikusumo, 2006). On par with other industries, construction work has three to six times more risks related with it (Gurcanli & Mungen, 2013). The major cause of fatal accidents occurring at construction sites is the fall of persons from heights and through openings (Shirur & Torgal, 2014). The construction industry in India has substantial economic importance as it provides employment opportunities to 33 million people and remains in the second position among all industries after agriculture (Kumar & Bansal, 2013). In the present scenario the Indian construction industry is quite large and complex because of the introduction of the latest technology and manpower. Although the Indian construction industry is developing, some of the drawbacks are seen in terms of occupational safety and health (OSH) issues (Kanchana et al., 2015).

A study by Dorji & Hadikusumo (2006) affirmed that the lack of safety awareness, safety training, and the lack of safety regulations and their enforcement are the major problems for occupational safety and health in construction. Inefficiency of safety laws and regulations are the main reasons for high injury and fatality rates in construction (Awwad et al., 2016). Accidents and fatalities are observed mostly in developing countries. Idubor and Osiamoje (2013) stated that regulations lacking appropriate enforcement are equivalent to no laws. For ensuring the effectiveness of the regulations, enforcement is very important. Lack of strict enforcement of OSH regulations is a key driver for non-compliance with OSH regulations. A study by Adeyemo and Smallwood (2017) affirmed that lack of enforcement officers, lack of knowledge regarding health and safety, lack of skilled personnel, lack of commitment, and insufficient funding are key barriers for implementing health and safety legislation.

In India most construction projects consist of contractual workers. These workers are often from an agricultural background, rural areas and other sectors and don't have adequate knowledge about construction safety (Singh, 2014). As a result they are unable to predict the risks involved in the construction process and face severe injuries at construction sites. The Building and Other Construction Workers (BOCW) Act, 1996 is the only legislation related to safety in the Indian construction industry. Construction safety in India is still in its early stages because safety laws are not strictly enforced. In India there are bad safety records due to improper enforcement of safety laws and regulations (Awwad et al., 2016). The main reason preventing India from developing a construction safety program is corruption. Most state governments in India are unable to fully implement the BOCW Act and enforcement from government is also lacking (Rajaprasad & Chalapathi, 2015).

In the Indian construction industry there is no adequate research on safety laws, regulations and organizations implementing safety laws. Apart from this research on construction firms' level of awareness, research on compliance and implementation of safety regulations is also lacking. Some of the research has been conducted about construction safety in India (Kanchana et al., 2015; Singh, 2014; Rajaprasad & Chalapathi, 2015; Shirur and Torgal, 2014; Kumar and Bansal, 2013). But none of this research addresses key issues such as

- Existing safety laws and regulations in the Indian construction industry
- The regulatory authority at the state level and central level regarding the implementation of safety laws

- Challenges faced by regulatory authorities while enforcing safety regulations at construction projects
- Construction firms' level of awareness, compliance and implementation with safety regulations

All the aforementioned issues are the existing knowledge gaps for this chapter and the current research addresses these issues. The main aim of the current research is to study the safety systems, laws and regulations influencing the Indian construction sector. The current study aims

1. To investigate safety laws, regulations and organizations implementing safety laws in the Indian construction sector
2. To investigate the level of awareness, implementation, compliance and enforcement of safety regulations at construction projects
3. To propose recommendations for improving construction safety in India to government agencies and construction firms

Scope of the research

As mentioned earlier, this research mainly focuses on safety laws, systems and regulations influencing the Indian construction sector. Semi-structured interviews were conducted with safety managers of construction firms and government inspectors in the Labor Department of the central and state government. This research is focused on building construction firms in India. A qualitative approach with case studies as the tool is used for data collection. In India there are 29 states and 7 union territories, and in each state there are many construction firms undertaking numerous construction activities. Apart from these there are labor departments at the state level and central level for carrying out the provisions of the BOCW Act. Since research has not been carried out in each state and their respective labor departments, there are certain limitations for this research:

1. Results of this research are confined to only nine construction companies visited in the states of Andhra Pradesh, Telangana and Maharashtra in India.
2. Results of this research are confined to the Deputy Chief Labor Commissioner Office, Hyderabad in the labor department at the central government level and the Andhra Pradesh labor department at the state government level.

Safety law and regulations related to the construction sector in India

Workplace safety and health in is broadly divided into four areas, namely factories, dock works, mines and construction sites. The Constitution of India, under the Directive Principles of State Policy, includes provisions relating to workplace safety and health (WSH) in all economic activities. The Factories Act, 1948 is the legislation dealing with workplace safety and health in factories. Safety in dock works and mines is ensured through the implementation of The Dock Workers (Safety, Health and Welfare) Act 1986 and Regulations 1990 and the Mines Act 1952, respectively. Workplace safety and health in construction sites are enforced by the Building and Other Construction Workers (Regulation of Employment and and Conditions of Service) Act, 1996 and the applicable central and state government rules (Planning Commission of India, 2011).

The BOCW Act was passed on 1 March 1996 and is the only act related to construction safety. The act is applicable to all establishments employing ten or more workers in any building and other construction works. The central government is the appropriate government for notifying the rules and regulations under the act as well as the enforcement of the provisions under the said rules, in respect of establishments in relation to which the central government is the appropriate government under the Industrial Disputes Act, 1947. Regarding the other establishments, the state government is the appropriate government for notifying the rules and enforcing the provisions.

The Building and Other Construction Workers (Regulation of Employment and Conditions of Service) Act, 1996

The preamble of the BOCW Act is as follows:

> An Act to regulate the employment and conditions of service of building and other construction workers and to provide for their safety, health and welfare measures and for other matters connected therewith or incidental thereto.

The BOCW Act extends to the whole India. The act contains 64 sections and these sections are further divided into chapters. In the BOCW Act, Chapter VII (Safety and Health Measures) is the only chapter related to occupational safety and health matters.

The Building and Other Construction Workers (Regulation of Employment and Conditions of Service) Act, 1996 at central and state level

After the president's assent was given on 20 August 1996, the act was published in the public gazette. In the exercise of the powers discussed under Sections 40 (related to occupational safety and health) and 62 of the BOCW Act, the central government should make the rules for carrying out the provisions of the BOCW Act. Finally, the central government made the rules in 1998 and named them the Building and Other Construction Workers (Regulation of Employment and Conditions of Service) Central Rules, 1998. The BOCW Central Rules, 1998 not only consists of rules related to occupational safety and health, but there are some other aspects involved (Government of India, 1996).

In the exercise of the powers discussed under Sections 40 (related to occupational safety and health) and 62 of the BOCW Act, the state government should make the rules for carrying out the provisions of the act. The government of Andhra Pradesh (AP) took almost 3 years to frame the rules as per sub-section (1) of 40 and sub-section (1) of 62 of the BOCW Act. Finally, the government of AP completed its rules in 1999 and named them the Andhra Pradesh Building and Other Construction Workers (Regulation of Employment and Conditions of Service) Rules, 1999 (Government of Andhra Pradesh Labor Department, 1999). Safety regulations in BOCW Central Rules, 1998 and AP BOCW Rules, 1999 comprise 17 chapters and they cover areas such as general provisions for safety and health; lifting appliances and gear; runways and ramps; transport and earth moving equipment; concrete work; demolition; excavation; ladders and step ladders; catch platform and hoardings, chutes, safety belts and nets; structural frame and frame work; stacking and unstacking; scaffold; safety organizations; explosives; piling; medical facilities; and information to the Bureau of Indian Standards (Government of Maharashtra, 1996). Apart from these there are 11 schedules and 11 forms related to occupational safety and health (see Figure 5.1).

Methodology

Research design

The research design for the present study starts with the study questions. Study questions mainly depend upon the objectives of the study. For the present research, the study questions are mainly comprised of what and why. Most of the research questions were related to safety systems, laws and regulations in the Indian construction sector and the organizations related to OSH in India. The research questions are

- What are the existing safety laws and regulations in the Indian construction sector?
- What organizations are responsible for the implementation of these safety laws?
- What are the challenges faced by government inspectors while enforcing the safety regulations at the construction projects?
- What is the firm's level of awareness, implementation and compliance with safety regulations at construction projects?

The main aim of the study was to investigate the safety systems, laws and regulations influencing the Indian construction sector. To achieve this aim, a qualitative research design, with a type 2 (multiple case design with single unit of analysis) case study design was adopted. To achieve the objective, data was collected from government agencies and construction companies. Semi-structured interviews and sites visits were conducted to investigate the companies' level of awareness, compliance and implementation of safety regulations. Semi-structured

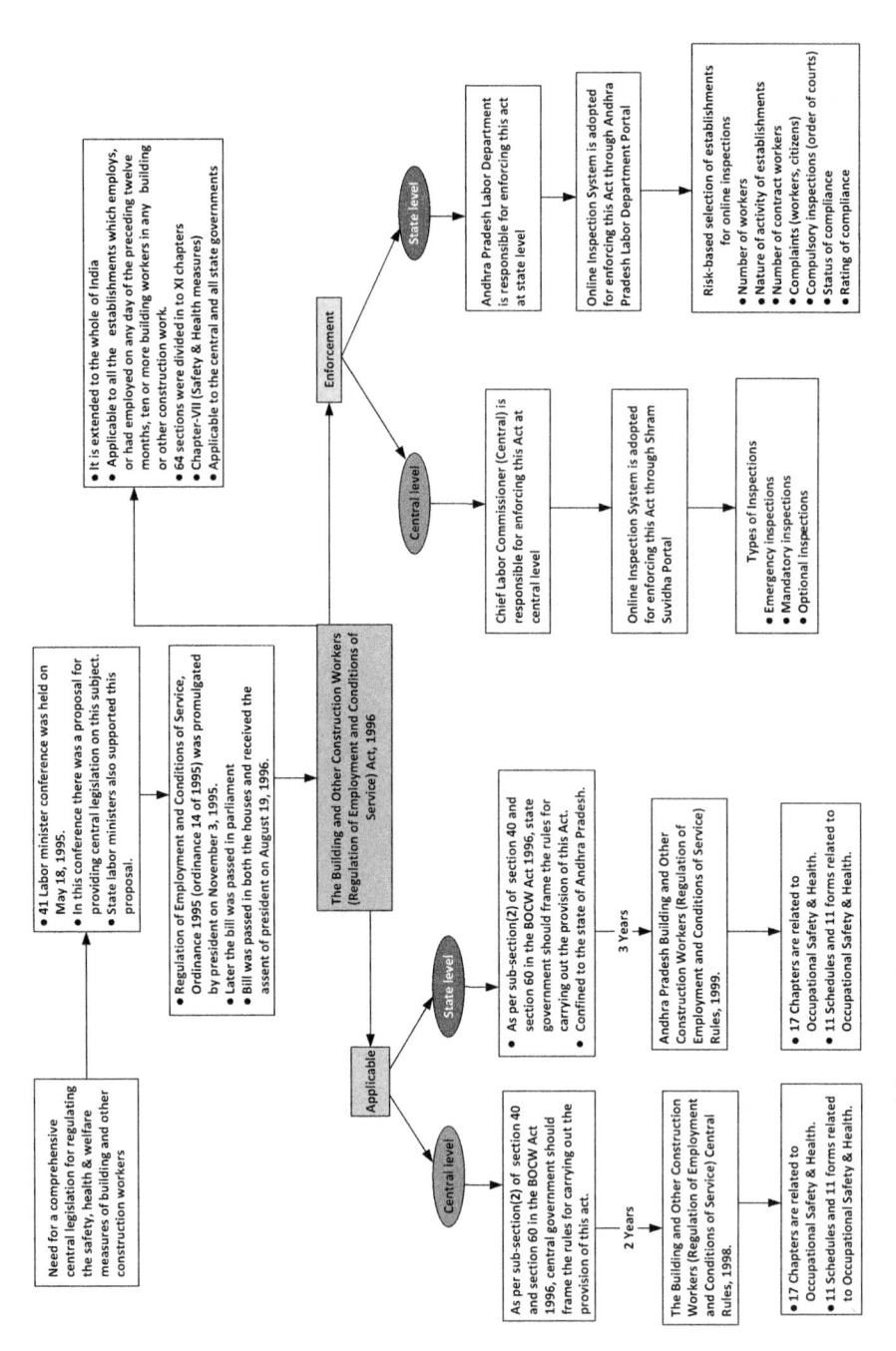

Figure 5.1 Overview of the BOCW Act.

interviews were also conducted with government inspectors regarding the enforcement of safety regulations in construction projects. Keeping all these constraints in mind, the researchers adopted a qualitative approach rather than a quantitative approach.

Data collection

Construction firms

For the first part of data collection, data was collected from three small, three medium and three large building construction firms in India. The targeted respondent was the safety manager of the firm. Some firms don't have a safety manager, so the person responsible for OSH matters was interviewed. Conducting interviews, site visits and gathering relevant documents were the methods chosen for data collection. The interview guidelines were comprised of three parts as follows:

- Part A: Introduction. This section contains information about the general instructions.
- Part B: Profile of the respondent and company. This section contains information about the role of the respondent, experience in OSH, number of employees, size of the company and company registration under the BOCW Act.
- Part C: Contains information about the firm's level of awareness, implementation and compliance with safety regulations.

After interviewing the respondent, a project under that firm is visited to verify whether the responses given by respondent during the interview are appropriate. As there are numerous construction activities, the researchers used convenience sampling for selection of construction companies based on the following criteria:

(i) Nature and scope of the construction:
Researchers considered only the companies that were dealing with building construction projects in the construction stage or later stages of the cycle. This ensures the researcher will observe whether these construction firms are implementing and complying with the safety regulations at the construction sites.

(ii) Company size:
Researchers considered small, medium and large size construction companies in India. The classification of construction companies regarding size is based on the number of employees (see Table 5.1).

(iii) Company and project location:
The study was carried out in the states of Andhra Pradesh, Telangana and Maharashtra. Researchers considered these places for the current study because there is a boom in the construction sector in these regions. Recently bifurcation of Andhra Pradesh and Telangana took place, so there are numerous construction activities going on.

Table 5.1 Classification of companies based on number of employees

Company size	Number of employees
Small Size	11 to 50 employees
Medium Size	51 to 150 employees
Large Size	More than 150 employees

(Source: Manu et al., 2018)

Government agencies

In the second part of data collection, data was collected from government agencies, specifically from the Labor Department at the central and state government levels. Data was collected in the form of interviews and gathering of relevant documents. Criteria regarding the selection of government officials were based on the officer's experience in the Labor Department and the officers responsible in the Labor Department for enforcing the BOCW Act. Interviews were conducted with two inspectors from central government and two inspectors from state government in the Labor Department. Interview guidelines consisted of two parts: Part A – profile of the respondent (i.e., experience, designation in the Labor Department); and Part B – the respondent's awareness regarding safety regulations, the BOCW Act and enforcement of safety regulations.

Data analysis

All the data collected was sorted for analysis. The obtained data was individually analyzed for each case first. Later, cross-cases analyses were executed among the small, medium and large companies. Cases A, B and C are small firms; Cases D, E and F are medium firms; and Cases G, H and I are large firms. During the site visit appropriate photos were taken as per the safety regulations in the BOCW Act. Data from the interviews and site visits were compared to investigate the firm's level of awareness, implementation and compliance with safety regulations. Regulations that were observed at the site were compared with the interview and then the appropriate response based on the scale adopted is given. For the rest of the regulations, the response was given based on the interview. Data obtained from government inspectors were individually analyzed first for all the four respondents. Later a comparison was done among the two state government and two central government inspectors based on interview guidelines.

Table 5.2 depicts information about the definitions of the scale used for the factors Awareness, Implementation, Compliance and Effectiveness in this research. Figure 5.2 depicts information about the research methodology framework.

Table 5.2 Definition of scale for the factors

Factors		Description
(i) Awareness	Not aware (NA)	Firm was not aware of this regulation
	Partially aware (PA)	Firm was partly aware of this regulation
	Fully aware (FA)	Firm was completely aware of this regulation
(ii) Implementation	Not implementing (NI)	Firm was not implementing this regulation
	Partially implementing (PI)	Firm was partly implementing this regulation
	Fully implementing (FI)	Firm was completely implementing this regulation
(iii) Compliance	Not complying (NC)	Firm was not complying with this regulation
	Partially complying (PC)	Firm was partly complying with this regulation
	Fully complying (FC)	Firm was completely complying with this regulation

Results and discussions

Results of the study are discussed in the following for construction firms and government agencies separately.

Construction firms

Results obtained from nine construction firms (three small, three medium and three large) are mentioned under two main headings: profile of the respondents and companies; firms' level of awareness, implementation and compliance with safety regulations.

Profile of respondents and companies

Table 5.3 provides information about the profile of the respondents and companies. The table shows that one small company employed safety personnel and the other two small companies did not. So when there was no safety manager, the researcher interviewed the person responsible for OSH matters. All the medium and large companies employed safety personnel; a safety manager was employed at all three large companies and but at only one medium company. Compared to the small companies, the medium and large companies employed safety personnel with adequate experience. Regarding the registration under the BOCW Act, all the medium and large companies were registered. Only one small company was registered. Researchers visited eight residential projects and one commercial project.

68 Duddukuru and Hadikusumo

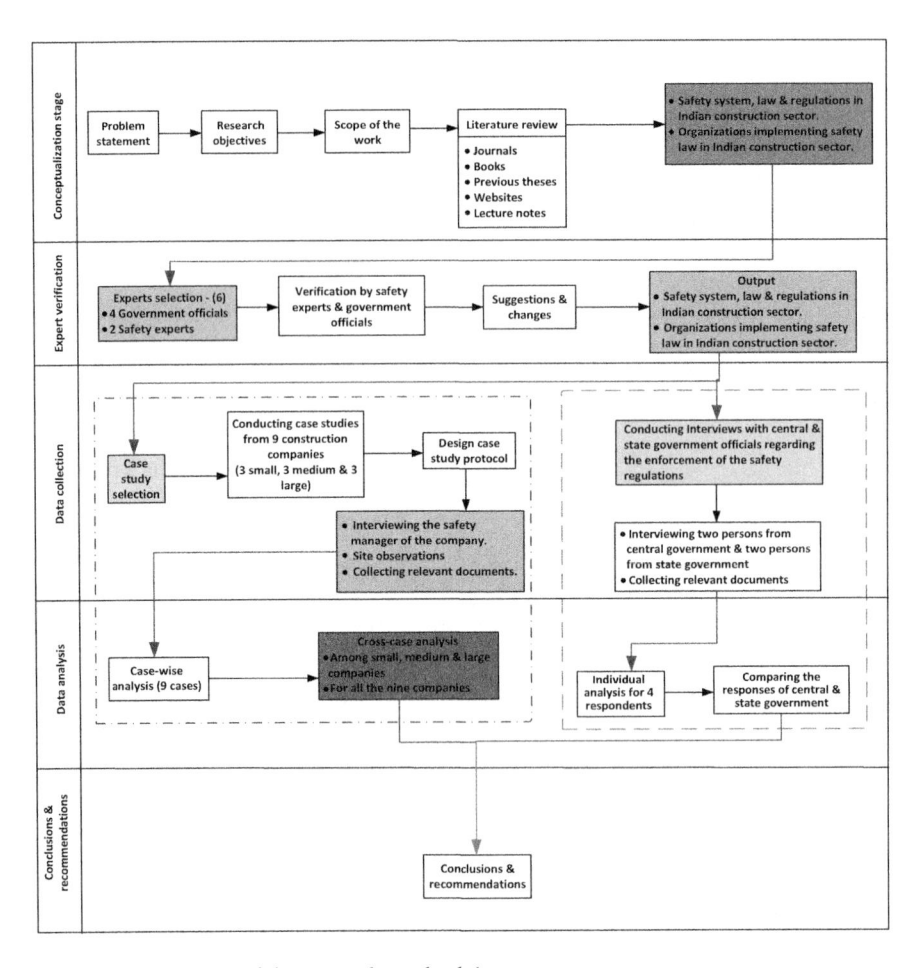

Figure 5.2 Overview of the research methodology.

Firms' level of awareness, implementation and compliance with safety regulations

SMALL FIRMS

Table 5.4 provides the information about the small firms' level of awareness, implementation and compliance with safety regulations of the APBOCW Rules, 1999. For a clear understanding of the notations used, please refer to Table 5.2. In Table 5.4 it is clearly shown that two companies are not aware of all the safety regulations and one company is fully aware of all the safety regulations. Two companies are not implementing and not complying with all the safety regulations,

Table 5.3 Profile of the company and respondents interviewed

Name of the case	Profile of the respondent interviewed		Profile of the company			
	Designation	Experience in OSH	Size	Number of employees	Type of project visited	Registration under BOCW Act
Case A	Safety officer	4 years	Small	45	Residential	Registered
Case B	Site engineer	—	Small	25	Residential	Not registered
Case C	Site engineer	—	Small	30	Residential	Not registered
Case D	Safety manager	10 years	Medium	150	Residential	Registered
Case E	Safety officer	6 years	Medium	100	Residential	Registered
Case F	Safety officer	7 years	Medium	130	Residential	Registered
Case G	Safety manager	16 years	Large	350	Residential	Registered
Case H	Safety manager	12 years	Large	250	Residential	Registered
Case I	Safety manager	13 years	Large	300	Commercial	Registered

Table 5.4 Awareness, implementation and compliance with safety regulations in the AP BOCW Rules 1999 among small companies

Safety regulations in the Andhra Pradesh Building and Other Construction Workers Rules, 1999	Small companies								
	Case A			Case B			Case C		
	Aw	Im	Co	Aw	Im	Co	Aw	Im	Co
1. General provisions for safety and health	FA	PI	PC	NA	NI	NC	NA	NI	NC
2. Lifting appliances and gear	FA	PI	PC	NA	NI	NC	NA	NI	NC
3. Runways and ramps	FA	PI	PC	NA	NI	NC	NA	NI	NC
4. Transport and earth moving equipment	FA	PI	PC	NA	NI	NC	NA	NI	NC
5. Concrete work	FA	PI	PC	NA	NI	NC	NA	NI	NC
6. Demolition	FA	PI	PC	NA	NI	NC	NA	NI	NC
7. Excavation	FA	PI	PC	NA	NI	NC	NA	NI	NC
8. Ladders and step ladders	FA	PI	PC	NA	NI	NC	NA	NI	NC
9. Catch platform and hoardings, chutes, safety belts and nets	FA	PI	PC	NA	NI	NC	NA	NI	NC
10. Structural frame and form work	FA	PI	PC	NA	NI	NC	NA	NI	NC
11. Stacking and unstacking	FA	PI	PC	NA	NI	NC	NA	NI	NC
12. Scaffold	FA	PI	PC	NA	NI	NC	NA	NI	NC
13. Safety organizations	FA	PI	PC	NA	NI	NC	NA	NI	NC
14. Explosives	FA	PI	PC	NA	NI	NC	NA	NI	NC
15. Piling	FA	PI	PC	NA	NI	NC	NA	NI	NC
16. Medical facilities	FA	PI	PC	NA	NI	NC	NA	NI	NC
17. Information to Bureau of Indian Standards	FA	FI	FC	NA	NI	NC	NA	NI	NC

and one company is partially implementing and partially complying with all the safety regulations.

MEDIUM FIRMS

Table 5.5 provides the information about the medium firms' level of awareness, implementation and compliance with safety regulations of the BOCW Central Rules, 1998. For a clear understanding of the notations used, please refer to Table 5.2. Table 5.5 clearly shows that all three companies are fully aware of all the safety regulations. One company is fully implementing and fully complying with safety regulations 1, 2, 3, 5, 6, 7, 8, 12, 14 and 15. Two companies are fully implementing and fully complying with safety regulations 4, 9 and 11. All three

Table 5.5 Awareness, implementation and compliance with safety regulations in the BOCW Central Rules 1998 among medium companies

| Safety regulations in the Building and Other Construction Workers Central Rules, 1998 | Medium companies |||||||||
| | Case D ||| Case E ||| Case F |||
	Aw	Im	Co	Aw	Im	Co	Aw	Im	Co
1. General provisions for safety and health	FA	FI	FC	FA	PI	PC	FA	PI	PC
2. Lifting appliances and fear	FA	FI	FC	FA	PI	PC	FA	PI	PC
3. Runways and ramps	FA	FI	FC	FA	PI	PC	FA	PI	PC
4. Transport and earth moving equipment	FA	FI	FC	FA	PI	PC	FA	FI	FC
5. Concrete work	FA	FI	FC	FA	PI	PC	FA	PI	PC
6. Demolition	FA	FI	FC	FA	PI	PC	FA	PI	PC
7. Excavation	FA	FI	FC	FA	PI	PC	FA	PI	PC
8. Ladders and step ladders	FA	FI	FC	FA	PI	PC	FA	PI	PC
9. Catch platform and hoardings, chutes, safety belts and nets	FA	FI	FC	FA	FI	FC	FA	PI	PC
10. Structural frame and form work	FA	FI	FC	FA	FI	FC	FA	FI	FC
11. Stacking and unstacking	FA	FI	FC	FA	PI	PC	FA	FI	PC
12. Scaffold	FA	FI	FC	FA	PI	PC	FA	PI	PC
13. Safety organizations	FA	PI	PC	FA	PI	PC	FA	PI	PC
14. Explosives	FA	FI	FC	FA	PI	PC	FA	PI	PC
15. Piling	FA	FI	FC	FA	PI	PC	FA	PI	PC
16. Medical facilities	FA	PI	PC	FA	NI	NC	FA	PI	PC
17. Information to Bureau of Indian Standards	FA	FI	FC	FA	FI	FC	FA	FI	FC

companies are fully implementing and fully complying with safety regulations 10 and 17. One company is partially implementing and partially complying with safety regulations 4, 9 and 11. Two companies are partially implementing and partially complying with safety regulations 1, 2, 3, 5, 6, 7, 8, 12, 14, 15 and 16. One company is not implementing and not complying with safety regulation 16.

LARGE FIRMS

Table 5.6 provides information about the large firms' level of awareness, implementation and compliance with safety regulations of the BOCW Central Rules, 1998. For a clear understanding of the notations used, please refer to Table 5.2. Table 5.6 clearly shows that all three companies are fully aware of

Table 5.6 Awareness, implementation and compliance with safety regulations in the BOCW Central Rules 1998 among large companies

Safety regulations in the Building and Other Construction Workers Central Rules, 1998	Large companies								
	Case G			Case H			Case I		
	Aw	Im	Co	Aw	Im	Co	Aw	Im	Co
1. General Provisions for safety and health	FA	FI	FC	FA	FI	FC	FA	FI	FC
2. Lifting appliances and gear	FA	FI	FC	FA	FI	FC	FA	FI	FC
3. Runways and ramps	FA	FI	FC	FA	FI	FC	FA	FI	FC
4. Transport and earth moving equipment	FA	FI	FC	FA	FI	FC	FA	FI	FC
5. Concrete work	FA	FI	FC	FA	FI	FC	FA	FI	FC
6. Demolition	FA	FI	FC	FA	FI	FC	FA	FI	FC
7. Excavation	FA	FI	FC	FA	FI	FC	FA	FI	FC
8. Ladders and step ladders	FA	FI	FC	FA	FI	FC	FA	FI	FC
9. Catch platform and hoardings, chutes, safety belts and nets	FA	FI	FC	FA	FI	FC	FA	FI	FC
10. Structural frame and form work	FA	FI	FC	FA	FI	FC	FA	FI	FC
11. Stacking and unstacking	FA	FI	FC	FA	FI	FC	FA	FI	FC
12. Scaffold	FA	FI	FC	FA	FI	FC	FA	FI	FC
13. Safety organizations	FA	PI	PC	FA	PI	PC	FA	PI	PC
14. Explosives	FA	FI	FC	FA	FI	FC	FA	FI	FC
15. Piling	FA	FI	FC	FA	FI	FC	FA	FI	FC
16. Medical facilities	FA	FI	FC	FA	FI	FC	FA	PI	PC
17. Information to Bureau of Indian Standards	FA	FI	FC	FA	FI	FC	FA	FI	FC

all the safety regulations. All three companies are fully implementing and fully complying with safety regulations 1, 2, 3, 5, 6, 7, 8, 12, 14 and 15. Two companies are fully implementing and fully complying with safety regulation 16. All three companies are partially implementing and partially complying with safety regulation 13. One company is partially implementing and partially complying with safety regulation 16.

Government agencies

Results from the two central government inspectors and two state government inspectors are discussed under two main headings: profile of the respondents; and respondents' awareness regarding safety regulations, the BOCW Act and enforcement of safety regulations.

Profile of the respondents

Table 5.7 provides information about the profile of the respondents interviewed by researchers. Researchers interviewed the assistant labor commissioner and labor enforcement officer of the Chief Labor Commissioner Office of the central government. From the state government, researchers interviewed the deputy chief labor commissioner and the assistant commissioner of labor in the Andhra Pradesh Labor Department. All the inspectors interviewed by researchers had adequate experience in the Labor Department, as shown in the Table 5.7.

Awareness regarding safety regulations, BOCW Act and enforcement of safety regulations

From the semi-structured interviews conducted by researchers, the following results were obtained. Regarding the awareness of the BOCW Act, central government inspectors were fully aware, whereas state government inspectors were partially aware. Central government inspectors are fully aware of the safety regulations in the BOCW Rules, whereas state government inspectors are partially aware of safety regulations in the BOCW Rules. Both state and central government inspectors are partially enforcing the safety regulations of the BOCW Rules. Both the central and state government inspectors suggested increasing the penalties 30–50%. Both central and state government inspectors conveyed that regular inspections are important for enforcing the safety regulations, but due to the online inspection system inspections are drastically reduced. Lack of inspectors, awareness, registrations, training and online inspection system are the challenges faced by central government inspectors when enforcing the safety regulations, and the lack of awareness, lack of inspectors, online inspection system, lack of training and lack of registrations are the challenges faced by state government inspectors when enforcing the safety regulations. Table 5.8 compares the responses of state and central government inspectors.

Table 5.7 Profile of the respondents interviewed in the Labor Department at central and state levels

Name of the government agency	Designation of the employee	Years of experience in the government agency	Years of experience in the current position
Chief Labor Commissioner office (Central Government)	1. Assistant Labor Commissioner	16 years	12 years
	2. Labor Enforcement Officer	10 years	6 years
Government of Andhra Pradesh Labor Department (State Government)	1. Deputy Chief Labor Commissioner	19 years	13 years
	2. Assistant Commissioner of Labor	14 years	8 years

Table 5.8 Comparison of central and state government respondents' responses regarding the factors considered in the study

Respondent	State level — Assistant Labor Commissioner	State level — Labor Enforcement Officer	Central level — Deputy Chief Labor Commissioner	Central level — Assistant Commissioner of Labor	Comparison
1. Awareness regarding BOCW Rules (at state and central level)	Partially aware	Partially aware	Fully aware	Fully aware	• State government inspectors are partially aware of BOCW Rules. • Central government inspectors are fully aware of BOCW Rules.
2. Awareness regarding safety regulations in BOCW Rules (at state and central level)	Partially aware	Partially aware	Fully aware	Fully aware	• State government inspectors are partially aware of safety regulations BOCW Rules. • Central government inspectors are fully aware of safety regulations in BOCW Rules.
3. Enforcement of safety regulations in BOCW Rules (at state and central level)	Partially enforcing Moderately effective	Partially enforcing Moderately effective	Partially enforcing Moderately effective	Partially enforcing Moderately effective	• Both state and central government inspectors are partially enforcing the safety regulations in BOCW Rules and being moderately effective while enforcing the regulations.
4. Severity of penalties	Increase by 30–40%	Increase by 35–40%	Increase by 50%	Increase by 45–50%	• Severity of penalty should be increased 30–50%.
5. Existing system	Existing system is adequate for enforcing the safety regulations.	Existing system is adequate for enforcing the safety regulations.	Existing system is adequate for enforcing the safety regulations.	Existing system is adequate for enforcing the safety regulations.	• Existing system is adequate for enforcing the safety regulations.

| 6. Regular inspections | Yes, regular inspections are important because of the online system inspections were drastically reduced | Yes, regular inspections are important, but due to online inspection system inspections were not taking place. | Yes, regular inspections are important, but due to online inspection system inspections were not taking place. | Yes, regular inspections are important, but due to online inspection system inspections were not taking place. | • Yes, regular inspections are important, but due to online inspection system inspections were not taking place. |
| 7. Challenges faced while enforcing the safety regulations | A. Lack of training construction safety
B. Awareness of safety regulations
C. Lack of inspectors
D. Online inspection system
E. Lack of registrations
F. Judicial system | A. Lack of awareness regarding safety regulations
B. Online inspection system
C. Lack of training regarding safety
D. Lack of registrations
E. Lack of inspectors | A. Lack of inspectors
B. Lack of registrations
C. Online inspection system
D. Judicial system | A. Judicial system
B. Lack of registrations
C. Online inspection system
D. Lack of inspectors | • Common challenges for both inspectors are
• Lack of inspectors
• Lack of registrations
• Online inspection |

Key issues identified by the researcher during site visits and interviews

Construction firms

Researchers conducted an extensive literature review, semi-structured interviews and site visits to investigate the firms' level of awareness, implementation and compliance with safety regulations. During this investigation some key issues were identified:

(i) From the extensive literature review concerning safety laws and regulations, researchers identified that regulations are not updated from the past 10 years.
(ii) The penalty for not complying with safety regulations is too little.
(iii) During the interviews with all the small companies and two of the medium companies, researchers identified that companies ignore safety regulations before starting any work.
(iv) During the interviews, researchers identified that two of the small companies did not have any knowledge regarding construction safety.
(v) During the interviews with three small companies and two medium companies, researchers identified that the companies focused on time, cost and quality and did not give any priority to safety.
(vi) Two of the small companies did not employ any safety personnel at the construction site due to lack of safety awareness and negative perception about safety.
(vii) During the interviews with small firms, researchers came to know that lack of financial capacity is a key issue for implementing and complying with safety regulations.
(viii) No training regarding construction safety is given in all three of the small firms and one of the medium firms.
(ix) Two of the medium companies employed safety personnel at the construction site, but they were unable to fully implement and comply with the safety regulations due to the lack of front-line management. This issue was also observed at one of the small companies.
(x) Two of the small companies did not have any management practices related to safety.
(xi) One of the small companies was trying to comply and implement the safety regulations, but it did not have appropriate management practices (for example tool box meeting, safety plan) related to construction safety. This issue was also observed at two of the medium companies.
(xii) The necessary budget for safety is allocated in all the three large companies and three medium companies. Two of the small companies didn't allocate any budget for safety and one small company allocated limited budget.

All the aforementioned key issues were truly observed by the researchers during the site visits and from interviews.

Government agencies

Researchers conducted an extensive literature review on safety systems, laws, regulations and organizations implementing safety laws in the Indian construction sector. Researchers conducted semi-structured interviews with central and state government inspectors. During the process, researchers came to know about some of the key issues, which are mentioned next:

(i) Government inspectors in state labor and central labor departments do not have adequate knowledge regarding construction safety and regulations associated with the BOCW Act.
(ii) There is a lack of data regarding accident statistics in the construction industry.
(iii) Due to the lack of training, central and state government level inspectors face problems when enforcing the safety regulations.
(iv) Government inspectors at the central and state government levels additionally enforce other Acts (e.g. Industrial Disputes Act 1947 and The Minimum Wages Act 1948) alongside the BOCW Act, and as a result the existing staff is not adequate to enforce the BOCW Act.
(v) During the interviews with government inspectors, researchers came to know that the priority for safety during inspections is low.
(vi) Due to online inspection system, regular inspections were drastically reduced and inspectors were unable to enforce safety regulations.
(vii) During the interviews with government inspectors, researchers came to know that small companies are not registering under the BOCW Act.

Implications for further research and recommendations

Implications for future research

Current research investigated two aspects: (1) the safety systems, laws, regulations and organizations implementing safety laws in the Indian construction sector; and (2) firms' level of awareness, implementation and compliance with safety regulations in the BOCW Act. Further research should be carried out to find why firms in India are not implementing and complying with safety regulations. Further research should also be carried out to find what challenges firms face when implementing and complying with safety regulations. So by conducting this type of future research, one can find out the ways to make firms in India implement and comply with the safety regulations.

From the government side, further research should be carried out into why government inspectors are unable to fully enforce the safety regulations at construction projects. Further research should also be carried out to find why government inspectors are giving low priority to safety during inspections. So by conducting this type of research, necessary recommendations can be identified for ensuring strict legal enforcement regarding safety regulations.

Recommendations for improving construction safety in India

- For government agencies
  1. Governments should conduct training programmes for government inspectors to create awareness regarding safety legislations, corresponding regulations and their enforcement.
  2. The existing staff is not adequate for enforcing the safety regulations. Therefore, a separate department or inspectors should be appointed to only enforce the BOCW Act.
  3. In India, small firms are unaware of construction safety, corresponding legislations and regulations associated with them. So to avoid this issue government should conduct safety campaigns, workshops and conferences for educating small firms regarding the importance of construction safety and benefits when they register under this act.
  4. In India most construction activities are done by small firms. But these small firms don't have any financial support for implementing and complying with this act. Although the fee for registration is low, small firms were not registered under this act. So government agencies should give subsidies to the small firms that register under this act.
  5. Currently there are no official records for construction safety in India. There are no accident statistics, profiles and reports available related to construction safety. So when there are no statistics related to accidents, then it is very difficult for the government to improve construction safety. The labor departments at the central and state levels should collect data related to accidents at construction sites in India.
  6. Since 1996 the legislation has not undergone any changes. As technology is rapidly increasing and new technologies are being adopted by various construction firms for improving productivity, new risks are also emerging. Regulations should be updated accordingly to overcome the risks. Like Occupational Safety and Health Administration (OSHA), research activities should also be conducted in this act.
  7. Penalties in the BOCW Act, 1996 are not adequate to make construction firms comply with safety regulations. So government should increase penalties 35–40%.
- For construction firms
  1. In small organizations management doesn't care about safety requirements. Allocation of funds for safety measures is out of scope for organization management. So small firms should allocate the necessary budget for safety.
  2. In India most small organizations do not have any safety personnel at the construction site for ensuring a safe workplace, so management should appoint safety personnel at the construction site.
  3. Small organizations ignore safety training, as it is expensive. Investing money in safety training is better than compensating injured individuals. Mandatory safety training should be given to the workers and employees in small organizations to create awareness regarding safety.

4. For implementing and complying with all the safety regulations, construction companies irrespective of size should adopt the following management practices:
   - Safety policy
   - Safety plan
   - Safety training
   - Safety committee
   - Tool box meeting
   - Recording of safety violations
   - Near-miss reports
   - Regular inspections on site
   - Penalties for non-compliance
   - Frontline management
   - Maintain safety records
   - Regular inspections
   - Reward system
   - External support from OSH organizations
   - Regular inspections at the sites
   - Checklists for all the activities
   - Medical examination of building workers
   - Safe work procedures

By adopting all the management practices, firms can implement and comply with safety regulations at worksites.

Conclusion

The current research started with a problem that firms in India ignore the basic safety rules and regulations, and enforcement from government is lacking. Based on this, objectives were developed and research was conducted. Construction safety in India is still in the early stages because safety laws are not strictly enforced. Based on three cases of small companies, three cases of medium companies and three cases of large companies, the results of the study revealed that most of the small firms were not registered under the BOCW Act, 1996, whereas most of the medium and large firms were registered. Most of the small firms were still not aware, not implementing and not complying with safety regulations. However, the medium and large firms were fully aware of all the safety regulations, but not fully implementing and fully complying with all the safety regulations. Further research should be carried out to find why firms in India are not implementing and complying with safety regulations. Further research should also be carried out to find the challenges firms face firms when implementing and complying with safety regulations.

Both central and state governments are not fully implementing the BOCW Act, 1996 and enforcement is also lacking. Compared to state government inspectors, central government inspectors are better at enforcement. In India, after introducing an online inspection system, regular inspections were drastically reduced. More liberalization has taken place after introducing the online inspection system. Lack of inspectors, awareness, registrations, training and online inspection system are the major challenges faced by government agencies when enforcing the safety regulations. From the results of the study, governments can find ways to improve construction safety in India and ensure that firms implement and comply with the safety regulations of the BOCW Act. Due to the lack of strict enforcement, contractors are ignoring safety regulations. From the government side, further research should be carried out into why government inspectors are unable to fully enforce the safety regulations in construction projects. Apart from these, further research should also be carried out to find why government inspectors are giving low priority to safety during the inspections.

Acknowledgements

The authors are grateful to the Asian Institute of Technology, Thailand for giving the necessary support throughout the research period. The authors express their sincere gratitude to all the construction companies and government agencies in India for spending their valuable time and giving their support during the research period.

References

Adeyemo, O., & Smallwood, J. (2017). Impact of occupational health and safety legislation on performance improvement in the Nigerian construction industry. *Procedia Engineering*, 196 (June), 785–791.

Awwad, R., El Souki, O., & Jabbour, M. (2016). Construction safety practices and challenges in a Middle Eastern developing country. *Safety Science*, 83, 1–11.

Dorji, K., & Hadikusumo, B. H. W. (2006). Safety management practices in the Bhutanese construction industry. *Journal of Construction in Developing Countries*, 11(2), 53–75.

Government of Andhra Pradesh Labor Department. (1999). Andhra Pradesh Building and Other Construction Workers (RE & CS) Rules 1999. Available from: http://labour.ap.gov.in/documents/APBuildingOtherConstructionWorkersRECSRul es1999.pdf.

Government of India, Chief Labor Commissioner (Central). (1996). The Building and Other Construction Workers (Regulation of Employment Conditions Service Act 1996 [Internet]. Available from: https://clc.gov.in/clc/clcold/Acts/shtm/bocw.php [accessed 20 April 2018].

Government of Maharashtra. (1996). The Building and Other Construction Workers (Regulation of Employment Conditions Service Act 1996 [Internet]. Available from: https://maitri.mahaonline.gov.in/pdf/building-and-other-construction-workers-act-1996.pdf [accessed 27 September 2017].

Gürcanli, G. E., & Müngen, U. (2013). Analysis of construction accidents in Turkey and responsible parties. *Industrial Health*, 51(6), 581–595.

Idubor, E. E., and Osiamoje, M. D. (2013). An Exploration of Health and Safety Management Issues in Nigeria's Efforts to industrialize. *European Scientific Journal*, 9 (12), pp 154–169.

Kanchana, S., Sivaprakash, P., & Joseph, S. (2015). Studies on labour safety in construction sites. *Scientific World Journal, 2015*, 590810.

Kumar, S., & Bansal, V. K. (2013). Construction safety knowledge for practitioners in the construction industry. *Journal of Frontiers in Construction Engineering, 2*(2), 34–42.

Manu, P., Mahamadu, A. M., Phung, V. M., Nguyen, T. T., Ath, C., Heng, A. Y. T., & Kit, S. C. (2018). Health and safety management practices of contractors in South East Asia: A multi country study of Cambodia, Vietnam, and Malaysia. *Safety Science, 107*, 188–201.

Planning Commission of India. (2011). Report of the working group on occupational safety and health [Internet]. Available from: http://planningcommission.nic.in/aboutus/committee/wrkgrp12/wg_occup_safety.pdf [accessed August 15 2017].

Rajaprasad, S. V. S., & Chalapathi, P. V. (2015). Factors influencing implementation of OHSAS 18001 in Indian construction organizations: Interpretive structural modeling approach. *Safety and Health at Work, 6*(3), 200–205.

Shirur, E. S., & Torgal, D. S. (2014). Enhancing safety and health management techniques in Indian construction industry. *International Journal of Engineering and Technical Research (IJETR), 2*(4), 52–56.

Singh, K. (2014). Safety in Indian construction. *Computing in Civil Engineering (2012), 3*(11), 1564–1566.

Part II
Occupational health and safety management

6 The strategies to enhance safety performance in the construction industry of Pakistan

Hafiz Zahoor and Albert P. C. Chan

Summary

Though large construction companies in Pakistan are following various safety management systems, unsafe behaviours and unsafe conditions are at times noticed resulting in fatal accidents. This necessitates exploring the root causes of such accidents as well as identifying the significant safety climate (SC) factors that can positively influence the safety performance (SP) in the Pakistani construction industry (CI). For this study, a mixed-method research strategy comprising concurrent and sequential research methods was adopted. The study concludes that accidents in the Pakistani CI can be reduced by strict enforcement of safety rules and regulations by government agencies; guaranteed provision of a safety budget and the introduction of safety incentives by the clients; adequate provision of safety training and resources by the contractors; and giving due importance to past SP while selecting the subcontractors. A 24-item SC scale, comprised of four SC factors, was developed and validated for the Pakistani CI. The study also verified the significant positive impact of SC on safety compliance and safety participation, and the negative impact on the number of self-reported accidents/injuries. The analysis of demographic variables confirmed that the employees in the age group of 20 years or below, and working for more than 40 hours per week, have a tendency towards unsafe behaviour. The study adds to the body of knowledge by highlighting the implications of adopting an existing SC scale across the region and culture, and revealing the deviations in SC dimensions. Caution should be exercised while generalizing the research findings to other sectors of CI. Nonetheless, the study's robust methodology can be replicated in other industries and regions for element prioritization and SC investigation.

Introduction

The construction industry (CI) of Pakistan has experienced a consistent increase in the percentage of injuries from 14.54% in 2008–2009 (PBS 2009) to 16.3% in 2014–2015 (PBS 2015, p. 38). From the statistics given in Figure 6.1, it can be inferred that construction is the second most injury-prone industry after agriculture, even though its employment rate is fifth amidst all other industries.

Large and medium-sized construction enterprises are employed on tall building projects in Pakistan, as main contractors and subcontractors.

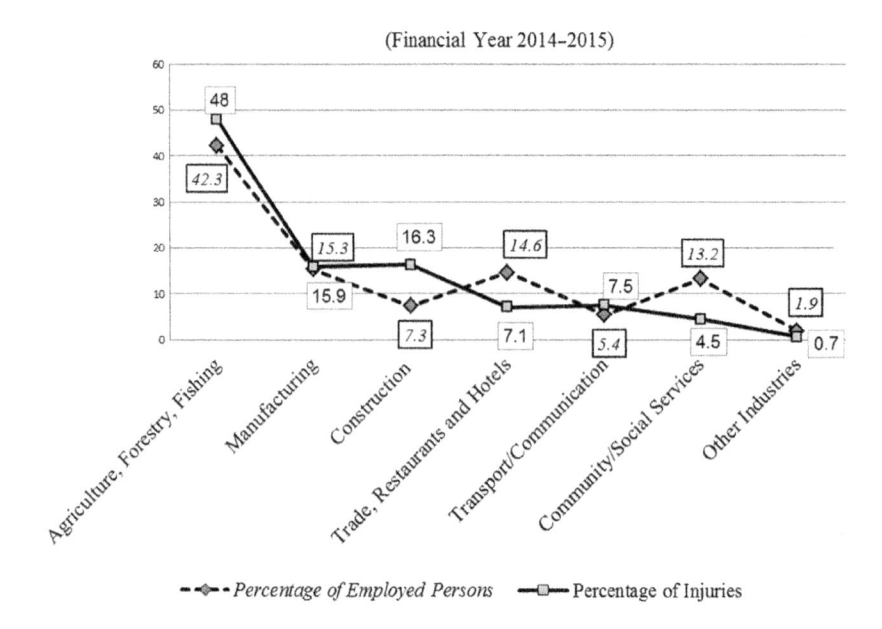

Figure 6.1 Percentage of employment vis-à-vis injuries in major industries of Pakistan. (Source: PBS, 2015)

Despite implementing various safety management systems and providing adequate safety training to employees, unsafe behaviours and unsafe conditions are at times noticed on these projects (Raheem and Issa 2016). Workers are sometimes found working under sizzling heat and extreme weather conditions. As a result, accidents are consistently increasing on these projects. Such a situation warrants investigating the worker's unsafe behaviour and unsafe conditions.

Accident causation factors

Forty under-construction tall building projects (at least 70 meters high) were considered to conduct this research. These projects were located in six major cities of Pakistan: Karachi (27), Lahore (7), Rawalpindi and Islamabad (3), Faisalabad (2), and Hyderabad (1) ('List of the tallest buildings' 2015). *Fall from height, followed by lifting activity and electrocutions were* observed as the leading causes of accidents on these projects. This fact was highlighted in past studies as well (Zahoor et al. 2015a, b; Choudhry et al. 2014; Hassan 2012). Fall accidents, in particular, and their contribution to the overall higher accident rate, in general, encouraged examining the flaws in existing safety practices, as well as, exploring the causes of such accidents. However, due to the non-availability of reliable accident statistics, only the underlying causes of accidents (leading indicators) were investigated. These underlying causes include unsafe behaviour and unsafe conditions, both of which together, contribute to 98% of construction accidents (Hassan 2012). De-chun et al. (2010) also concluded that out of 90% of all construction fatalities, 70% could be prevented through effective commitment on part of the management. Given this, the study evaluated those factors that contributed to accidents on building projects in Pakistan.

Safety climate and safety performance

The most frequently documented safety climate (SC) factors in the CI include perception of managerial commitment and employees' involvement in safety, safety communication, safety training, safety systems and procedures, and workers' attitude to safety and risk (Barbaranelli et al. 2015; Hon et al. 2013; Reader et al. 2015; Wu et al. 2015). As most of the SC studies were conducted in developed countries, they could not be successfully cross-validated across the regions and cultures (Bahari and Clarke 2013; Barbaranelli et al. 2015; Cigularov et al. 2013). Likewise, consensus could not be developed either on the number of SC factors or which factors are the most effective in measuring the SC (Milijic et al. 2013; Zohar 2010; Zhou et al. 2011). Reader et al. (2015) argued to investigate the differences in the dimensions of SC from a more global perspective to determine whether the meaning of SC and its causes and determinants are invariant or not.

Research has shown a strong association between SC and safety performance (SP) (Barbaranelli et al. 2015; Kleiner et al. 2015). The SC is considered to be a strong predictor of SP (Pousette et al. 2008; Zohar 2010). The SP can be measured using safety compliance, safety participation and number of accidents. According to DeArmond et al. (2011), safety participation refers to the behaviours that are voluntary in nature, while safety compliance refers to the behaviours that are compulsory. It is worth mentioning that there is no single measure of SP that can be said to be superior to others, however, safety participation has developed a stronger positive relationship with SP as compared to safety compliance (Clarke 2006). Considering this, the study has developed an exhaustive approach to identify the SC factors that can significantly influence the SP of building projects in Pakistan. In addition, the study evaluated the personal attributes of the employees that can significantly influence SC and SP (Zhou et al. 2008).

Research aim and objectives

This study was aimed at developing the strategies for enhancing the SP on building projects in Pakistan. It was achieved through the following objectives:

1. Identifying the contributory factors of accident causation in the CI of Pakistan
2. Determining the SC factors that can significantly enhance SP

3. Examining the causal relationship between SC factors and SP indicators
4. Examining the effects of demographic variables on SC
5. Recommending the strategies to enhance SP in the CI of Pakistan

Methodology

The study started with an all-inclusive literature review of the safety research work conducted in the CI of Pakistan. In addition, the causes of accidents, the concept of SC and its impact on SP were comprehensively reviewed. A mix-method research strategy, comprising concurrent and sequential mix methods (see Figure 6.2), was adopted for the data collection and analysis (Creswell 2009, p. 16). The quantitative research methods, such as a questionnaire survey and a Delphi survey, and the qualitative research method, such as semi-structured interviews, were adopted as key data collection instruments. Questionnaires for the field survey, Delphi survey and interviews were designed separately. The frontline workers, supervisors, managers and safety officials of key stakeholder organizations, such as clients, consultants and contractors/subcontractors, were targeted for the collection of SC data in Pakistan. The results of the semi-structured interviews were integrated with the quantitative analysis while interpreting the overall findings.

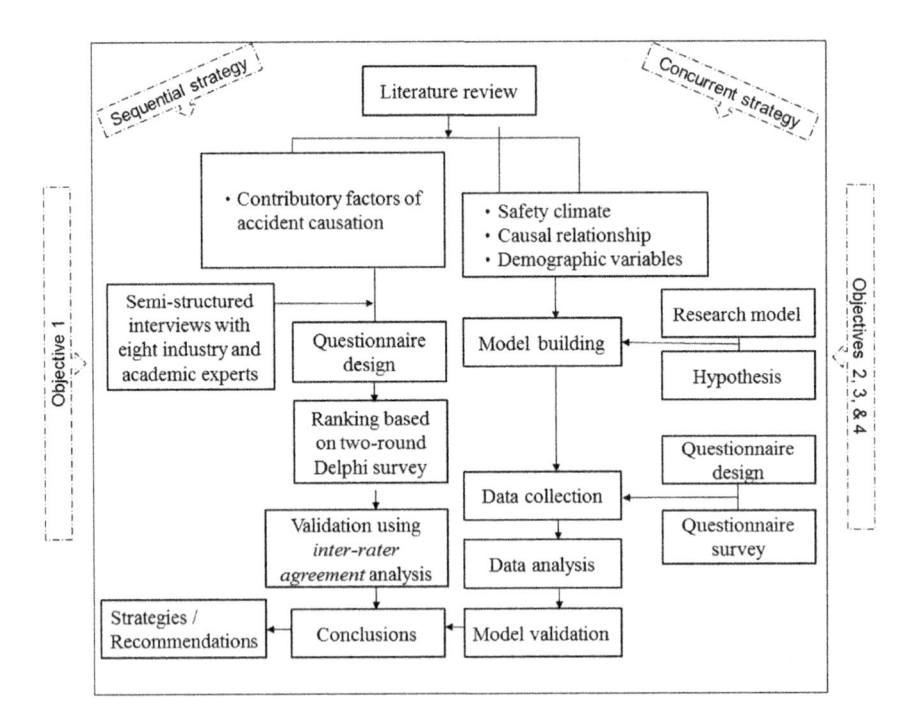

Figure 6.2 Research framework of the study.

The contributory factors of accident causations (Objective 1) were shortlisted through a triangulation strategy between 8 face-to-face semi-structured interviews with academic and industry experts, and a comprehensive literature review of 58 related studies. The targeted interviewees (namely, A to H) were the senior representatives of their respective organizations and had a rich experience of working on building projects in Pakistan. They were also registered members of the Pakistan Engineering Council (PEC). Each interview lasted for approximately an hour. Interviews were tape-recorded and transcribed for interpretation. Two of the interviewees provided printed responses, as they were too busy to be available for an interview. Interviewee A was the professor of OSH in a public university, whereas interviewees B and C represented the client and consultant organizations in the public sector. Three of the interviewees (D, E, and F) belonged to large construction companies/contractors, whereas two of interviewees (G and H) represented the medium-sized construction enterprises employed as subcontractors. The Delphi survey and *inter-rater agreement analysis* were conducted among four respondent groups (clients, contractors, safety officials and academic experts) to prioritize the shortlisted 32 accident causation factors (Okoli and Pawlowski 2004). A two-round Delphi exercise was adopted as it reduces the fatigue and attrition of the industry experts, compared with the repeated rounds (Ameyaw and Chan 2015), and at the same time, it also provides trustworthy feedback and reliable revision of the responses (Mullen 2003). The agreement analysis among respondents' groups was conducted using various statistical methods including mean score ranking, Kendall's coefficient of concordance (W), chi-square analysis, and inter-rater agreement analysis. The details regarding the identification of thirty-two accident causation factors from the literature review, their analysis, prioritization and validation can be read from Zahoor et al. (2017c).

To achieve Objectives 2 to 4, a research model (see Figure 6.3) was hypothesized based on the literature review. The model comprised of four constructs displaying the relationship between SC and three common indicators of SP, namely safety compliance, safety participation, and 'self-reported accidents/injuries and near-misses.' The following five hypotheses were established to examine and validate the influence of SC on the indicators of SP, in the construction of building projects in Pakistan.

1. Hypothesis 1 (H1): SC has a significant positive relationship with safety compliance.
2. Hypothesis 2 (H2): SC has a significant positive relationship with safety participation.
3. Hypothesis 3 (H3): SC has a significant negative relationship with self-reported accidents/injuries and near-misses.
4. Hypothesis 4 (H4): Safety participation has a stronger positive relationship with SP than safety compliance.
5. Hypothesis 5 (H5): SC factors can measure SP significantly.

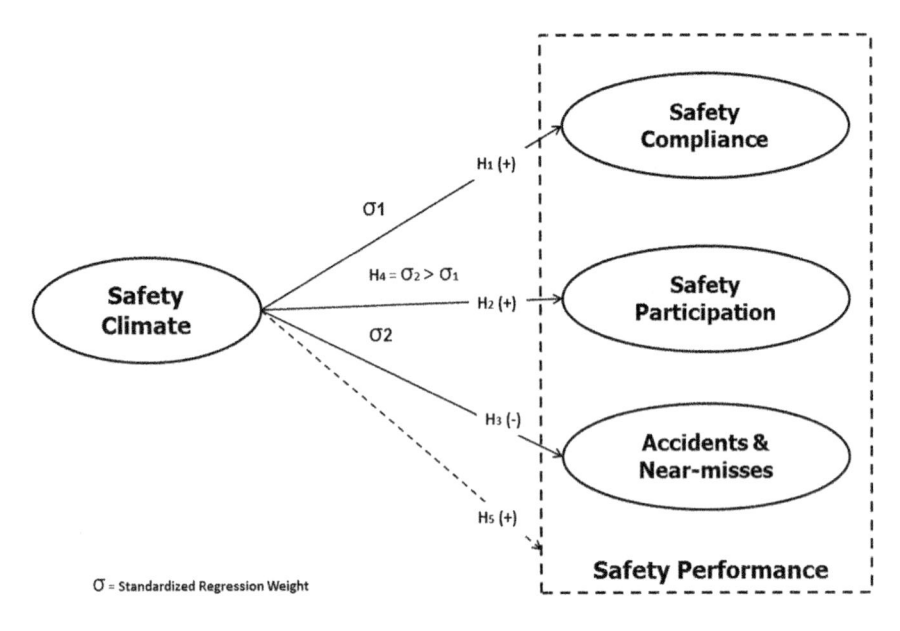

Figure 6.3 Research model and research hypotheses.

The 38-item SC index (SCI) questionnaire of the Occupational Safety and Health Council of Hong Kong (OSHC 2008) was adopted for measuring SC (Hon et al. 2013). The SCI questionnaire, however, was integrated with the suggested changes based on the current safety situation, cultural and regional values, and pilot study. Thus, the finalized SC questionnaire was comprised of 45 items. As the sampling of the questionnaire can significantly affect the generalizability of the findings, an effort was made to collect the responses in a similar environment, from the employees working at different levels and representing various stakeholder organizations including the clients/owners, main contractors, subcontractors, consultants and experts from academia. To reduce the data variability and minimize the bias due to outliers and non-serious responses, treatment of the collected data ($N = 426$) was carried out (Seo et al. 2004). The data were randomly split into calibration and validation sub-samples (Hon et al. 2014). The Statistical Package for the Social Science (SPSS, ver. 19) was utilized to examine the influence of ten demographic variables on SC, and conduct exploratory factor analysis (EFA) for identifying the SC factors and SP indicators using a calibration sub-sample. Confirmatory factor analysis (CFA) was then conducted using Analysis of Moment Structures (AMOS, ver. 20) on a validation sub-sample to validate the results achieved through EFA (Pousette et al. 2008; Seo et al. 2004). The details regarding the literature review, hypotheses development, data suitability for analysis, reliability and validity tests for EFA, model fit indices, composite reliability, convergent validity and discriminant validity for CFA can be read from Zahoor et al. (2017a, 2017b). The identified SC constructs and the effects of demographic variables were then compared with other studies to unveil the variances (Bahari and Clarke 2013; Pousette et al. 2008). Based on the results of EFA, a causal relationship between SC factors and SP indicators was investigated by conducting CFA on the calibration sub-sample (Le et al. 2014). The results were also verified by repeating CFA on the validation sub-sample (Hon et al. 2014). Finally, strategies were recommended for enhancing the SP in the CI of Pakistan (Objective 5).

The research objectives along with their respective methods and analytical techniques are summarized in Table 6.1.

Results and discussion

Contributory factors of accident causation

The study has shortlisted and prioritized the 32 contributory factors of accident causation. The most significant category of accident causation (i.e. *inadequate setting of safety level by the contractor*) can be improved by (1) sufficient provision of safety training and resources by the *contractor* (including personal protective equipment [PPE]), (2) considering past safety performance while selecting the subcontractors, and (3) developing a mechanism at company level for reporting and investigating the accidents and near-misses. Likewise, the second important category of accident causation (i.e. *lack of leadership from the 'Government' as a key client*) can be enhanced by (1) strict enforcement of safety rules and regulations by the government agencies, (2) developing a mechanism to maintain the accident statistics at industry level, and (3) establishing a safety certification system for all the stakeholders at the industry level. Similarly, the third most important accident causation category (i.e. *lack of safety ownership and appreciation by the client*) can be improved by the sufficient allocation of safety budget and safety incentives by the clients.

Identification of significant SC factors

The developed scale (as shown in Figure 6.4) was comprised of 24 SC statements, clustered into following four SC factors: (1) management commitment and employees' involvement in the provision of communication, resources and training (MC&EI); (2) safety enforcement and promotion (SE&P); (3) applicability of safety rules and safe work practices (SR&WP); and (4) amplifying the safety consciousness and responsibility (SC&R).

The factor of MC&EI developed the strongest relationship with the perceived SC, followed by SE&P, SC&R and SR&WP. The factor of SE&P was discovered as the second most influential SC factor, whereas SR&WP was noticed as the most ignored SC factor. The existence of correlation among the error variables of SE&P and SR&WP factors dictated developing a synergy in the enhancement efforts. It implies that the efforts towards safety enforcement and promotion must

Table 6.1 Research methods and analytical techniques to achieve five objectives of the study

Research objectives	Research methods			Analytical techniques				Remarks
	Literature review	Questionnaire survey	Interview	Ranking	Delphi survey	EFA/ SPSS	CFA/ AMOS	
1. Identifying the contributory factors of accident causation in the CI of Pakistan	✓	✓	✓	✓	✓			a. Groups: client, contractor, safety official, academics. b. Two-round Delphi survey. c. Ranking based on mean values. d. Kendall coefficient of concordance and chi-square analysis for inter-group comparison. e. Validation using inter-rater agreement analysis technique.
2. Determining the SC factors that can significantly enhance SP	✓	✓				✓	✓	a. SC data is split into calibration and validation sub-samples for conducting EFA and CFA. b. Reliability and validity tests.
3. Examining the causal relationship between SC factors and SP indicators	✓	✓					✓	a. Hypothesis development. b. Model testing and validation using AMOS on calibration and validation sub-samples.

4. Examining the effects of demographic variables on SC		✓	a. The analysis is based on mean SC scores of all categories of demographic variables.
b. Cross-tabulation.
c. Spearman's rho correlation.
d. Mann-Whitney U test.
e. Kruskal-Wallis test.
5. Recommending the strategies to enhance SP in the CI of Pakistan		✓	Summarized on the basis of the findings of Objectives 1 to 4.

Abbreviations: CI, construction industry; SC, safety climate; SP, safety performance; SPSS, Statistical Package for the Social Sciences; EFA, exploratory factor analysis; AMOS, analysis of moment structures; CFA, confirmatory factor analysis.

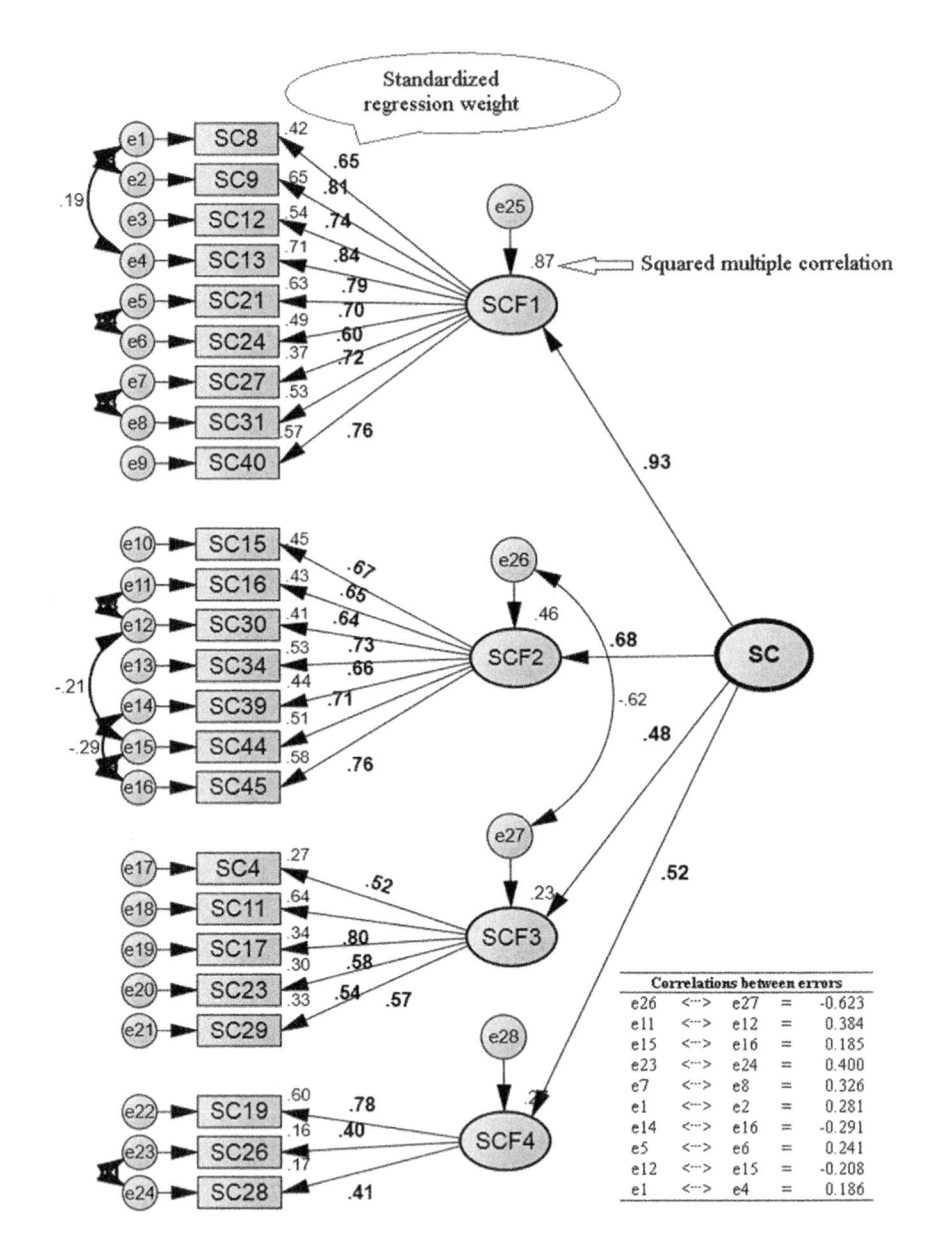

Figure 6.4 Twenty-four items SC scale.

Safety performance 95

be coupled with revising the safety regulations and work practices. As expected, the study could not replicate the SC measurement scale of the Hong Kong CI in the cross-cultural setting of the Pakistani CI. Hence, it has supported the need for considering the region-specific and culture-specific aspects while designing the SC scale. The findings are expected to help construction stakeholders to streamline their safety enhancement initiatives.

The most influential SC factor of MC&EI can be improved by focusing on the four most neglected aspects: (1) ensuring that all the persons on the project site wear their PPE when they are required to (SC12); (2) making adequate arrangements for periodic safety training of the workers (SC9); (3) providing sufficient resources and equipment to all the employees, especially to the subcontractors (SC24); and (4) developing good communication between the higher management and workers (SC21). Similarly, the most overlooked SC factor of SR&WP can be improved by updating and implementing the obsolete safety rules and regulations in the CI (SC4), and streamlining the safe work procedures as per the varying work site conditions (SC29). Likewise, the factor of SE&P can be enhanced by giving due consideration to the suggestions made by the site staff for improving the safe work procedures (SC15), encouraging the employees to work safely by announcing monetary incentives (SC34), and carrying out job hazard analysis regularly before commencing each activity (SC45).

Influence of SC on SP

Based on the hypothesized SP measurement model (see Figure 6.5), complex associations between the identified 4-factor structure of SC encompassing 24

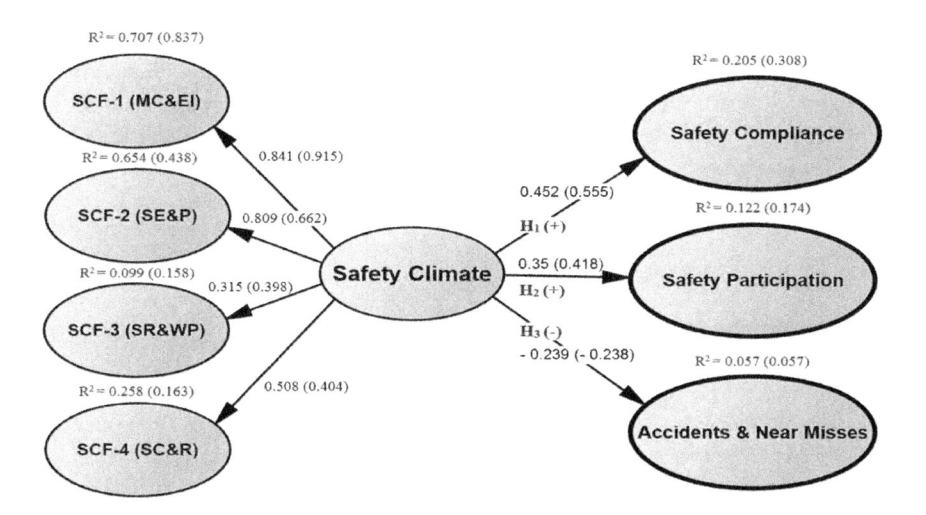

Figure 6.5 Hypothesized model showing the relationship between SC and SP.
(Note: Values in the brackets are the standardized path coefficients for the validation sub-sample.)

observed variables and 3 indicators of SP were examined using AMOS. The SP indicators included safety compliance, safety participation, and the number of self-reported accidents/injuries.

The results, as displayed in Figure 6.6, offered a perfect understanding of the interdependence and causal relationship among various constructs of SC and the indicators of SP. The study has established the significant positive impact of SC on *safety compliance* and *safety participation*, and significant negative impact on the *number of self-reported accidents/injuries*. Noticeably, the hypothesized model could not retain one of the observed variables of SP, i.e. *number of near-misses*. Hence, further investigation is needed to check the applicability of this variable in other regions and cultures, so as to eventually include or exclude it from the SP measurement model. Deviating from past studies (Christian et al. 2009; Clarke 2006; Hon et al. 2014; Hu et al. 2016), the study has established the strongest positive impact of *safety compliance* on SP as compared to *safety participation*. Likewise, it has quantified the *number of accidents/injuries* to have the weakest influence on SC among all the three indicators of SP. These variances in the impact of various SP indicators confirmed the necessity of investigating the safety behaviour in the cross-cultural and cross-regional environment, especially in the developing countries.

Therefore, construction stakeholders are suggested to focus more on *safety compliance* than the *safety participation* for achieving a better performance level. Employees also need to be motivated to comply with obligatory safe procedures and work practices. Such efforts can be augmented by inculcating a higher degree of self-motivation among the employees through various incentives to enhance their *safety participation*. Though safety participation will not contribute directly to personal safety, it will definitely nurture the culture of voluntary participation in the organization. The lower mean value of *safety rules and work practices* signified the inadequate enactment and incompatibility of safety rules to the worksite requirements. Hence, there is a need to update and implement the safety rules and regulations, matching with the rapid technological advancement in the industry. Similarly, the mean performance level for the factor of *management commitment & employees' involvement* was relatively low; however, it could attain the strongest impact on SC. This implies that a little attention by the senior management on this factor will significantly enhance SP.

Effect of demographic variables on SC

Ten demographic variables were analyzed for their associations with each other, and for their influence on the overall SC. The results have pointed out that the employees who are married, aged *over 50 years*, supporting *7 or more* family members, working for *up to 40 hours* per week, having *6–10 years* of service in the current company and/or *more than 15 years* of industry experience attained a relatively higher SC level. In contrast, employees in the age group of *20 years or below*, working for *more than 40 hours* per week, having a *low* level of education, and/or *less than 1 year* of experience in the current company, achieved a relatively lower SC level. This group can be targeted for the enhancement of their safety

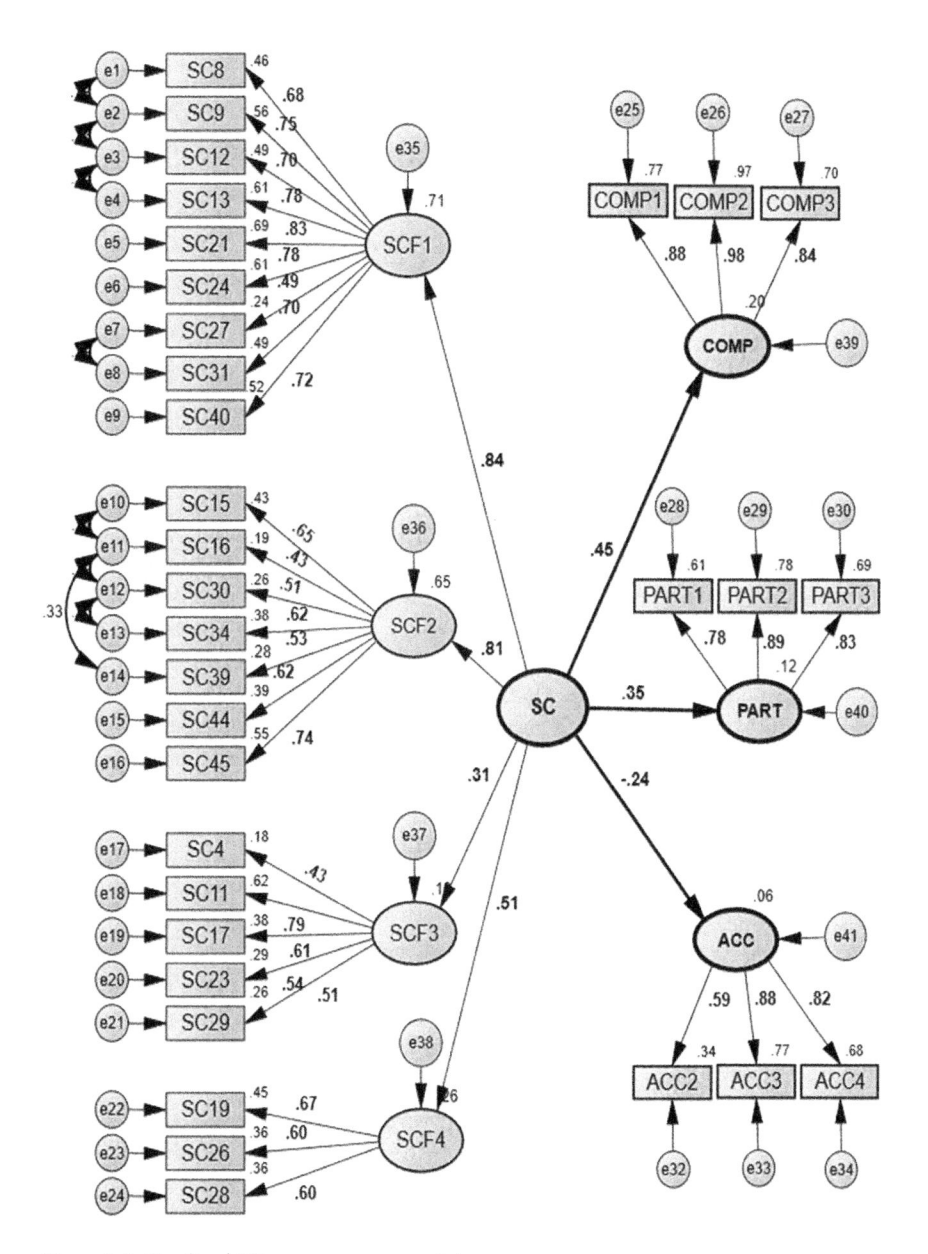

Figure 6.6 Finalized SP measurement model.

behaviour through safety education, promotion, and training. It is believed that a joint focus on both the highlighted personal attributes and the significant SC factors will enhance the safety behaviour of workers in the CI.

Strategies for safety enhancement on building projects in Pakistan

Based on the study's results, strategies are proposed that can significantly enhance the SP on building projects in the CI of Pakistan.

1. The Pakistan Engineering Council (PEC) is recommended to (1) institutionalize the accident reporting and examination mechanism at the industry and company levels; (2) revise the contracting and bidding documents so as to ensure the allocation of the safety budget by the client at the time of contracting; (3) incorporate the safety credit points in the process of contractor licensing, renewal and selection; and (4) establish a safety certification system for employees' induction in the CI.
2. Construction accidents can be reduced by (1) strict enforcement of safety rules and regulations by government agencies, (2) guaranteed provision of safety budget and safety incentives by the clients, (3) adequate provision of safety training and resources by contractors, and (4) giving due importance to past safety performance while selecting the subcontractors.
3. A focus on the four identified SC factors can also enhance SP on building projects. *Management commitment and employees' involvement* can ensure safety communication; and the provision of safety resources, equipment and training. Likewise, *safety enforcement and promotion* can be guaranteed by implementing the site staff's suggestions, announcing monetary incentives and carrying out job hazard analysis. *Applicability of safety rules and safe work practices* can be fortified by updating obsolescent rules and regulations, and streamlining the safe work procedures. Last, paying attention to *safety consciousness and responsibility* among all the employees and stakeholders can also enhance SP.
4. Last but not the least, employees in the age group of *20 years or below*, working for *more than 40 hours* per week, having a *low* level of education, and/or *less than 1 year* of experience in the current company need to be targeted for enhancing their safety behaviour through safety education, promotion, and training.

Conclusion

The study has prioritized the contributory factors of accident causation, and developed a 24-item SC scale comprised of four SC factors (MC&EI, SE&P, SR&WP, and SC&R), which can concisely measure the SC on building projects in the CI of Pakistan. It has also pointed out the variations in the interrelationships among the constructs of SC and SP indicators. Besides, the study has recommended the strategies for safety enhancement on building projects in Pakistan. Validation of the study's results has substantiated the diligence of the results and robustness of the adopted methodology. The methodologies adopted for the Delphi survey data analysis, the validation of the results using inter-rater agreement analysis technique, identification of the SC factor structure, and investigation of the causal relationship can be replicated in any industry and region.

Theoretical and practical implications

The identified accident causation factors can guide key stakeholders in prioritizing their efforts towards eliminating the root causes of construction accidents. The study has highlighted the repercussions of adopting an existing SC questionnaire in an entirely different region, and confirmed the necessity of integrating the region and culture-specific dimensions while measuring SC. Besides, *Safety enforcement and promotion* has been discovered as a new SC factor. It is believed that dimensions related to this factor might have been split among other SC factors in previous studies. The study's results are anticipated to assist construction professionals and safety practitioners to measure, benchmark, monitor, and enhance the SP of their construction enterprises.

Future research directions

This research offers a valuable starting point for future safety interventions in the Pakistani CI. As the employees' perceptions can change over time, a longitudinal study is advocated to examine the steadiness of the developed SC scale, and obtain a more comprehensive and coherent picture of the antecedents of SC in the CI of developing countries, in general, and Pakistan, in particular. Realizing the varied influence of cultural and regional aspects on SC, the designed SC scale can be tested in other developing regions and cultures, especially in the non-English speaking countries. It will help to further probe into the consequences of cross-validation. It will also assist in investigating the applicability of the observed variable of *number of near-misses* in other regions and cultures, as it could not be retained in the SP measurement model of this study. There is also a need to revise the safety regulations and update the safety clauses in the contract documents. The development of a mechanism at the industry level needs to be established for the maintenance of accident statistics so that the database can be used to quantitatively analyze the causes of construction accidents. Most importantly, a cost-benefit analysis between the safety investment and the cost of accidents, in the regional context, may be conducted to convince the construction stakeholders regarding the professed benefits of investing in safety.

Acknowledgement

The work described in this chapter was fully funded by the International Postgraduate Scholarship of the Hong Kong Polytechnic University. The suggestions made by the interviewees and the respondents to the questionnaire surveys are gratefully appreciated.

References

Ameyaw, E. E., and Chan, A. P. C. (2015). 'Evaluating key risk factors for PPP water projects in Ghana: A Delphi study.' *Journal of Facilities Management*, 13(2), 133–155.

Bahari, S. F., and Clarke, S. (2013). 'Cross-validation of an employee safety climate model in Malaysia.' *Journal of Safety Research*, 45, 1–6.

Barbaranelli, C., Petitta, L., and Probst, T. M. (2015). 'Does safety climate predict safety performance in Italy and the USA? Cross-cultural validation of a theoretical model of safety climate.' *Accident Analysis and Prevention*, 77, 35–44.

Choudhry, R. M., Tariq, B., and Gabriel, H. F. (2014). 'Investigation of fall protection practices in the construction industry of Pakistan.' *CIB W099 International H&S Conference, Achieving Sustainable Construction Health & Safety*, 2–3 June, Sweden, pp. 211–220.

Christian, M. S., Bradley, J. C., Wallace, J. C., and Burke, M. J. (2009). 'Workplace safety: A meta-analysis of the roles of person and situation factors.' *Journal of Applied Psychology*, 94(5), 1103–1127.

Cigularov, K., Lancaster, P. G., Chen, P. Y., Gittleman, J., and Haile, E. (2013). 'Measurement equivalence of a safety climate measure among Hispanic and White Non-Hispanic construction workers.' *Safety Science*, 54, 58–68.

Clarke, S. (2006). 'The relationship between safety climate and safety performance: A meta-analytic review.' *Journal of Occupational Health Psychology*, 11(4), 315–327.

Creswell, J. W. (2009). ' Chapter 1: A Framework for Design'. In *Research Design: Qualitative, Quantitative, and Mixed Methods Approaches*. Sage, London.

DeArmond, S., Smith, A. E., Wilson, C. L., Chen, P. Y., and Cigularov, K. P. (2011). 'Individual safety performance in the construction industry: Development and validation of two short scales.' *Accident Analysis and Prevention*, 43, 948–954.

De-chun, N., Jian-ping, W., and Guo-dong, N. (2010). 'Analysis of factors affecting safety management in construction projects.' *International Conference of IEEE on Management and Service Science (MASS)*, pp. 1–5. https://ieeexplore.ieee.org/document/5576911 (accessed on 7 September 2018).

Hassan, S. A. (2012). 'Health, safety and environmental practices in the construction sector of Pakistan.' Masters dissertation, Uppsala University.

Hon, C. K. H., Chan, A. P. C., and Yam, M. C. H. (2013). 'Determining safety climate factors in the repair, maintenance, minor alteration, and addition sector of Hong Kong.' *ASCE's Journal of Construction Engineering and Management*, 139(5), 519–528.

Hon, C. K. H., Chan, A. P. C., and Yam, M. C. H. (2014). 'Relationships between safety climate and safety performance of building repair, maintenance, minor alteration, and addition (RMAA) works.' *Safety Science*, 65, 10–19.

Hu, X., Griffin, M., and Bertuleit, M. (2016). 'Modelling antecedents of safety compliance: Incorporating theory from the technological acceptance model.' *Safety Science*, 87, 292–298.

Kleiner, B. M., Hettinger, L. J., DeJoy, D. M., Huang, Y., and Love, P. E. D. (2015). 'Sociotechnical attributes of safe and unsafe work systems.' *Ergonomics*, 58(4), 635–649.

Le, Y., Shan, M., Chan, A. P. C., and Hu, Y. (2014). 'Investigating the causal relationships between causes of and vulnerabilities to corruption in the Chinese public construction sector.' *ASCE's Journal of Construction Engineering and Management*, 140(9), 1–12.

'List of the tallest buildings in Pakistan – Under construction buildings.' (2015). http://goo.gl/o3RLX4; https://goo.gl/uphSVU (accessed on 25 August 2018).

Milijic, N., Mihajlovic, I., Strbac, N., and Zivkovic, Z. (2013). 'Developing a questionnaire for measuring safety climate in the workplace in Serbia.' *International Journal of Occupational Safety and Ergonomics*, 19(4), 631–645.

Mullen, P. M. (2003). 'Delphi: Myths and reality.' *Journal of Health Organization and Management*, 17(1), 37–52.

Okoli, C., and Pawlowski, S. D. (2004). 'The Delphi method as a research tool: An example, design considerations and applications.' *Information and Management*, 42(1), 15–29.

OSHC. (2008). 'Construction industry safety climate index questionnaire.' Hong Kong Occupational Safety and Health Council (OSHC). Hong Kong. https://goo.gl/C1dOax (accessed on 31 October 2018).

PBS. (2009). Pakistan Bureau of Statistics – Labour Force Statistics (2008–2009). Islamabad. http://goo.gl/xpfRKF (accessed on 27 August 2018).

PBS. (2015). Pakistan Bureau of Statistics – Labour Force Statistics (2014–2015). Islamabad. http://goo.gl/ZHsbrX (accessed on 18 September 2018).

Pousette, A., Larsson, S., and Torner, M. (2008). 'Safety climate cross-validation, strength and prediction of safety behavior.' *Safety Science*, 46, 398–404.

Raheem, A., and Issa, R. (2016). 'Safety implementation framework for Pakistani construction industry.' *Safety Science*, 82, 301–314.

Reader, T., Noort, M., Shorrock, S., and Kirwan, B. (2015). 'Safety sans frontières: An international safety culture model.' *Risk Analysis*, 35(5), 770–789.

Seo, D. C., Torabi, M. R., Blair, E. H., and Ellis, N. T. (2004). 'A cross-validation of safety climate scale using confirmatory factor analytic approach.' *Journal of Safety Research*, 35, 427–445.

Wu, C., Song, X., Wang, T., and Fang, D. P. (2015). 'Core dimensions of the construction safety climate for a standardized safety-climate measurement.' *ASCE's Journal of Construction Engineering and Management*, 141, 1–12.

Zahoor, H., Chan, A. P. C., Choudhry, R. M., Utama, W. P., and Gao, R. (2015a). 'Construction safety research in Pakistan: A review and future research direction.' *7th International Civil Engineering Congress (ICEC-2015)*, 12–13 June, Karachi, pp. 1–8. https://goo.gl/WxyQvN (accessed on 19 July 2018).

Zahoor, H., Chan, A. P. C., Utama, W. P., and Gao, R. (2015b). 'A research framework for investigating the relationship between safety climate and safety performance in the construction of multi-storey buildings in Pakistan.' *Procedia Engineering*, 118, 581–589.

Zahoor, H., Chan, A. P. C., Utama, W. P., Gao, R., and Memon, S. A. (2017a). 'Determinants of safety climate for building projects: SEM-based cross-validation study.' *ASCE's Journal of Construction Engineering and Management*, 143(6), 05017005.

Zahoor, H., Chan, A. P. C., Utama, W. P., Gao, R., and Zafar, I. (2017b). 'Modeling the relationship between safety climate and safety performance in developing countries: A cross-cultural validation study.' *International Journal of Environmental Research and Public Health*, 14(4), 351, 1–19.

Zahoor, H., Chan, A. P. C., Gao, R., and Utama, W. P. (2017c). 'The factors contributing to construction accidents in Pakistan: Their prioritization using the Delphi technique.' *Engineering, Construction and Architectural Management*, 24(3), 463–485.

Zhou, Q., Fang, D. P., and Wang, X. (2008). 'A method to identify strategies for the improvement of human safety behavior by considering safety climate and personal experience.' *Safety Science*, 46(10), 1406–1419.

Zhou, Q., Fang, D. P., and Mohamed, S. (2011). 'Safety climate improvement: Case study in a Chinese construction company.' *ASCE's Journal of Construction Engineering and Management,* 137(1), 86–95.

Zohar, D. (2010). 'Thirty years of safety climate research: Reflections and future directions.' *Accident Analysis and Prevention,* 42(5), 1517–1522.

7 Health and safety performance in the Ugandan construction industry

Moses Okwel, Henry Alinaitwe and Denis Kalumba

Summary

The construction industry is possibly the most hazardous industry with regard to the health and safety (H&S) of workers. In Uganda, there has been poor management of H&S at construction sites and, as a result, the industry registers increased frequencies of safety incidents. These incidents have led to loss of lives, injuries, damage to property and equipment, reduced productivity and loss of revenue, amongst other pitfalls. Unfortunately, these H&S incidents are reported after the occurrence and do not show preventive measures undertaken. The aim of this study was to analyse the H&S performance in Uganda's construction industry in order to propose measures for effective preventive management. Construction firms that had active construction projects within the Kampala, Mukono and Wakiso Districts during the study period were assessed. The data on performance were collected using questionnaires and observation checklists. Data on accident records were collected using documented accident reports. The results showed that contracting firms were generally aware of the need to uphold good H&S practices, but that only 35% of the H&S programmes were implemented at a good level or above. On average, 40% of the construction site practices were generally unsafe. Objectively, H&S performance was characterised by a high accident injury rate (20.2), non-fatal injury rate (18.2) and fatal injury rate (2.0) per 100 equivalent full-time workers (EFTW). The Ugandan government should incorporate the identified 17 H&S programmes into a formal regulation and strengthen the laws governing H&S in the construction industry. Construction firms should train their workers on H&S requirements in order to improve their H&S regimes.

The high rate of fatal accidents in construction

Uganda's construction sector grew at 13% per annum for the period 2005–2015 (Uganda Bureau of Statistics [UBOS], 2016), twice the national economy average growth rate of 6.5%. Despite its important contribution of 12.9% of the national gross domestic product (GDP) in year 2015 (UBOS, 2016), the construction industry experienced very high numbers of fatal accidents during the period 1996–2006. Over this period, collapses at building construction sites were major occurrences (Mwakali, 2006). The main incidents identified included the

Bwebajja building construction accident that occurred on 1 September 2004 with 11 deaths and over 26 serious injuries (Ministry of Works and Transport [MoWT], 2004); the St. Peter's Naalya school building construction accident with 11 deaths and over 10 serious injuries (Ssempogo et al., 2008); and the National Social Security Fund (NSSF) building construction accident with 8 deaths and 1 serious injury (Musoke et al., 2008; MoWT, 2008). Globally, the construction industry is perceivably the most hazardous industry with regard to the health and safety (H&S) of workers (Edmonds & Nicholas, 2002).

Health and safety at construction sites

The health and safety of individuals who are associated with the construction process should be of major concern. The occupational safety and health (OSH) aspect should not be disregarded in the process of pursuing project completion or improving the economy (Jafaar et al., 2017). Handling of OSH depends on the risk assessment process and the coherence of decisions taken to eliminate or reduce risk. Kartam (1997) developed a system to integrate H&S incident prevention issues into all phases of a construction project from design and planning through to construction and maintenance. The system was based on the 'three E's' of H&S: engineering, education and enforcement. However, the main obstacle to the implementation of construction safety improvements is that both the contractor and the client/employer often perceive H&S as a cost rather than a benefit. Mwakali (2006) noted that serious accidents and injuries had occurred with alarming frequency at construction sites in Uganda. Alinaitwe et al. (2007) asserted that accidents at construction sites occurred as a result of lack of knowledge or training; poor supervision; lack of proper means to execute tasks safely; or because of carelessness, apathy or downright recklessness. Abdul et al. (2008) was of the view that accidents do not just happen; they are caused by unsafe acts, unsafe conditions or both. Over time, many theories have evolved regarding the cause of accidents. Some of these theories include the single factor theory, multiple causation theory and psychological accident causation theory (Ahmed & Adnan, 2005). Arguably, all accidents are multi-causal, with a combination of factors needing to coincide to give rise to an accident (Mwakali, 2006).

The construction industry is arguably one of the worst performing with respect to occupational health. A report from Health and Safety Executive (HSE, 2007) shows that for 2006–2007, respiratory diseases in construction have the highest incidence rate across all other industries in the United Kingdom (UK). Occupational cancer deaths amongst construction workers and tradesmen reflect the risks inherent within particular construction work processes, environments and materials. They also reflect the failure by the industry to prevent or control exposures and to manage health problems adequately (Donaghy, 2009). For too long, health has had minimal attention when compared with safety (Lin & Mills, 2001). To secure improvement, the industry needs to manage both H&S issues as an integral part of its day-to-day business management – to manage the risk, not the symptom (Donaghy, 2009).

Research has indicated that H&S on construction sites has remained a formidable problem for construction management (Vatin, Gumayunova & Petrosova, 2014)). Mohamed (2003) asserted that the construction industry suffers from the general inability to manage workplace H&S at a level where a pro-active, zero-accident culture prevails. On the other hand, Aksorn and Hadikusumo (2008, 2007) and Idoro (2008) argued that continuous improvement of H&S can be attained through performance evaluations to guide management on future undertakings. These performance evaluations can be done through reactive and pro-active approaches. Previous studies in Uganda's construction industry have focused mainly on accidents and injuries (Alinaitwe et al., 2007; Lubega et al., 2000; Mwakali, 2006). However, these accident studies have had limitations in that they reflect a negative image of contractors and reactive information that might not bring about improvement (Marosszeky et al., 2004). This shortcoming presented a gap for this study of H&S performance in the construction industry in Uganda. An alternative approach is to focus on pro-active efforts dealing with the factors responsible for such accidents, injuries and ill-health, and how to control them (Mohamed, 2003). This approach involves both objective and subjective measurement of H&S preventive performance. Objective measurement of H&S performance is through the evaluation of accident statistics, while subjective measurement is through evaluation of the implementation of H&S programmes as well as construction practices (unsafe acts and conditions). Research studies (Mohamed, 2003; Musonda, 2012) brought out H&S programmes that were effective in the improvement of H&S performance in the construction industry. These key H&S programmes included H&S policy, safety committee, safety induction, H&S training, H&S inspection, accident investigation, first aid programme, in-house safety rules, safety incentive, control of sub-contractors and selection of employees. Others are personal protection programmes (PPP), emergency preparedness planning (EPP), safety-related promotion, safety auditing, safety record keeping and job hazard analysis (JHA). These H&S programmes were thus used in the current study as part of the pro-active measurement of H&S performance.

Methodology

H&S preventive performance in Uganda's construction industry was analysed using two main approaches, namely subjective measurement and objective measurement. A total of 138 construction firms that had active construction projects were purposively selected for study within the Kampala, Mukono and Wakiso Districts. Data for the subjective measurement of H&S performance were obtained through the use of questionnaires for assessing the implementation of 17 H&S programmes and observation forms for assessing unsafe acts and unsafe conditions at construction sites. Data for the objective measurement of H&S performance were obtained through accident forms from the reported accident records at various relevant government authorities and recorded accidents at the construction sites of the projects under study. The accident report form was developed with guidance from similar previous studies including those by Alinaitwe et al. (2007) and Mwakali (2006), for example, and 'Form 300' of the Oregon Occupational Safety and Health Division from the United States (Oregon OSHA, 2015).

Observation forms included observed items and criteria for unsafe scores developed in compliance with the Occupational Safety and Health Act (2006) and verified by a form used in a similar previous study in Thailand (Aksorn & Hadikusumo, 2008). The observation of workers' practices at a construction site was conducted first. Results for each construction project were tallied, analysed and combined with results of the whole study. Data collection through questionnaires and observations at construction sites was executed during a 4-month period from February to June 2015.

The accident injury rate (AIR), fatal injury rate (FIR) and non-fatal injury rate (NIR) were calculated per 100 equivalent full-time workers (EFTW) in accordance with the formulae used by the United States (US) Department of Labour (BLS, 2011):

$$AIR = \frac{N_a}{EH} \times 200{,}000 \qquad (7.1)$$

$$FIR = \frac{N_f}{EH} \times 200{,}000 \qquad (7.2)$$

$$NIR = \frac{N_n}{EH} \times 200{,}000 \qquad (7.3)$$

where
N_a = Total number of persons that sustained injuries (fatal + non-fatal)
N_f = Total number of persons that sustained fatal injuries (death)
N_n = Total number of persons that sustained non-fatal injuries
E_H = Total number of hours worked by all employees during the calendar year
200,000 = Base for 100 EFTW (working 40 hours per week, 50 weeks per year)

The unsafe acts (UA) and unsafe conditions (UC) observation indices were calculated using the following equations and reported as percentages (Aksorn & Hadikusumo, 2008):

$$UA\ Observation\ Index = \frac{Unsafe\ acts}{Unsafe\ acts + Safe\ acts} \times 100\% \qquad (7.4)$$

$$UC\ Observation\ Index = \frac{Unsafe\ acts}{Unsafe\ conditions + Safe\ conditions} \times 100\% \qquad (7.5)$$

Data were analysed using the Statistical Package for Social Sciences (SPSS) version 17 and Microsoft Excel software, 2007. The actual status of implementation of the H&S programmes was analysed and their rankings were determined using the coefficient of variation (CV). Multiple regression analysis was done to investigate the relationships of H&S programmes with performance. Cronbach's alpha (Bonnet & Wright, 2015) and randomly split sample reliability tests (two parts) were conducted to determine the internal consistency of the data. The values of Cronbach's alpha were 0.94 for the whole data set, and 0.95 and 0.88 for parts 1 and 2 of the split data respectively. All alpha values were greater than the recommended minimum value of 0.7 (Bonnet & Wright, 2015).

Results and discussions

A total of 125 responses were received from the expected 138, of which 84.8% were found valid and 15.2% found invalid owing to incomplete data for analyses. For the unsafe scores, a total of 117 construction sites corresponding to firms participating in the questionnaire survey were visited. This was to allow inferential analysis of H&S performance across safety programmes being implemented.

Overall ranking of the H&S programmes

The overall ranking of H&S programmes according to their CV is given in Table 7.1. The six most highly rated H&S programmes are circled with CV less than 0.50. Three of the five best-ranked H&S programmes differ from those found out by Aksorn & Hadikusumo (2008) in their study of the Thai construction industry.

Unsafe acts and unsafe conditions

The totals obtained from the observations were unsafe acts = 5105; safe acts = 8796; unsafe conditions = 513; and safe conditions = 685. Hence, from Equations 7.4 and 7.5:

$$\text{Unsafe act observation Index} = \frac{5105}{5105+8796} \times 100\% = 36.7\%$$

The unsafe act observation index was 36.7 percent. This implied that 36.7% of the methods and actions of workers used in the execution of the construction activities were unsafe, potentially putting the lives of workers and the safety of property in danger.

$$\text{Unsafe conditions observation Index} = \frac{513}{513+685} \times 100\% = 42.8\%$$

Table 7.1 Ranking of the status of H&S programmes according to coefficient of variation (CV)

H&S programme	Mean (μ)	Standard deviation (σ)	$V = \dfrac{\sigma}{\mu}$	Ranking*
P1: H&S policy	2.81	1.016	0.362	②
P2: Safety committees	2.94	1.206	0.410	③
P3: Safety inductions	2.32	1.334	0.575	11
P4: H&S training	2.49	1.436	0.577	12
P5: Safety inspections	2.45	1.885	0.560	10
P6: Accident investigation	2.93	1.031	0.352	①
P7: First-aid programmes	2.57	1.225	0.477	⑥
P8: In-house safety rules	1.88	1.098	0.584	13
P9: Safety incentives	1.43	0.880	0.615	15
P10: Control of subcontractors	2.72	1.154	0.424	④
P11: Selection of employees	2.76	1.171	0.424	④
P12: PPP	2.03	1.118	0.551	9
P13: EPP	1.89	1.406	0.744	17
P14: Safety promotions	1.3	0.713	0.548	8
P15: Safety auditing	2.1	1.242	0.592	14
P16: Safety record keeping	2.41	1.252	0.520	7
P17: JHA	1.86	1.249	0.672	16

* The most highly rated H&S programmes are circled.

The unsafe condition observation index was 42.8%. It depicts that 42.8% of the working conditions at the construction sites were unsafe for workers.

Accidents at construction sites

The numbers of accident cases from the three districts of Uganda (study area) for both on-site and reported records are shown in Table 7.2. Comparison of on-site accident data and those obtained from the labour offices in this study indicates that only 24% of accidents on site were actually reported to the government authorities. Based on the on-site accident records and Equations 7.1, 7.2 and 7.3, the AIR (20.2), FIR (2.0) and NIR (18.2) were calculated per 100 EFTW. The results showed an unacceptable objective level of H&S performance. This trend contrasts with those in developed economies like the UK, US, Australia and Thailand where there has been a tremendous decline in accident cases. Over the same period (2006–2010), the non-fatal work injury rate in the US's private construction sector declined by over 20% from 4.4 to 3.5 per 100 EFTW (BLS, 2011).

The results in Table 7.2 show the distribution of tradesmen and causes of recorded accidents during the period 2006–2010. The findings show that most victims of accidents were labourers, followed by painters, carpenters, electricians and supervisors.

Table 7.2 Distribution of tradesmen, nature and causes of recorded accidents (2011) and reported accidents (2006–2010) at construction sites

Indicators		On-site recorded accidents (2011)	Percent (%)	Reported accidents (2006–2011)	Percent (%)
SUM		628		150	
AIR		20.0		—	
Type of worker involved	Labourer	292	46.5	59	39.3
	Carpenter	54	8.6	13	8.7
	Concreter	18	2.9	4	2.7
	Electrician	20	3.2	12	8.0
	Painter	96	15.3	28	18.7
	Plaster	11	1.8	6	4.0
	Steel fixer	19	3.0	3	2.0
	Plumber	17	2.7	2	1.3
	Masons	13	2.1	1	0.7
	Welder	22	3.5	8	5.3
	Technician	27	4.3	6	4.0
	Engineers	18	2.9	5	3.3
	Foremen	21	3.3	3	2.0
	Total	628	100.0	150	100
Nature of injury incurred	Fatal injury	63	10.0	16	10.7
	Major injury	453	72.2	101	67.3
	Minor injury	112	17.8	33	22.0
	Total	628	100	150	100
Cause of accident	Fall (from height)	94	15.0	24	16
	Cut	41	6.5	13	8.7
	Hit by object	154	24.5	43	28.7
	Vehicle	12	1.9	3	2.0
	Machine	81	12.9	16	10.7
	Electricity	31	4.9	13	8.7
	Falling debris	55	8.8	8	5.3
	Chemical	60	9.6	4	2.7
	Suffocation	44	7.0	14	9.3
	Burnt	56	8.9	12	8.0
	Total	628	100	150	100

The results were consistent with the outcome of previous studies and accident statistics in other developed countries (Alinaitwe et al., 2007; BLS, 2010; HSE, 2007). Alinaitwe et al. (2007) suggested that labourers are more vulnerable to construction site accidents because they are mostly unskilled and are the predominant labour component at construction sites, especially where there are labour-intensive activities such as building sites. The three major causes of both the reported and on-site recorded accidents were hit by object, fall (from height) and machine. The manual performance of most construction site tasks and the poor implementation status of some H&S programmes like PPP, H&S inspections and JHA could be the major reason. Comparatively, fall from height is the major cause of accidents even in developed countries like the UK and US.

For over a decade (1996–2008), fall from height accounted for about 50% of all fatal accidents in the UK's construction industry (HSE, 2009).

H&S performance in the construction projects

The relationship of implementation between H&S preventative programmes and H&S performance was investigated using multiple regression analysis. The H&S programmes were the independent variables, while unsafe scores for AIR calculated from on-site recorded accident cases were used as the dependent variables.

The results of regression analysis indicated that the models reflected a set of the most effective H&S programmes for H&S performance improvement. The results of the analysis are tabulated in Table 7.3. The regression analysis created three models. First, the model signifying that 5 of the 17 H&S programmes, namely H&S inspection (P5), in-house safety rules (P8), control of sub-contractors (P10), safety auditing (P15) and job hazard analysis (P17), were the most effective in reducing the AIR at the construction sites. The model can be expressed as follows:

$$AIR = 39.234 - 9.171P5 - 7.271P8 - 7.011P10 - 7.917P17 \qquad (7.6)$$

The coefficient of determination, R^2, of the model was 0.74 indicating that approximately 75% of the accident injury rates could be accounted for by these five H&S programmes.

The second and third models were identical and reflected the most effective programmes regarding the occurrence of unsafe practices (unsafe acts and unsafe conditions). Four of the 17 H&S programmes, namely H&S policy (P1), H&S inspection (P5), PPP (P12) and safety auditing (P15), were found to be most effective in minimising the occurrences of unsafe acts (UA) and unsafe conditions (UC). The model can be represented mathematically as follows:

$$UA = 25.22 - 6.032P1 - 2.777P5 - 5.511P12 - 6.508P15 \qquad (7.7)$$

$$UC = 26.237 - 7.528P1 - 3.333P5 - 3.981P12 - 7.248P15 \qquad (7.8)$$

The value of R^2 of the UA model was 0.56, indicating that 56% of the variation in the level of occurrences of unsafe acts could be explained by these four H&S programmes. The value of R^2 of the UC model was 0.68, indicating that 68% of the variation in the occurrences of unsafe conditions could be explained or controlled by these four H&S programmes. UA and UC were associated with the same H&S programmes, signifying that the same efforts directed at alleviating the status of either unsafe practice at the construction site would result in double success. The models further posited that UA and UC practices at construction sites were inter-related. Therefore, management efforts have to focus on both equally in the quest to improve H&S performance (Aksorn & Hadikusumo, 2007).

Table 7.3 Regression analysis of the effects of H&S programmes on H&S performance

Independent variables (H&S programmes)	Dependent variables									
	AIR (accident injury rate)			UA (unsafe acts)			UC (unsafe conditions)		
	Unstandardized coefficients (β)	Standardized coefficients (α)	Ranking	Unstandardized coefficients (β)	Standardized coefficients (α)	Ranking	Unstandardized coefficients (β)	Standardized coefficients (α)	Ranking
P1: H&S policy	—	—	—	−6.032	−0.451	2	−7.538	−0.462	2
P2: Safety committees	—	—	—	—	—	—	—	—	—
P3: Safety inductions	—	—	—	—	—	—	—	—	—
P4: H&S training	−9.171	−0.706	1	—	—	—	—	—	—
P5: H&S inspections	—	—	—	−2.777	−0.273	4	−3.333	−0.277	3
P6: Accident investigation	—	—	—	—	—	—	—	—	—
P7: First aid programmes	—	—	—	—	—	—	—	—	—
P8: In-house safety rules	−7.271	−0.426	5	—	—	—	—	—	—
P9: Safety incentives	—	—	—	—	—	—	—	—	—
P10: control of subcontractors	−7.011	−0.469	3	—	—	—	—	—	—
P11: Selection of employees	—	—	—	—	—	—	—	—	—
P12: PPP	—	—	—	−5.511	−0.449	3	−3.981	−0.274	4
P13: EPP	—	—	—	—	—	—	—	—	—

(Continued)

Table 7.3 Continued

Independent variables (H&S programmes)	Dependent variables									
	AIR (accident injury rate)			UA (unsafe acts)			UC (unsafe conditions)		
	Unstandardized coefficients (β)	Standardized coefficients (α)	Ranking	Unstandardized coefficients (β)	Standardized coefficients (α)	Ranking	Unstandardized coefficients (β)	Standardized coefficients (α)	Ranking
P14: Safety promotions	—	—	—	—	—	—	—	—	—
P15: Safety auditing	−6.678	−.445	4	−6.508	−.575	1	−7.248	−.543	1
P16: Safety record keeping	—	—	—	—	—	—	—	—	—
P17: Job hazard analysis	−7.917	−0.514	2	—	—	—	—	—	—
Constant	39.234			25.219			26.237		
Adjusted R^2	0.74			0.56			0.68		

Conclusions and recommendations

Conclusions

The results showed that, overall, five H&S programmes, namely accident investigation, H&S policy, safety committees, first-aid programmes, selection of employees and control of sub-contractors, had the best actual achievement status. Generally, there was a general acceptance of H&S programmes in principle, but low levels of implementation. Only six H&S programmes (35%) were implemented at a good level. The remaining 65% were either fairly or poorly implemented H&S programmes and they are of great concern if H&S performance is to improve. The unsafe acts and unsafe conditions performance indices were calculated to be 36.7% and 42.8% respectively. Subjectively, H&S performance was measured to be inadequate with 40% of the construction site practices unsafe.

Furthermore, the objective measurement of H&S performance revealed high frequency of H&S incidents at Uganda's construction sites. Reporting of accidents to the authorities was found to be considerably low (about 24%). This indicated that there was lack of enforcement of H&S regulations in reporting of occupational injuries and illness. The NIR and FIR of 18.2 and 2.0, respectively, per 100 EFTW were also clear evidence that Uganda's construction sites were generally unsafe.

Recommendations

It is recommended that the Ugandan government mandates the implementation of the 17 H&S programmes into a formal regulation, especially the key programmes P1, P5, P7, P8, P10, P12, P15 and P17. The laws governing H&S within the workplace specific to the construction industry should be strengthened. The government should also develop guidelines on occupational health and safety hazards at construction sites and regularly update them. Moreover, consideration should be given to constituting a dedicated regulatory authority such as a construction industry commission (CICO). H&S performance should be incorporated as one of the criteria for the evaluation of tenders for construction projects.

It is recommended that contracting firms adopt H&S programmes as a corporate responsibility and ensure their implementation to achieve better H&S performance and hence improved productivity. It is also recommended that designers, such as architects, engineers and environmentalists, adopt a 'design for safety' culture in all infrastructure projects and append H&S plans as part of deliverables.

Further research is recommended in the areas of analysis of the costs of accidents in Uganda's construction industry; identification of critical factors that affect the implementation of H&S programmes for improved H&S performance; and illnesses related to construction site health problems.

Acknowledgement

The authors acknowledge the contribution made by the contractors on which the study was based.

References

Abdul, R. A. H., Muhd, Z. A. M. & Bachan, S. (2008), Causes of accidents at construction sites, *Malaysian Journal of Civil Engineering*, Vol. 20(2), pp. 242–259.

Ahmed, M. H. & Adnan, E. (2005), Improving safety performance in construction projects in Gaza Strip. Unpublished master's thesis, The Islamic University of Gaza. [Online]. Available at: http://library.iugaza.edu.ps/Thesis/65061.pdf.

Aksorn, T. & Hadikusumo, B. H. W. (2008), Measuring effectiveness of safety programmes in Thai construction industry, *Journal of Construction Management and Economics*, Vol. 26, pp. 409–421.

Aksorn, T. & Hadikusumo, B. H. W. (2007), Gap analysis approach for construction safety programme improvement, *Journal of Construction in Developing Countries*, Vol. 12(1), pp. 77–97.

Alinaitwe, H., Mwakali, J. & Hansson, B. (2007), An analysis of accidents on building construction sites reported in Uganda during 2001–2005, in T. C. Hampt & R. Milford (Eds), *CIB World Building Congress 2007, Construction for Development*, Published by CIB World Building Congress, pp. 1208–1221.

Bonnet, D. G. & Wright, T. A. (2015), Cronbach's alpha reliability: Interval estimation, hypothesis testing, and sample size planning, *Journal of Organizational Behaviour*, Vol. 36, pp. 3–15.

Bureau of Labour statistics. (2011), Construction: Fatalities, injuries and illnesses. [Online]. Available at: http://www.bls.gov/iag/tgs/iag23.htm.

Bureau of Labour statistics. (2010), Construction: Fatalities, injuries and illnesses. [Online]. Available at: http://www.bls.gov/iag/tgs/iag23.htm.

Donaghy, R. (2009), One death is too many: An inquiry into the underlying causes of construction fatal accidents. Report to the Secretary of State for Work and Compensation, Norwich, UK.

Edmonds, D. J. & Nicholas, J. (2002), The state of health and safety in the UK construction industry with a focus on plant operators, *Structural Survey*, Vol. 20(2), pp. 78–87.

Health and Safety Executive. (2007), The annual statistics reports. [Online]. Available at: www.hse.gov.uk/statistics/index.htm.

Health and Safety Executive (2009), The annual statistics reports. [Online]. Available at: www.hse.gov.uk/statistics/index.htm.

Idoro, G. I. (2008), Health and safety management efforts as correlates of performance in the Nigerian Construction Industry, *Journal of Civil Engineering and Management*, Vol. 14(4), pp. 277–285.

Jaafar, M. H., Arifin, K., Aiyub, K. Razman, M. R., Ishak, M. S. & Samsurijan, M. S. (2017), Occupational safety and health management in the construction industry: A review, *International Journal of Occupational Safety and Ergonomic*, Vol. 23, pp. 1–12.

Kartam, N. A. (1997), Integrating health and safety performance into construction project management, *Journal of Construction Engineering and Management*, Vol. 123(2), pp. 121–126.

Lin, J. & Mills, A. (2001), Measuring the occupational health and safety performance of construction companies in Australia, *Facilities*, Vol. 19(3/4), pp. 131–139.

Lubega, H. A., Kiggundu, B. M. & Tindiwensi, D. (2000), An investigation into the causes of accidents in the construction industry in Uganda, in A. B. Ngowi & J. Ssegawa (Eds), *Proceedings of 2nd International Conference on Construction in Developing Countries*, 15–17 November 2000, Gaborone, Botswana, pp. 1–12.

Mohamed, S. (2003), Scorecard approach to benchmarking organizational safety culture in construction, *Journal of Construction Engineering and Management*, Vol. 129(1), pp. 80–88.

Marosszeky, M., Karim, K., Davis, S. & Naik, N. (2004), Lessons learnt in developing effective performance measures for construction safety management, International Group on Lean Construction (IGLC, 2004) Conference.

Ministry of Works and Transport. (2008), Report of the Technical Investigation into the NSSF Pension Towers Accident of 14th October, 2008 on Plots 15A, 15B and 17 Lumumba Avenue Kampala (Main Report). Entebbe, Uganda.

Ministry of Works and Transport. (2004), Bwebajja Building Accident Report (Main Report). Entebbe, Uganda.

Musoke, C., Muwanga, D. & Ssempogo, H. (2008), Eight crushed at NSSF building site. *The New Vision Newspaper*, Vol. 23(207), p. 1.

Musonda, I. (2012), *Construction Health and Safety (H&S) Performance Improvement – A Client Based Model*, PhD thesis, University of Johannesburg.

Mwakali, J. A. (2006), A review of the causes and remedies of construction related accidents: The Uganda experience, in J. A. Mwakali & G. Tabani-Wani, (Eds), *Advances in Engineering and Technology*, Entebbe: Elsevier, pp. 285–300.

Occupational Safety and Health Act. (2006), *The Uganda Gazette*, No. 36 Volume XCVIX.

Oregon OSHA (2015) OSHA forms for recording work-related injuries and illnesses. Oregon OSHA. Available at https://osha.oregon.gov/OSHAPubs/3353.pdf

Ssempogo, H., Candia, S., Kajoba, N., Ouma, F. & Wasike, A. (2008), Survivors rescued after 24 hours, *The New Vision Newspaper*, Vol. 23(23), p. 1.

Uganda Bureau of Statistics. (2016), Statistical Abstracts, Kampala, Uganda.

Vatin, N., Gamayunova, O. & Petrosova, D., (2014) Relevance of education in construction safety area, *Applied Mechanics and Materials*, Vols. 635–637, pp. 2085–2089.

8 Towards the development of an integrated safety, health and environmental management capability maturity model (SHEM-CMM) for uptake by construction companies in Ghana

Millicent Asah-Kissiedu, Patrick Manu, Colin Booth and Abdul-Majeed Mahamadu

Summary

With construction activities and processes accounting for several types of work-related fatalities and negative environmental impacts, it has become imperative for construction companies to effectively implement management tools that will address these problems. Whilst safety, health and environmental (SHE) management systems offer a useful framework for SHE management and have been adopted by companies, especially in developed countries, the situation is different in many developing countries, especially those in sub-Saharan Africa such as Ghana. A critique of the literature on the subject reveals that whilst the low level of adoption of international SHE management systems by construction companies in developing countries is attributable to a myriad of challenges, the high cost of adoption of these systems as stand-alone management functions remains a major inhibitor. Therefore, it is argued that integrated management of SHE through a single system could be less costly and onerous, and yet effective in delivering desired SHE outcomes that could be useful in stimulating greater adoption in developing countries. Drawing on this, as well as literature on continuous process improvement, the need for the development of an integrated SHE management capability maturity model (SHEM-CMM) is explored in this chapter for use by Ghanaian contractors to help improve their SHE management processes. The key knowledge gaps and research questions that need to be addressed in order to enable the development of a SHEM-CMM are also highlighted in this chapter.

The benefits of integrated management systems in construction

The construction sector in many countries is one of the most significant industries in terms of its contribution to gross domestic product (GDP). However, construction continues to contribute significantly to occupational related accidents, injuries and illness as well as having an adverse effect on the environment (Kheni, 2009; Geipele & Tambovceva, 2011). Over the years, considerable efforts have been made to address the poor status of safety, health and environmental

(SHE) issues in construction. However, the industry today, in both developed and developing countries, is still characterised by higher numbers of occupational fatalities, injuries and illnesses affecting construction workers as well as having detrimental effects on the environment (Zou, 2011; Enshassi et al., 2014). The situation is even more dire in developing countries, particularly those in sub-Saharan Africa, where SHE issues are often neglected and rarely managed, partly owing to the cost involved in implementing safety and environmental measures (Kheni et al., 2008).

With the rate of accidents and illnesses, negative environmental impacts and other issues of well-being recorded in construction through conventional practices, and the social and economic impacts arising from these incidents, there have been genuine concerns and increased acknowledgment of the need to adopt systematic approaches to address the poor status of SHE management in construction. As a result, the construction industry over the last few decades has embraced management systems (MSs), particularly environmental management systems (EMS) and safety and health management systems (SHMS), as one of the important approaches to help construction companies to manage and control the key management functions of safety and environment effectively in a systematic way (Griffith, 2011). Recent research shows that the EMS and SHMS can play a key role in improving the health, safety and well-being of workers and in tackling adverse environmental impacts (Podgorski, 2015). However, the adoption and subsequent implementation of these MSs in the construction sector in developing countries have been slow and generally low, mainly because of the cost and the unnecessary bureaucracy that comes with the parallel implementation of various stand-alone management systems (Zeng et al., 2008; Owolana & Booth, 2016). This situation is exacerbated by the limited financial resources and expertise within developing countries (Ayarkwa et al., 2010; Owolana & Booth, 2016). This has led to some industry stakeholders and researchers advocating the integration of management systems since such a single system could generate substantial benefits such as streamlining activities to achieve greater organisational efficiency and effectiveness (Rebelo et al., 2016).

Since construction safety issues are closely connected to environmental problems, and initiatives aimed at improving safety during construction could lead to enhanced environmental management and vice versa (Zutshi & Creed, 2015), efficient management of SHE through an integrated framework could be a single, consistent and simple approach to planning and management of SHE risks with maximum effectiveness and minimum bureaucracy (Griffith, 2011). This could be beneficial in reducing the number of fatalities, injuries, illnesses and the potentially negative impacts of construction operations on the environment, leading to better SHE performance outcomes. However, there is still no, single, integrated SHE management framework for construction organisations to use, especially those within developing countries. Consequently, there are also no tools or systematic mechanisms that enable construction companies to ascertain the maturity of their SHE management practices based on an integrated SHE management framework. However, organisations being able to ascertain the maturity of their processes in delivering a function is important in ensuring continuous process improvement as organisations are able to identify the strengths and weaknesses within their processes and practices. Maturity models, which are management-oriented tools that have proven to be valuable in assessing organisational processes in delivering performance in many management-related disciplines (Becker et al., 2009), can offer such a mechanism.

Based on a review of relevant literature, the aim in this chapter is to illustrate the need to develop an integrated SHE management capability maturity model for uptake by construction companies in the context of a developing country – Ghana – where SHE performance is poor. To achieve this, the chapter begins with an overview of the status of SHE performance in the construction industry in developing countries, especially in Ghana. Subsequently, SHE management systems are examined, as well as literature on integrated management systems. Following this, the need for an integrated safety health and environment management capability maturity model (SHEM-CMM) is presented together with concluding remarks.

SHE performance in the construction industry of developing countries

There is well-documented evidence of a high infrastructure deficit in developing countries (Fay et al., 2011; Bhattacharya & Kharas, 2012). In order to fuel economic growth and improve the quality of lives of the people within these countries, governments have been steadily increasing their infrastructure investments resulting in high levels of construction activities (Global Construction Perspectives and Oxford Economics, 2013). Thus, the impact of construction activities on the environment as well as on the safety and health of workers has become much greater and detrimental (Yahaya & Abidin, 2013). Unfortunately, owing to the lack of concern, statutory regulations on safety and health, and accurate records, developing countries do not put much effort into ensuring that construction has minimal effects on the safety and health (S&H) of both workers and society (Farooqui et al., 2008). Therefore, it is common to hear of occupational accidents and incidents that result in tragic loss of life, illness and some bodily harm to workers. For instance, in Tanzania, Matico and Naidoo (2015) recorded that the construction sector was responsible for approximately 10% of all occupational accidents, whilst Idoro (2011) recorded five injuries per worker and two accidents per hundred workers in Nigeria. Mosanawe (2013) reported that 55% of workplace accidents recorded in Botswana occurred in the construction industry.

Apart from the poor S&H situation in construction, its activities and operations consume over 50% of all raw materials extracted globally and 16% of water; they produce considerable volumes of waste and dust and 20–30% of greenhouse emissions; and they are responsible for 36% of global final energy use and contribute to global warming (Gupta & Deshmukh, 2016; UN Environment, 2017), making the construction industry one of the least sustainable industries globally.

These adverse environmental impacts and work-related injuries, ill health and fatalities have significant socio-economic implications. Yet, not much is done by developing countries to curb this increasing assault on the environment. Safety and environmental considerations in the delivery process of building projects are not given priority and, therefore, SHE issues are often neglected (Kheni et al., 2008; Ametepey & Ansah, 2015).

The SHE performance situation in the Ghanaian construction sector

Existing literature indicates that the SHE performance in the Ghanaian construction sector is poor (Laryea & Mensah, 2010; Ametepey & Ansah, 2015). The poor performance is largely attributed to an inefficient institutional and legal framework; lack of knowledge of SHE issues; absence of SHE ethics; low senior management commitment; inadequate SHE regulations specifically for the construction sector; and the laxity in the enforcement of existing regulations, all of which point to a poor SHE management culture (Kheni et al., 2008; Ameyaw et al., 2014). Notably, this realisation comes at a time when there seemed to be a resurgence of growth in the construction sector within sub-Saharan Africa because of the increases in investments especially from China (Pigato & Tang, 2015). Though the momentous growth in construction output has several socio-economic benefits, it also raises grave concerns as a result of potentially detrimental SHE and associated cost consequences. Certainly, there is an urgency to improve SHE management in the Ghanaian construction sector. Therefore, a systematic implementation of SHE management practices stipulated in the management systems could address the SHE management issues in construction (Fewings, 2013).

Safety and health management systems (SHMS)

SHMS is designed to help organisations reduce work-related injuries, ill health and fatalities, and continually improve overall S&H performance by demonstrating conformity to established requirements (Abad et al., 2013). Different variants of SMHS, including ILO-OHS-2001, AS/NZS 4801, BS 8800 and OHSAS 18001, are available. However, the Occupational Health and Safety Assessment Series (OHSAS) 18001:2007, developed by the British Standard Institute, is the most widely used standard for SHMS, albeit a new international certifiable standard, ISO 45001, has recently been published to replace OHSAS 18001. Both standards offer a useful framework for safety management in construction operations, which can inspire a positive safety culture. Nonetheless, SHMS is still not commonly adopted and implemented in the construction industry (Zeng et al., 2008). However, it has been argued that a well-designed, effectively implemented and managed SHMS contributes to improvements in a company's working conditions and management practices and the prevention of injuries. It also increases productivity and improves internal safety communication (Ghahramani, 2016).

Environmental management systems (EMS)

An EMS is a management tool that provides an integrated management framework that assists organisations to control and improve their environmental performance on a voluntary basis and in a systematic manner through the comprehensible allocation of resources, assignment of responsibilities, continuing evaluation of practice, management of its legal compliance and a focus on continuous improvement (ISO 14001, 2015). Several standards applying to an EMS have been established over the years. These include British Standard 7750 (BS7750), the Eco-Management and Auditing Scheme (EMAS) and the ISO 14001 by the International Standard Organisation (ISO). However, the ISO 14001 standard is the most accepted certification and widely used standard for an EMS (Campos et al., 2016). EMS adoption and implementation within the construction industry is low, but has been implemented in some countries because of the compulsion of the market demands within those countries and the benefits derived, such as competitive advantage, prevention of pollution and waste (which result also in cost reduction), reduction in environmental incidents and improvements on site and project safety (Oliveira et al., 2016). Companies in developing countries have encountered a deterrent to implementation in the high cost of resources required for EMS implementation (Ayarkwa et al., 2010; Owolana & Booth, 2016).

From individual systems to integration

Living up to the various requirements of several stakeholders, the development of different MSs, based on management system standards, has proliferated making separate management practices more difficult and expensive (Abad et al., 2014). Implementation and use of stand-alone MSs, such as EMS, QMS and SHMS, have been criticised for being bureaucratic, costly, paper-driven and arduous (Simon et al., 2013; Rebelo et al., 2014). This is because management of stand-alone MSs in a parallel manner could involve extensive documentation, duplication of efforts, waste of time and resources, cause complexity of internal management and increase the probability of faults and errors (Abad et al., 2014; Nunhes et al., 2016). Evidence from Ghana shows that adoption and implementation of separate MSs in the construction industry are generally low because of the cost and bureaucratic implementation and maintenance (Ayarkwa et al., 2010). As the cost of implementing stand-alone MSs remains a major inhibitor, there is good reason to consider the integration of MSs.

Over the last decade, the various MS standards, such as ISO14001, ISO 9001, OSHAS 18001 and ISO 45001, have become more aligned since their structure is based on the Deming Cycle of continual improvement (i.e. Plan–Do–Check–Act) and, as a result, they have similar methodologies for their creation, structure and implementation processes (Rebelo et al., 2014). These similarities make them synergistic and amenable to integration. Therefore, an integrated management system (IMS) has been strongly advocated (Abad et al., 2014).

This advocacy has been overtaken by a more practical approach in recent years, based on empirical evidence that indicates that the integration of MSs is vital to managing the requirements and expectations of stakeholders in improving organisational efficiency in use of resources (Simon et al., 2013; Rebelo et al., 2014). As a result, the subject of IMS in terms of safety, health, quality and environmental management has become increasingly part of an organisation's management portfolio (Asif et al., 2010). Therefore, the integration of MSs is one of the most important strategies for organisations to ensure survival and savings (in time, cost and resources) in today's competitive and stringently regulated business environment (Simon et al., 2013).

Furthermore, the relevant literature on IMS across different industrial sectors has been focused more on the fusion of the three-standardised management systems (EMS, QMS and SHMS) and, whenever possible, the fusion of two systems (either QMS and EMS, or EMS and SHMS) (von Ahsen, 2014). However, the fusion of EMS and SHMS in construction is scarce (Zeng et al., 2008). The integration of an EMS and SHMS is proposed because of the similarities that the structure of ISO 14001 EMS and OHSAS 18001 OHSMS share and also the fact that both are based on the control of risks and on the Plan–Do–Check–Act management structure (Rebelo et al., 2014). Construction health and safety and the negative effect of construction operations on the environment are the major challenges facing the construction sector. Thus, integrating EMS and SHMS could enable construction companies to use similar practices to help manage SHE issues jointly in a sustainable and cost-effective way to ensure that negative effects on the environment are minimised and workers' safety and health are safeguarded (Muzaimi et al., 2017).

Managing the challenges of integration

As no process is without obstacles, the existing body of research on IMS suggests the process of integration and implementation of IMS is fraught with challenges (Simon et al., 2012). These challenges include the lack of qualified personnel to cover all system requirements, technical guidance and support; the lack of support from governments, unions and regulatory bodies; time delays in the integration process; predictable cultural resistance; and people's attitudes towards change (Simon et al., 2012). However, Lopez-Fresno (2010) acknowledged there are potential challenges to implementation and the organisation, but argued that if organisations are able to address these challenges early in the process of integration, several benefits could be realised by all stakeholders. Some of the main improvements and advantages include elimination of conflicts between separate MSs with better utilisation of resources; cost and time savings; elimination of several waste types; improvement in the management of safety, health and environmental risks; and improved public image. In fact, Olaru and colleagues (2014) summarised 40 benefits that could be gained from integrating MSs. Some of these benefits have been corroborated by researchers such as Simon and colleagues (2013) and Rebello and colleagues (2016). To gain the potential benefits of IMS and sustain a competitive advantage and sustainability in the business market, several organisations, especially in developed countries, have implemented IMS or are planning to integrate their MSs (Rebello et al., 2016). This, according to Salomone (2008), explains the cultural shift that is underway, where many companies are advancing towards integration.

Towards an integrated SHEM capability maturity model

The key element of SHE management systems is continuous improvement of SHE performance. However, MSs often provide performance criteria and targets based on outcomes (e.g. number of injuries), and are not based on the operational methods or processes needed to achieve continuous improvement in the outcomes. Thus, whilst EMS and SHMS highlight management areas and processes or practices that need to be implemented to achieve positive outcomes, they do not offer a mechanism for ascertaining how well a company is performing in implementing those practices, i.e. the level of maturity in performing a practice (Zobel, 2008). However, the premise of EMS and SHMS is that if they are well established and implemented effectively, they will reduce or eliminate negative environmental impacts and S&H risks to move a company toward better SHE performance. In this vein, implementing companies should be able to establish their current level of SHE management performance maturity and identify actions to improve their SHE management practices and processes continuously.

Various process improvement models and approaches are available to enable organisations to improve their performance continuously. These include Lean, Six Sigma, Excellence models and capability maturity models (Sun et al., 2009). However, apart from the maturity models, the others do not really assess the effectiveness of the processes involved and show no evidence of the capability improvements of organisation processes (Sun et al., 2009). Maturity models (MMs), on the other hand, show the sequence of levels that describes how practices, processes and actions of an organisation can consistently show an expected or desired progressive path of improvement that could produce essential and desired outcomes (Paulk et al., 1993; Curry & Donnallen, 2012; Manu et al., 2018). Over the last few years, maturity models have been recognised as widely used tools that have proved valuable for performance improvement in organisation business processes in many disciplines (Proença & Borbinha, 2016). However, MMs have been criticised for oversimplifying reality and being based mostly on espoused best practices with their reliability not justified empirically in most cases, lacking rigour in their model development process and limited in scope (Becker et al., 2009). Nevertheless, they do offer a framework with a methodical approach for assessing the capability and capacity of an organisation to manage its business processes in the best way (Becker et al., 2009). In view of this, an integrated safety, health and environmental management capability maturity model (SHEM-CMM) could be a useful process improvement tool for assessing the maturity of a construction company's SHE management practices and help it to improve its processes to achieve better performance outcomes.

While an integrated SHEM-CMM would be beneficial, especially for contractors to enable them to improve on their SHE management, there is none existent at present and very limited research has been done to inform their development. The closest to date are the integrated models such as SHEQ-MS (Hamid et al., 2004) and IMS-QES (Rebelo et al., 2014), which do not enable assessment of SHE management capability maturity in order to pave way for process improvement. Moreover, SHE management studies in the construction industry in Ghana have largely covered areas such as environmental impacts of construction activities, perceptions of adoption and implementation of an EMS, on-site S&H management issues, legislation and procurement (Kheni et al., 2008; Laryea & Mensah, 2010; Ametepey & Ansah, 2015; Ayarkwa et al., 2014). Consequently, knowledge gaps remain regarding the key attributes or elements in terms of SHE management that should be incorporated in an integrated SHEM-CMM; the relative importance/priorities of such attributes so as to enable prioritisation of process improvement actions; and the levels of capability maturity that are appropriate for capturing stages of maturation in those attributes. Against this backdrop, the following research questions need to be addressed:

1. What organisational attributes regarding SHE management are required for the development of an integrated SHEM-CMM?
2. What are the relative priorities of those attributes?
3. What levels of maturity are appropriate for capturing maturity on the capability attributes?

The preceding questions offer an opportunity for research aimed at the development of a capability maturity model focused on integrated SHE management in construction, especially for a developing country, with the dual purpose of easing the financial and resource burden associated with the implementation of separate stand-alone MSs by contractors, and also making it possible to ascertain the maturity of their MS to guide efforts to improve processes.

Conclusions

In this chapter, the safety, health and environmental (SHE) performance status of the construction industry has been examined. It has been shown that SHE problems are a global issue, but are more prevalent in developing economies, such as Ghana. The poor state of SHE management in the Ghanaian construction sector could deteriorate owing to the growth in construction output. While the study has shown SHMS and EMS could be useful in addressing, SHE challenges, their adoption, particularly in developing countries, is rather low because of several factors including the cost of implementation. Considering the identified benefits and challenges of integrated management systems, it has been argued that an integrated management of SHE through a single system could be useful in stimulating greater adoption and operation of well-structured SHE management systems in the Ghanaian construction industry since the cost of implementing stand-alone MSs for S&H and environmental management is a major inhibitor. Additionally, to enable construction companies to assess and improve their SHE management processes and practices continuously, a case for research aimed at the development of an integrated SHEM-CMM has been presented outlining the relevant research gaps and questions that need to be addressed. It is anticipated that undertaking such research could have far-reaching benefits in terms of construction organisations in developing countries effectively managing S&H and environmental challenges.

Acknowledgement

The study presented in this chapter was funded by the Commonwealth Scholarship Commission (Scholarship Reference: GHCS - 2016 - 147).

References

Ametepey, S.O. & Ansah, S.K. (2015), Impacts of construction activities on the environment: The case of Ghana. *Journal of Environment and Earth Science*, Vol. 5(3), 18–26.

Abad, J., Dalmau, I. & Vilajosana, J. (2014), Taxonomic proposal for integration levels of management systems based on empirical evidence and derived corporate benefits, *Journal of Cleaner Production*, Vol. 78, pp. 164–173.

Abad, J., Lafuente, E. & Vilajosana, J. (2013), An assessment of the OHSAS 18001 certification process: Objective drivers and consequences on safety performance and labour productivity, *Safety Science*, Vol. 60, pp. 47–56.

Asif, M., Fisscher, O. A. M., Bruijn, E. J. & Pagell, M. (2010), Integration of management systems: A methodology for operational excellence and strategic flexibility, *Operations Management Research*, Vol. 3(3–4), pp. 146–160.

Ayarkwa, J., Ayirebi-Dansoh & Amoah, P. (2010), Barriers to implementation of EMS in construction industry in Ghana, *International Journal of Engineering Science*, Vol. 2(4), pp. 37–45.

Ayarkwa, J., Acheampong, A., Hackman, J. K. & Agyekum, K. (2014), Environmental Impact of Construction Site Activities in Ghana, *Journal of Africa Development and Resources Research Institute*, Vol. 9(2), 1–19.

Becker, J., Knackstedt, R. & Pöppelbuß, J. (2009), Developing maturity models for IT management, *Business and Information Systems Engineering*, Vol. 1(3), pp. 213–222.

Bhattacharya, A. & Homi, K. (2012), *Infrastructure for Development: Meeting the Challenge*, Centre for Climate Change Economics and Policy, London.

Campos, L. M. S., Trierweiller, A. C., Nunes De Carvalho, D. & Šelih, J. (2016), Environmental management systems in the construction industry: A review, *Journal of Environmental Engineering and Management*, Vol. 16(2), pp. 453–460.

Curry, E., & Donnellan, B. (2012), Understanding the Maturity of Sustainable ICT. vom Brocke, J., Seidel, S., and Recker, J., (Eds) *Green Business Process Management -Towards the Sustainable Enterprise*, Springer, 2012, 203–216.

Enhhassi, A., Kochendoerfer, B. & Rizq, E. (2014), Evaluacion de los impactos medioambientales de los proyectos de construccion, *Revista Ingenieria de Construccion*, Vol. 29(3), pp. 234–254.

Farooqui, R. & Ahmed, S. (2008), Assessment of Pakistani construction industry – Current performance and the way forward, *Journal for the Advancement of Performance Information and Value*, Vol. 1(1), pp. 51–72.

Fay, M., Michael, T., Daniel, B. & Stefan, C. (2011), Infrastructure and sustainable development. In: Fardoust, S., Yongbeom, K., and Claudia, P.S. (Eds.) *Postcrisis Growth and Development: A development agenda for the G-20*, Washington, DC: World Bank.

Fewings, P. (2013), *Construction Project Management – An Integrated Approach*. 2nd ed. Oxon: Routledge Publishing.

Geipele, I. & Tambovceva, T. (2011), Environmental management systems experience among Latvian construction companies, *Technological and Economic Development of Economy*, Vol. 17, pp. 595–610.

Ghahramani, A., (2016), Factors that influence the maintenance and improvement of OHSAS 18001 in adopting companies: A qualitative study, *Journal of Cleaner Productions*, Vol. 137, pp. 283–290.

Global Construction Perspectives and Oxford Economics (2013), Global Construction 2025. [Online]. Available at: www.globalconstruction2025.com/ (accessed 3 April 2017).

Griffith, A. (2011), *Integrated Management Systems for Construction: Quality, Environment and Safety*. Harlow, UK: Prentice Hall.

Gupta, A. R. & Deshmukh, S. K. (2016), Energy efficient construction materials, *Engineering Materials*, Vol. 678, pp. 35–49.

Hamid, A. R. A., Wan, Z. W., Singh, B. & Yusof, A. K. T. Y. (2004), Integration of safety, health, environment and quality management system in construction: A review, *Journal Kejuruteraan Awam*, Vol. 16(1), pp. 24–37.

Idoro, G. I. (2011), Effect of mechanisation on occupational health and safety performance in the Nigerian construction industry, *Journal of Construction in Developing Countries*, Vol. 16(2), pp. 27–45.

ISO 14001 (2015), *Environmental Management Systems – Requirements with Guidance for Use*. Geneva, Switzerland: International Standards Organization, p. 32.

Kheni, N. A. (2009), Impact of health and safety management on safety performance of small and medium-sized construction businesses in Ghana. Unpublished PhD thesis, Department of Civil and Building Engineering, Loughborough University.

Kheni, N. A., Dainty, A. R. J. & Gibb, A. (2008), Health and safety management in developing countries: A study of construction SMEs in Ghana, *Construction Management and Economics*, Vol. 26(11), pp. 1159–1169.

Laryea, S. & Mensah, S. (2010), Health and safety on construction sites in Ghana. In: *The Construction, Building and Real Estate Research Conference of the Royal Institution of Chartered Surveyors*, 2–3 September 2010. Paris, Dauphine Université.

López-Fresno, P. (2010), Implementation of an integrated management system in an airline: A case study, *The TQM Journal*, Vol. 22(6), pp. 629–647.

Manu, P., Poghosyan, A., Mahamadu, A.-M., Mahdjoubi, L., Gibb, A., Behm, M. & Akinade, O. (2018), Development of a design for occupational safety and health capability maturity model. In: Saurin, T. A., Costa, D. B., Behm, M. & Emuze, F. (Eds.), *Proceedings of the Joint CIB W099 and TG59 Conference*, Brazil, 1–3 August 2018. Salvador, Marketing Aumentado.

Matiko, J. M. & Naidoo, R. N. (2015), Work-related fatalities and severe injuries in the Dar-es-Salaam region, Tanzania: A comparison of risk factors between the construction and non-construction sectors, 1980–2009, *Occupational Health Southern Africa*, Vol. 21(3), pp. 22–29.

Mosanawe, J.O (2013), Construction Safety and Occupational Safety and Health in Botswana. *African Newsletter on Occupational health and safety*, Vol. 23(3), pp 66–67.

Muzaimi, H., Chew, B. C. & Hamid, S. R. (2017), Integrated management system: The integration of ISO 9001, ISO 14001, OHSAS 18001 and ISO 31000, *AIP Conference Proceedings*, Vol. 1818, 020034.

Nunhes, T. V., Ferreira Motta, L. C. & de Oliveira, O. J. (2016), Evolution of integrated management systems research: Identification of contributions and gaps in the literature, *Journal of Cleaner Production*, Vol. 139(September), pp. 1234–1244.

Olaru, M., Maier, D., Nicoara, D. & Maier, A. (2014), Establishing the basis for development of an organization by adopting the integrated management systems: Comparative study of various models and concepts of integration, *Procedia – Social and Behavioral Sciences*, Vol. 109, pp. 693–697.

Oliveira, J.A., Oliveira, O.J., Ometto, A.R., Ferraudo, A.S., & Salgado, M.H. (2016), Environmental Management System ISO 14001 factors for promoting the adoption of Cleaner Production practices, *Journal of Cleaner Production*, Vol.133, 1384–1394.

Owolana, V. O. & Booth, C. A. (2016), Stakeholder perceptions of the benefits and barriers of implementing environmental management systems in the Nigerian construction industry, *Journal of Environmental Engineering and Landscape Management*, Vol. 24(2), pp. 79–89.

Paulk, M. C., Chrissis, C. & Weber, M. B. (1993), Capability maturity model, version 1.1, *IEEE Software*, Vol. 10(4), pp. 18–27.

Pigato, M. & Tang, W. (2015), *China and Africa: Expanding Economic Ties in an Evolving Global Context*. Washington, DC: World Bank, p. 2.

Podgorski, D. (2015), Measuring operational performance of OSH management system – A demonstration of AHP-based selection of leading key performance indicators, *Safety Science*, Vol. 73, pp. 146–166.

Proença, D. & Borbinha, J. (2016), Maturity models for information systems – A state of the art, *Procedia Computer Science*, Vol. 100(2), pp. 1042–1049.

Rebelo, M., Santos, G. & Silva, R. (2014), Conception of a flexible integrator and lean model for integrated management systems, *Total Quality Management*, Vol. 25(6), pp. 683–701.

Rebelo, M. F., Santos, G. & Silva, R. (2016), Integration of management systems: Towards a sustained success and development of organizations, *Journal of Cleaner Production*, Vol. 127, pp. 96–111.

Salomone, R. (2008), Integrated management systems: Experiences in Italian organisations, *Journal of Cleaner Production*, Vol. 16, pp. 1786–1806.

Simon, A., Bernardo, M., Karapetrovic, S. & Casadesus, M. (2013), Implementing integrated management systems in chemical firms, *Total Quality Management and Business Excellence*, Vol. 24, pp. 294–309.

Simon, A., Karapetrovic, S. & Casadesús, M. (2012), Difficulties and benefits of integrated management systems, *Industrial Management and Data Systems*, Vol. 112(5), pp. 828–846.

Sun, M., Vidalakis, C. & Oza, T. (2009), A change management maturity model for construction projects. In: Dainty, A. (Ed.), *25th Annual ARCOM Conference*, 7–9 September 2009. Nottingham, UK, Association of Researchers in Construction Management, pp. 803–812.

UN Environment (2017), Towards a zero-emission, efficient, and resilient buildings and construction sector: Global status report. [Online]. Available at: www.worldgbc.org/

sites/default/files/UNEP%20188_GABC_en%20(web).pdf (accessed 3 November 2018).

von Ahsen, A. (2014), The Integration of quality, environmental and health and safety management by car manufacturers - A long-term empirical study, *Business Strategy and the Environment*, Vol. 23(6), 395–416.

Yahaya, I. & Abidin, N. (2013), Commitment of Malaysian contractors for environmental management practices at construction site, *International Journal of Sustainable Development*, Vol. 1(3), pp. 119–127.

Zeng, S. X., Tam, V. W. Y. & Tam, C. M. (2008), Towards occupational health and safety systems in the construction industry of China, *Safety Science*, Vol. 46, pp. 1155–1168.

Zobel, T. (2008), Characterisation of environmental policy implementation in an EMS context: A multiple case study in Sweden, *Journal of Cleaner Production*, Vol. 16(1), pp. 37–50.

Zou, P. X. (2011), Fostering a strong construction safety culture, *Leadership Management Engineer*, Vol. 11(1), pp. 11–22.

Zutshi, A. & Creed, A. (2015), An international review of environmental initiatives in the construction sector, *Journal of Cleaner Production*, Vol. 98, pp. 92–106.

9 Ad hoc and post hoc analysis of contractors' safety risks during procurement in Nigeria

Oluwole Olatunji, Abimbola Windapo and Nnedinma Umeokafor

Summary

Poor safety culture is a systemic issue for construction workers in Nigeria. Evidence suggests workers' rights to safe work and dignity are abused frequently. Although extant Nigerian laws compel contractors to maximise work safety, media reports are rife with incidents of collapse of structures, site accidents and hazards. Fatalities, and consequential losses from these, are significant. In context, the Nigerian procurement law requires contractors to be pre-qualified before receiving contract awards. Through a systematic literature review (SLR), this study examines pre-award assessment processes and standards relating to contractors' compliance with workers' health and safety. In addition, pre-award assessment objectives are compared with normative objectives of health and safety standards during construction. A significant gap is found between the two: pre-award assessment is incapable of translating into considerable safety outcomes for workers and projects. As a result, a new framework for assessing contractors' safety capability is proposed. The study also argues the significance of the proposed framework to extant pre-qualification frameworks used in Nigeria. Rather than being prescriptive, the framework can measure health and safety capabilities quantitatively. Conclusions are elicited from these on how to reform the Nigerian procurement landscape in terms of health and safety standards, and the cost benefits therefrom.

Purpose of this study

Safety is a paramount objective of a construction project. A clear consensus amongst construction management researchers is that there is no success in projects unless completion and operations are achieved and certified as truly safe – that is, safe for the builders and their workers, as well as the owners, end users and the environment (Chan et al., 2004; Gido & Clements, 2003; Gunduz & Yahya, 2015; Prabhakar, 2009). The works of Love and colleagues (2015) and Wanberg and colleagues (2013) are instructive on the relationship between project quality and safety. There is little to argue against the import of their conclusions: poor quality of work is unsafe, wasteful and unsatisfactory; and unsatisfactory works are just what they are – project owners do not pay large amounts for projects to be dissatisfied. Thus, a significant challenge to every project owner is how to ingrain success in its true meaning into their decisions across project development processes and post-construction operations. A way to do this is to ensure true success is enunciated clearly as the primary objective of the project owners' projects right from conception, and that this is further defined clearly by safety values rather than speed and short-sighted economic gains.

The cost of doing otherwise is dire. Unsafe projects involve costs to all stakeholders. In most developing economies, workers' fatalities and bodily injuries on construction sites have little or no recompense. Affected families (wives, children, parents and dependent relatives) often suffer from such irreparable losses without help. Also enormous are psychological costs and costs related to remediation (such as rehabilitations of work and persons, re-work and social re-engineering), disruptions to work, interventions by public administrators and legal costs arising from dysfunctional relationships. Despite the obvious, construction industries in developing countries do not have records. Where records exist, they are awful, reflecting only a fraction of actual occurrences and, most times, wrongly so. For example, Okoye (2018) is quite clear about the poor health and safety culture and lack of safety records in the Nigerian construction industry. The seriousness of this cannot be overestimated. Issues do occur, but the industry has had a culture of getting away with its poor records. Only major incidents, such as building collapses, are reported in the media, and these occur several times in a year. Actual costs beyond fatality figures are never known. Other major observables, such as performance issues relating to shoddy workmanship and debacles of poor material application, are commonplace occurrences which seldom appear in the media or government's official records (if any). Project owners are not interested in such records either (this will become clearer in the course of this study). Meanwhile, ignoring the issues does not improve the outcomes of projects nor the safety reputation of the construction sector.

Most construction workers in Nigeria do not wear protective gear (Olatunji et al., 2007). Whilst work is largely primitive and risk exposure is huge, high-risk items, such as scaffolds, are of the lowest quality possible. This is because contractors often view the commercial reality of safety as adding to project costs – costs which are unrewarded, unwarranted and avoidable (El-Rayes & Khalafallah, 2005). Nonetheless, the factual, objective reality is that safety cost is sublime (Goetsch & Goetsch, 2003). It involves more than the simplistic cost of materials and labour. Project owners who are keen for their projects to succeed must consider this carefully (Egan, 1998).

Pre-qualification is a legal requirement for all public projects in Nigeria (Aje, 2012; Olatunji, 2008). It is a mechanism enforced by regulators to ensure projects are executed only by competent bidders, and that project owners are able to achieve value for their investments through competitive bidding processes. Safety is one crucial criterion on which project owners focus during pre-qualification. Project stakeholders often want to be sure that their contractors are able to deliver projects safely, within budget, on time and at the appropriate quality. This approach is in line with previous arguments in literature where project owners have been challenged to be involved in the evolution of their projects rather than leaving important primary decisions about their project outcomes to other parties (Kometa et al., 1995). There are two key questions regarding this.

First, what is in pre-qualification regarding construction safety and the safety culture of the Nigerian construction industry as a whole? In spite of the obvious, how has pre-qualification policy performed in ingraining the appropriate safety outcomes in construction projects in Nigeria?

The aim of this study was to ameliorate the endemic nature of poor safety considerations in Nigerian procurement systems. Ad hoc considerations that precede contractor selection were reviewed. A framework for actual safety performance measurement was developed by synthesising empirical evidence from literature regarding what stakeholders from the Nigerian construction industry should be doing to meet global standards. In addition, the relationship between ad hoc safety considerations and post hoc (post-award) safety performance of construction projects were examined. Recommendations are drawn for closing the gap between world's best practices and the praxis gap between ad hoc and post hoc safety considerations in the Nigerian construction industry. Reviews aligning with these objectives are presented in three parts. First, there is a review of literature on the variables of ad hoc safety considerations. Second is a review of literature on the variables of post hoc safety considerations. Third, the study delineates the relationship between ad hoc and post hoc considerations towards eliciting the doctrine of absolute safety and global best practices. Implications of these will explain measurements and outcomes of construction projects regarding project safety.

Ad hoc safety considerations

There are several initiatives in construction management research regarding planning considerations that are focused on safety prior to commencing construction. An example of this is design for safety (Behm, 2005; Hadikusumo & Rowlinson, 2002). Gambatese and colleagues (2005) conclude such an initiative motivates positive safety outcomes during project implementation. If design motivates positive safety outcomes, procurement and operational cultures have significant roles to play. This is where most research in construction management about safety considerations during procurement has failed. Findings that support health and safety as a critical component of contractors' competence, and how it should be assessed, are few and exclusive (Idoro, 2004). For example, Ogunsemi and Aje (2006) surveyed 74 participants to identify key selection criteria used by project owners for selecting construction contractors. Of such criteria, 22, including safety, were identified, and were ranked accordingly. The analysis presented in the study is inconclusive and can only be applied with great care. This is because *health and safety policy of contractors*, the only variable relating to safety in the study, was ranked 11th, though most participants rated the criterion as being significant (3.86/5). Surprisingly, when the variables were revised for use in a regression model, an absolute selection model was created without any consideration for health and safety. The caution in applying this model is premised on the fact that a measured variable of considerable significance should not be discarded from a process without noticeable impact.

Similarly, Aje (2012) compares the views of project owners, consultants and contractors regarding pre-qualification criteria, where 194 respondents and 77 construction projects were analysed. Findings from the study suggested health and safety was ranked lowest by all the respondents across domains. Despite this, the criterion was one of the two considerations that was significant statistically in the model (p-value = 0.009) – after past performance only (p-value = 0.001). Other criteria proposed as being significant in a model by Ogunsemi and Aje's (2006) were found to be statistically insignificant (p-value > 0.05) by Aje (2012). Although obscured in the study, other findings established a strong correlation between safety and past performance, work quality and time performance of projects.

One approach is to examine how health and safety are measured, and see how this defines people's perception about their importance. For example, where financial capability is measured in such ad hoc considerations, contractors are judged on the basis of their turnover, working capital, audited accounts, financial statements, bank balance, bonding arrangements and records of project completion (Hatush & Skitmore, 1997). Similarly, technical competence is measured by the strength of experience of key personnel, quality of equipment and past performance history (Ng & Skitmore, 1999). The effect of these variables is that contractors who are able to provide convincing documentation about each criterion are assumed by project owners to be appropriate and competent to achieve project objectives accordingly. Holt (2018) argues that such an assumption is misleading, spurious and superficial because often the criteria are not linked to actual project objectives. For example, it is a commonplace assumption that contractors are only able to deliver projects within budget if they have a considerable bank balance in relation to the proposed project cost. Meanwhile, if at all, the bank balance of contractors often does not show the true financial situation of their business. Proceeds reported in such accounts could have been payments made in advance for projects not yet completed, supplies and subcontractors' work not yet paid, unpaid wages of workers, loans and debentures and unacquainted work.

Even if misleading, very limited analysis of contractors' safety capability as assessed by project owners is available in research literature about Nigerian construction. Ogunsemi and Aje (2006) mention *health and safety policy* only vaguely. Aje (2012) has no such breakdown. However, Ajayi (2010) provides an instructive list of sub-criteria for assessing contractors' safety capabilities before awarding a contract. The variables listed in Ajayi's study include

- Safety: perhaps use of personal safety equipment
- Experience modification rating: perhaps value engineered and earned through organisational learning in the context of safety
- Administration of occupational health and safety: perhaps policy settings in terms of self-regulation and legislative compliance
- Incidence rate: in terms of previous records of injuries and fatalities
- Management safety accountability: perhaps whether management has been accountable for reported safety issues in the past

There is no record of validation of these variables. They were not defined in any particular context either. However, it is partly logical to assume they are applied somewhere in Nigeria. Without bringing the popularity of such applications into question, it is rational to assume that having a metric to measure safety during construction is always good practice, and that it is possible to learn from such a framework.

Post hoc safety considerations

Safety objectives are difficult to measure unless there is a clear understanding of the goals from which such measurements are derived. For example, what do project owners want to achieve with their safety considerations before, during and after construction? Do clients really care about a safe artefact, built deliberately to be safe by a safe workforce, and that their contractors and workmen are able to leave a safe environment behind, and that they are able to possess a safe artefact that will enhance the safety of occupants and users? Answers to these questions are best imagined. However, it is often difficult to include all these factors in an achievable overall objective in developing countries, especially Nigeria. However, the appropriate thing to do is to ensure that safety objectives are well-rounded systemically and to consider the interest of all stakeholders. Davis (2014) and Littau and colleagues (2010) identify such stakeholders to include project owners, workers and their families, contractors, suppliers, people in the neighbourhood and the general public. Minimal safety considerations for each of these stakeholder groups could be defined in legislations, regulations and contract forms according to the nature of the project. Regardless, safety provisions should be taken beyond basic requirements, together and systemically.

As Olatunji (2005) reports, the single most important requirement of a typical Nigerian project owner is contractors' performance. The details of this expectation are often ill-defined. As a minimum, it means project owners expect contractors to deliver projects with little consideration for how they achieve their outcomes. Safety is neither measured nor paid for but is assumed to be embodied by performance. Furthermore, there is little, if any, compensation for injuries and dehumanisation underlying the achievement of such outcomes. This entrenched culture of feigned ignorance and indifference has no justification and should be discarded. The only option is to prioritise safety and to enhance the dignity of the humans involved in construction processes and the environment where projects are situated.

Safety management is measurable. Lin and Mills (2001) write about a continuous improvement matrix published by Australia's Construction Industry Development Agency (CIDA) for benchmarking the occupational health and safety (OHS) performance of construction contractors (see Table 9.1). The overarching goal of the CIDA tool is to ensure stakeholders are able to measure safety management during construction in a form that is assessable during contractor selection. Accordingly, contractors who are unable to demonstrate adequate commitment to safety, both in their work histories and in their proposals for

Table 9.1 CIDA's benchmark of OHS performance of construction contractors

CIDA system element description	Measurement
Management responsibility	Safety inspections, safety compliance assessment and integrity of test equipment
Health and safety system	Inspection regime and test status
Contract review	Control of non-compliance
Design control	Corrective and preventive safety actions
Document control	Safety integrity of handling, storage, packaging and delivery
Purchasing	Health and safety records
Purchaser supplied product	Health and safety auditing
Product identification and traceability	Training and servicing
Work method control	Statistical control, inspection and testing

Source: Adapted from Lin, J. & Mills, A., 2001, Measuring the occupational health and safety performance of construction companies in Australia, *Facilities*, Vol. 19(3/4), pp. 131–139.

future jobs, are unworthy to remain in business. CIDA's model, as adapted by Lin and Mills (2001), suggests safety is measurable through management responsibility, in that sites should be inspected and assessed regularly for safety integrity. In addition, contractors' health and safety systems can be assessed on the basis of the nature and context of inspection regimes and outcomes of safety examinations. For example, where a project fails on safety integrity tests, project owners can assume that such a project is of low quality and will only trigger unsatisfactory outcomes.

Non-compliance with safety regulations and work quality are also measurable elements of the CIDA safety assessment model. Besides governance provided by government inspectorates, safety compliance in construction contracts is also policed by labour unions and contractors' self-regulation mechanisms (Umeokafor, 2016). According to Behm (2005) and Gambatese and colleagues (2005), safety initiatives that are specified and designed for construction projects can be monitored for corrective and preventive outcomes. This objective can be measured to determine the appropriateness of design options and the benefits of realisation. In addition, CIDA's model also measures the integrity of material handling, storage, packaging and delivery. For example, the findings of Stern and colleagues (2001) regarding the harmful effects of cement on construction workers show that workers who are exposed to cement-based materials often have elevated risks of lung and stomach cancer. Mohler and colleagues (1998) report on how workers' exposure to polyvinyl chloride (PVC) and cement also causes peculiar liver and vascular diseases. Fatima and colleagues (2001) found that workers who are exposed to cement dust and particulate matters are at a high risk of DNA deformation. Evidence from the works of Akanbi and colleagues (2009) and Faremi and colleagues (2014) suggested that workers are poorly protected against these harmful materials and the environments that exacerbate them in Nigeria.

Furthermore, CIDA's model suggests health and safety can be measured through contactors' purchasing records. For example, contractors who are averse to procurement of safety materials and human capital are unlikely to have an ethos of concern for safety on and off their projects. If such contractors are unable to imbue an appropriate safety culture in their procurements, their suppliers are unlikely to do better. Manufacturers that produce to such contractors' and suppliers' demands are unlikely to do much better either. CIDA's model suggests safety can be traced from the source and that someone can be held responsible for unsafe materials and how they find their way onto construction sites. In particular, contractors can be held responsible for unsafe resources on their projects and in the supply chains they have incorporated into the development processes of a project.

In summary, it is possible to measure contractors' safety compliance through control mechanisms in designs and contract documentations, organisational leadership and the safety integrity of contractors' material handling and supply chains. Safety does not end with these. A significant number of incidents is often caused by workers' state of mind. Gillen and colleagues (2002) conclude that a dissatisfied worker is potentially an accident waiting to happen. According to Ajayi and Olatunji (2017), worker's job satisfaction is defined by their motivation to succeed, the quality of their relationship with their colleagues, organisational leadership, the reward and benefits they receive, their work–life balance and the impact of their job on their physical and mental health and well-being. These are often poorly researched, if at all, in relation to construction workers, their safety and the safety of their project outcomes.

Implications for practice and research

Contractors and their workers are the most important agents of outcome safety in construction projects. Both of them are regarded in normative literature as project stakeholders. However, contractors are responsible for the safety of their workers and the projects they deliver. As a result, project owners and contractors are the most influential determinants of project safety. Where owners fail to pay for a safe project but expect performance, contractors are unlikely to deliver safe projects. Indeed, this is an endemic problem in developing countries, Nigeria in particular. Project owners often want to achieve more in their projects than premised in the value of their projects; they often underestimate the role of safety in their considerations – or the lack thereof. Even where incidents are not reported, an unsafe site is unlikely to be a high-quality site, and owners tend to pay for their lack of safety discretion during construction throughout the life cycle of their projects.

Objectives of construction safety are not vague – and are measurable. An instrument that is potentially effective for developing countries is shown in Table 9.2. The instrument was synthesised from the extensive review of literature on which this current study is based. Clients desire safe projects for themselves, end users and the environment of their built assets throughout the project life. One way to make this happen is to ensure they select contractors with appropriate commitment to safety and to ensure that such commitment is policed sufficiently during construction. Unlike other ad hoc selection criteria that often do not deliver their objectives during construction (Holt, 2018), safety can be measured before and during construction, and the impact of such measurement is such that projects benefit in quality and in life-cycle costs.

As shown in Table 9.2, contractors can be assessed through the safety attributes of their workers, whether on-site or management workers. In essence, project owners must ensure contractors' workers are adequately trained and that they are well motivated to work safely. This is measurable by their experience, training, personal values, self-leadership, the safety integrity of the equipment and insurances. No worker or project is safe without these (Langford et al., 2000). In line with Aje (2012), an assessment of contractors' management capability is insufficient without some assurances that they are safe, and their sense of responsibility can be taken as reliable. This is measurable by contractors' compliance with specified safety standards, and that they have been tested with appropriate instruments. In addition, their proposed work method can be tested for safety issues. Akanmu and colleagues (2016) have developed an autonomous system that combines digital design and construction planning platforms by using building information modelling, smart tags and genetic algorithms to model on-site movements. Safety could be modelled the same way (Zhao & Lucas, 2015).

Table 9.2 Assessment instrument for construction contractors on project safety

Measures	Descriptors
A. Workers	
Safety qualifications of key personnel	Formal and ongoing training
Motivation to commit to safety ethos	Experience
Job satisfaction	Relationship with employers
Safety leadership	Self-regulation
Safety responsibility	Outcomes of inspection assessments
	Equipment integrity tests
Securitisation	Personal safety
	Insurance
B. Contractor's management responsibility	
Safety inspections	Safety compliance assessment
	Integrity of test equipment
Health and safety system	Inspection regime and test status
Work method	Control statistics
	Inspection and testing
Safety design	Corrective safety actions
	Preventive safety actions
Document control	Integrity of material handling and storage
	Resource packaging and delivery
Contract review	Control of non-compliance
C. Supply chain management	
Deliveries	Health and safety records of purchases
Safety integrity of suppliers	Health and safety audit of suppliers
Safety integrity of manufacturers	Product identification and traceability
	Training and servicing

Rather than being prescriptive, these variables are measurable on a Likert scale. For example, as in other pre-qualification models, some assessment variables could be assessed as 'Not Applicable'. This means there is no basis for their measurement relative to the specific project situation. Alternatively, a candidate contractor could be assessed in line with the variables as to whether they have 'Demonstrated' or 'Not Demonstrated' satisfactory compliance on the basis of the documentation they have supplied for assessment. This could be in the form of 'Not evident', 'Evident only in trace', 'Developing' and 'Established'. Users of the model are also able to apply weightings to the variables in relation to their safety objectives.

Conclusion

The Nigerian construction industry has had a poor safety record. Construction clients often do not demonstrate appropriate commitment to project safety. They have often mistaken contractors' performance as encapsulating self-regulation regarding safety by default. It has been argued in this study that the issue of low-quality projects that is evident in media reports about the Nigerian construction industry could be attributed to poor commitment to safety. Both clients and contractors have a role in this. Clients often do not assess contractors' safety credentials before they are appointed. Where safety has been reported in literature as a selection criterion, studies have shown that clients' application of safety knowledge is ignoble. Similarly, the efficacy of contractors' safety records during construction often is not monitored. One key constraint in this is that there is growing concern in recent literature regarding the relationship between ad hoc and post hoc considerations of selection criteria. In essence, researchers have pointed out that extant studies on pre-qualification criteria are inconclusive, misleading and often do not reflect the objectives for which they were planned. The instrument proposed in this study bridges this gap. It adapts an established model by Australia's CIDA by integrating clients' assessment of contractors' personnel as well as contractors' safety systems.

The model has not been validated. Further studies can be built around this. For example, an empirical study could be dedicated to understanding the relative importance of safety factors, the correlation between them and project outcomes. Alternatively, future studies can look into the relationship between policy impact and safety outcome in construction industries in developing countries. Apparently, there is a significant number of laws and policies. However, their final effect, whether combined or analysed individually, does not suggest a remarkable improvement in the industry.

References

Ajayi, M. O. (2010), Multi-criteria decision making model for contractor selection in construction projects in Nigeria. In: S. Laryea, R. Leiringer & W. Hughes (Eds.), *West Africa Built Environment Research Conference*, pp. 423–444. University of Reading, Accra, Ghana.

Ajayi, S. O. & Olatunji, O. A. (2017), Demographic analysis of turnover intentions amongst Nigerian high school teachers, *Australian and International Journal of Rural Education*, Vol. 27(1), pp. 62–87.

Aje, I. (2012), The impact of contractors' prequalification on construction project delivery in Nigeria, *Engineering, Construction and Architectural Management*, Vol. 19(2), pp. 159–172.

Akanbi, M., Ukoli, C., Erhabor, G., Akanbi, F. & Gordon, S. (2009), The burden of respiratory disease in Nigeria, *African Journal of Respiratory Medicine*, Vol. 4(March), pp. 10–17.

Akanmu, A., Olatunji, O. A., Love, P. E. D., Nguyen, D. & Matthews, J. (2016), Auto-generated site layout: An integrated approach to real-time sensing of temporary facilities in infrastructure projects, *Structure and Infrastructure Engineering*, Vol. 12(10), pp. 1243–1255.

Behm, M. (2005), Linking construction fatalities to the design for construction safety concept, *Safety Science*, Vol. 43(8), pp. 589–611.

Chan, A. P., Scott, D. & Chan, A. P. (2004), Factors affecting the success of a construction project, *Journal of Construction Engineering and Management*, Vol. 130(1), pp. 153–155.

Davis, K. (2014), Different stakeholder groups and their perceptions of project success, *International Journal of Project Management*, Vol. 32(2), pp. 189–201.

Egan, J. (1998), *Rethinking Construction*, Department of the Environment Transport and the Regions, HMSO, London.

El-Rayes, K. & Khalafallah, A. (2005), Trade-off between safety and cost in planning construction site layouts, *Journal of Construction Engineering and Management*, Vol. 131(11), pp. 1186–1195.

Faremi, F. A., Ogunfowokan, A. A., Mbada, C., Olatubi, M. I. & Ogungbemi, A. V. (2014), Occupational hazard awareness and safety practices among Nigerian sawmill workers, *International Journal of Medical Science and Public Health*, Vol. 3(10), pp. 1244–1248.

Fatima, S. K., Prabhavathi, P. A., Padmavathi, P. & Reddy, P. P. (2001), Analysis of chromosomal aberrations in men occupationally exposed to cement dust, *Mutation Research/Genetic Toxicology and Environmental Mutagenesis*, Vol. 490(2), pp. 179–186.

Gambatese, J. A., Behm, M. & Hinze, J. W. (2005), Viability of designing for construction worker safety, *Journal of Construction Engineering and Management*, Vol. 131(9), pp. 1029–1036.

Gido, J. & Clements, J. P. (2003), *Successful Project Management*, South-Western, New York.

Gillen, M., Baltz, D., Gassel, M., Kirsch, L. & Vaccaro, D. (2002), Perceived safety climate, job demands, and coworker support among union and nonunion injured construction workers, *Journal of Safety Research*, Vol. 33(1), pp. 33–51.

Goetsch, D. L. & Goetsch, D. D. L. (2003), *Construction Safety and Health*, Prentice Hall, Upper Saddle River, NJ.

Gunduz, M. & Yahya, A. M. A. (2015), Analysis of project success factors in construction industry, *Technological and Economic Development of Economy*, Vol. 24(1), pp. 67–80.

Hadikusumo, B. & Rowlinson, S. (2002), Integration of virtually real construction model and design-for-safety-process database, *Automation in Construction*, Vol. 11(5), pp. 501–509.

Hatush, Z. & Skitmore, M. (1997), Criteria for contractor selection, *Construction Management and Economics*, Vol. 15(1), pp. 19–38.

Holt, G. D. (2018), Let's change tack, not wind direction: A response to Kog and Yaman (2016), *Engineering, Construction and Architectural Management*, Vol. 25(3), pp. 335–341.

Idoro, G. I. (2004), The effect of globalization on safety in the construction industry in Nigeria. In: S. Ogunlana, C. Charoenngam, P. Herabat & B. H. W. Hadikusumo (Eds.), *International Joint Symposium of CIB W107 and CIB TG23 on Globalisation and Construction*, pp. 817–826. The International Council on Innovation and Research in Building and Construction (CIB), Rotterdam, The Netherlands.

Kometa, S. T., Olomolaiye, P. O., and Harris, F. C. (1995). An evaluation of clients' needs and responsibilities in the construction process. *Engineering, Construction and Architectural Management*, 2(1), 57–76.

Langford, D., Rowlinson, S. & Sawacha, E. (2000), Safety behaviour and safety management: Its influence on the attitudes of workers in the UK construction industry, *Engineering, Construction and Architectural Management*, Vol. 7(2), pp. 133–140.

Lin, J. & Mills, A. (2001), Measuring the occupational health and safety performance of construction companies in Australia, *Facilities*, Vol. 19(3/4), pp. 131–139.

Littau, P., Jujagiri, N. J. & Adlbrecht, G. (2010), 25 years of stakeholder theory in project management literature (1984–2009), *Project Management Journal*, Vol. 41(4), pp. 17–29.

Love, P. E. D., Teo, P., Carey, B., Sing, C.-P. & Ackermann, F. (2015), The symbiotic nature of safety and quality in construction: Incidents and rework non-conformances, *Safety Science*, Vol. 79(1), pp. 55–62.

Mohler, D. G., Chen, W. W. & Bloom, H. (1998), Angiosarcoma of the hand associated with chronic exposure to polyvinyl chloride pipes and cement: A case report, *Journal of Bone and Joint Surgery (JBJS)*, Vol. 80-A(9), pp. 1349–1354.

Ng, S. T. & Skitmore, R. M. (1999), Client and consultant perspectives of prequalification criteria, *Building and Environment*, Vol. 34(5), pp. 607–621.

Ogunsemi, D. & Aje, I. (2006), A model for contractors' selection in Nigeria, *Journal of Financial Management of Property and Construction*, Vol. 11(1), pp. 33–44.

Okoye, P. U. (2018), Occupational health and safety risk levels of building construction trades in Nigeria, *Construction Economics and Building*, Vol. 18(2), pp. 92–109.

Olatunji, O., Aje, O. & Odugboye, F. (2007), Evaluating health and safety performance of Nigerian construction site. In: *CIB World Building Congress*, Rotterdam, Netherlands, pp. 1176–1190.

Olatunji, O. A. (2005), The impact of prequalification on contractors' performance in Nigerian construction projects, Unpublished honours dissertation, Federal University of Technology Akure, Nigeria.

Olatunji, O. A. (2008), Due process and contractor selection for public works in Nigeria. In: *CIB Conference on Building Abroad: Procurement of Construction and Reconstruction Projects in the International Context*, pp. 385–396, Montreal, Canada.

Prabhakar, G. P. (2009), What is project success: A literature review, *International Journal of Business and Management*, Vol. 3(9), pp. 3–10.

Stern, F., Lehman, E. & Ruder, A. (2001), Mortality among unionized construction plasterers and cement masons, *American Journal of Industrial Medicine*, Vol. 39(4), pp. 373–388.

Umeokafor, N. (2016), Approaches, drivers and motivators of health and safety self-regulation in the Nigerian construction industry: A scoping study, *Architectural Engineering and Design Management*, Vol. 12(6), pp. 460–475.

Wanberg, J., Harper, C., Hallowell, M. R. & Rajendran, S. (2013), Relationship between construction safety and quality performance, *Journal of Construction Engineering and Management*, Vol. 139(10), pp. 04013003-1–10.

Zhao, D. & Lucas, J. (2015), Virtual reality simulation for construction safety promotion, *International Journal of Injury Control and Safety Promotion*, Vol. 22(1), pp. 57–67.

10 Integrating health and safety into labour-only procurement system

Opportunities, barriers and strategies

Nnedinma Umeokafor, Abimbola Windapo and Oluwole Olatunji

Summary

Integration of health and safety (H&S) into procurement is a proactive approach to improving H&S culture in project development processes, and more effective than reactive and active strategies. Nonetheless, procurement approaches that support H&S culture have received limited attention in construction research. This chapter examines a labour-only procurement system (LoPS) towards eliciting key barriers, opportunities and strategies for incorporating H&S into this growing procurement approach. A systematic review of literature is the main method of the study. Findings suggest clients are highly involved in LoPS. They control and monitor project objectives including H&S. We argue that the extant clients' approach presents a mixture of opportunities and barriers to facilitating robust H&S outcomes in construction projects. This is because in LoPS, inexperienced clients have unreserved rights to undertake leadership responsibilities and this often compromises H&S outcomes of their projects and project environments. We also found that traditional contract documents, in the ways they define stakeholders' roles and responsibilities, remain barriers to integrating H&S into LoPS. This is because these documents have been the tradition long before the industry commenced the renewal of its efforts at promoting H&S culture. One the other hand, incorporating H&S in LoPS triggers insights from cost-effective H&S strategies and H&S culture also benefits from risk–benefit transfer between construction contract parties. LoPS encourages early-stage collaboration and effective negotiation that could lead to appropriate revisions to the definitions of H&S responsibilities of contract parties. Conclusions of the study are drawn from robust conceptual theories in a way that inspires empirical studies.

Introduction

Health and safety (H&S) records of the construction industry is very poor and can be seen from industry reports. For example, industry data published by the Health and Safety Executive (HSE) (2018) show that the construction industry of Britain recorded 38 fatal injuries in 2017–2018, the highest across all the industries. The annual average of fatal injuries between 2013–2014 and 2017–2018 is 39. Again, this is the highest across all the industries. Although many developing countries (DCs) lack transparency and adequate data to trigger definitive conclusions on the state of H&S issues in their construction industries (Umeokafor 2018a), research studies have reported higher safety incidences in DCs than in developed countries. In particular, a study by Tadesse and Israel (2016) on 504 construction workers in Ethiopia shows the most prevalent causes of construction injuries are cutting (66.3 per cent) and falling (28.5 per cent), other causes of accidents account for the remaining 5.2 percent. The study also shows that majority of workers (83.9 per cent) who worked for up to 2 years did not use personal protective equipment.

According to Eriksson and Westerberg (2011), strategic use of procurement often significantly influences project delivery outcomes. This can be extended to mean project success, including project's H&S outturn situation, which can be shaped by steps taken by project stakeholders at the preconstruction stage. Extant studies have shown how this can happen in various ways. These include ingraining H&S into procurement of construction projects through evidence-based approaches (Deacon and Smallwood 2016; Wells and Hawkins 2010). Some studies have identified designing for safety (DfS) as another appropriate approach (see Gambatese 2013). In particular, Wells and Hawkins (2010) argue that where stakeholders' H&S measures are inadequate, appropriate procurements strategies and contract documentations have the potential to improve H&S outcomes in construction projects. In addition, according to Gambatese (2013), proactive measures such as DfS and other approaches aimed at integrating H&S into procurement and overall management of organisations are more beneficial to projects than reactive approaches where stakeholders attempt to correct impacts of post-incidents. The author argues that when H&S strategies are integrated early into the project life cycle, efforts to prevent and reduce risks are likely to be efficient and cost-effective. Smallwood (1998) concludes that health and environmental practices support situations where procurement systems and practices influence safety.

Despite a seeming consensus between researchers that procurement improves H&S outcomes in construction projects, the H&S records of the industry have remained relatively poor. Many studies have attributed this to inadequate research. For example, Wells and Hawkins (2010:5) claim that 'the use of procurement as an instrument to promote improved (H&S) practices among suppliers has received little attention'. Deacon and Smallwood (2016) identify paucity of normative literature relating to 'procurement and H&S' as another issue. In Umeokafor's (2018b) review of construction H&S research in Nigeria over a 36-year period, no study on 'procurement and H&S' is evident. Limited studies available on procurement and H&S include Deacon and Smallwood (2016), Mahamadu and colleagues (2015), Smallwood (1998), and Wells and Hawkins (2010). Their studies have mainly covered traditional procurement. Consequently, it can be argued that, at least, there is limited research on H&S and procurement practices from the sub-Saharan African perspective.

Several procurement methods exist for a reason: procurement approaches influence project success in various ways. In addition to traditional procurement systems used in building construction, other variants such as design and build, construction management, LoPS, direct labour and management contracting have been shown to impact project outcomes differently (Ogunsanmi 2015).

LoPS is a unique approach in which clients provide materials whilst contractors provide labour (Akinkunmi et al. 2018; Hardy 2013; Ogunde 2011; Ogunsanmi 2013a, 2013b). Clients are responsible for project coordination, supervision and quality control (Hardy, 2013). Evidence in Ogunsanmi (2013a, 2013b) suggests that LoPS is popular in DCs, including in sub-Saharan African countries such as Botswana, Kenya, South Africa, Uganda and Zimbabwe. Akinkunmi and colleagues (2018) and Ogunde (2011) have reported about LoPS in Nigeria, while Hardy (2013) has reported about LoPS in New Zealand. A key finding by Ogunde (2011) suggests a growing popularity of LoPS amongst construction clients, in that 58 per cent of respondents in the study preferred LoPS ahead of other procurement approaches such as the traditional procurement method, design and build, project management, construction management and direct labour.

While LoPS is commonly practised in sub-Saharan Africa, its impact on construction H&S management in the region is less known. In addressing this research problem and gap in knowledge established so far, this study examines LoPS by identifying the challenges, opportunities and strategies for optimising H&S outcomes in LoPS. Architecture and characteristics of LoPS are reviewed. Merits and demerits of LoPS are also discussed as well as key opportunities and barriers to H&S. Conclusions are drawn on the implications of integrating H&S into LoPS and towards maximising H&S outcomes in project outturns.

Conceptual framework

Health and safety in DCs

In Africa, H&S discourse has continued to emerge. The situation is similar in most DCs. While H&S adoption rates in many DCs are considerable but episodic, construction H&S in DCs is emerging. A low level of compliance with H&S laws has been reported in other studies also. For example, Annan and colleagues (2015) found that in Ghana, despite several regulatory provisions regarding incidence reporting, victims seldom report incidents as. Tadesse and Israel (2016) reported that injury information in Ethiopia is rare. Poor culture and regulation are rife; however, other studies have reported issues relating to institutions such as regulatory issues and inadequate policies in DCs such as Nigeria, Ghana and Ethiopia (Annan et al., 2015; Tadesse and Israel, 2016). Despite these, appropriate discussions and stakeholders' commitment to affirmative actions on the need to improve construction H&S continues to advance in many DCs (Umeokafor 2017, 2018a). For example, whilst the Nigerian construction industry has had no local H&S legislation, contractors often deploy H&S regulations and standards that they have adopted from countries such as Germany, United States, China and United Kingdom (Umeokafor 2017). Research also reveals communities' influence and clients' contribution to H&S outcome in construction projects.

Procurement and LoPS

Variants to traditional procurement systems are triggered by different situations. For example, according to Akinkunmi and colleagues (2018) and Ogunsanmi (2015), LoPS became prominent due to clients' quest to save on construction costs during an economic downturn. In particular, Ogunsanmi (2013b) reports how Nigeria's economic downturns in the 1980s and 1990s forced project owners to expand the scope of the LoPS from repairs, maintenance and refurbishments to new projects. Akinkunmi and colleagues (2018) claim persistent dissatisfaction of clients around projects' target costs, durations and stipulated quality led to increased desires to save cost through LoPS and other procurement systems such as direct-labour. A similar finding is reported in Hardy (2013) where in New Zealand, LoPS is used to save costs and to maximize utility and value for money.

How LoPS works has been described in the introduction. Further to this, LoPS involves some robust collaboration between designers, clients, contractors and subcontractors (Ogunsanmi 2015). LoPS's tender procedures can be open, selective and negotiated. Ogunsanmi (2015) found that negotiated tendering of LoPS is most prominent. The author found that negotiated tendering was used in 83 per cent of the LoPS projects surveyed in the study. Eighty-eight per cent of these LoPS projects was for new projects, 12 per cent was for refurbishment projects, 96 per cent was for building projects and 4 per cent was for civil engineering. Akinkunmi and colleagues (2018) show that LoPS suits small private projects such as residential buildings and minor alteration or modification works rather than large projects which require intense capital and high technical inputs. Private clients own these small projects and Umeokafor (2018a) found that they are less involved in H&S than public clients. They are, however, more involved in informal procurement activities than public clients (Umeokafor 2018a). This underpins the case of improving client involvement in H&S through integrating H&S into LoPS.

Methodology

The research adopts a systematic review of literature. 'Labour only procurement' and 'labour only contract' were the key themes of the review. When these keywords were searched on Google Scholar in June 2018 as exact keywords in the title of the article but excluding citations, the search returned only four relevant papers. When the search scope was expanded to the exact keywords anywhere in the article excluding citations, 19 papers were found on 'labour only procurement' and 85 on 'labour only contract'. Only one of these is relevant to this study. The article was added to the four articles found through the initial process. These were complemented by the citation search approach where the references of papers and books are searched for leads to articles that can be used. In all cases, the title and abstracts were scrutinised for relevant papers. The relevant ones were reviewed for manifest and latent meanings in line with the objectives of this study.

Akinkunmi and colleagues (2018), Ogunsanmi (2015) and Ogunsanmi (2013a, 2013b) are the relevant four papers from the first literature search. Ogunde (2011) was added from the mentions search, whilst Hardy (2013) and

Fagbenle (2010) were added after the citation search and scrutiny. All the articles were on Nigeria except one. None of the articles was indexed in Scopus or the Social Science Citation Index. None of the papers cover the interconnection between LoPS and H&S. The few studies that cover 'Procurement and H&S' have been outlined in the introduction of this study. Little attention to LoPS in normative research is evident. An additional search of other databases may have provided more results, however, results from this may be insignificant as most crucial scholarly databases are indexed in Google Scholar.

Results and discussion

Characteristics, merits and demerits of LoPS and opportunities for H&S

Results of the content analysis are presented in Tables 10.1 to 10.3. Table 10.1 shows the characteristics of LoPS. A 'them and us' attitude is a common characteristic of fragmented procurement methods. The phenomenon is a common cause of disputes between project owners and contractors in fragmented procurement methods (Ogunsanmi 2015). In LoPS, however, it improves the relationship amongst members of the project. Another attribute of LoPS is that clients are in firm control. The high level of their involvement and control means clients are assumed to possess appropriate experience and competence in material sourcing and handling. When clients lack experience and competence yet provide pivotal leadership roles, the projects' outturn situations (including H&S) are likely to suffer. While designers may design for safety, client's quest to maximise value for money may compromise H&S standards due to optimism bias and a deficient confidence.

It is prudent that clients and designers are able to make considerable inputs regarding the strategic roles of H&S in procurement (Umeokafor 2018a). However, H&S may attract a cost burden that clients often underestimate. Smallwood (1996) argues that contractors who make generous provision for H&S in their bids often stand the risk of losing to a competitor who is less committed to H&S standards. With the high level of clients' involvement and control that characterise LoPS, a logical question to ask is how H&S can be incorporated in LoPS for maximum outcomes. In countries like South Africa where clients have statutory duties relating to H&S, there are straightforward answers to this. However, in Nigeria where there are no local H&S laws, strategies have to be soft and creative.

Clients' high level of *control* in LoPS (see Table 10.1) implies that H&S is overly dependent on clients' willingness and knowledge depth. As explained earlier, this can be a barrier. On the other hand, a high level of client *involvement* in the development processes of projects and their desire to optimise value for money imply that clients are keen to share risks. This is in contrast to the traditional procurement method where clients transfer all the financial risks for construction to the contractor. The position of normative literature is instructive regarding the opportunities presented by clients' willingness to situate their

Table 10.1 Characteristics of LoPS

Characteristic	Description
• Moderate or higher level on diversity of responsibility	When clients engage in robust collaborative relationships with contractors, they are able to diversify their responsibilities and accrue value-adding savings (Akinkunmi et al. 2018; Ogunsanmi 2015).
• Varied level of client involvement	Clients can accrue benefits from varying degree of project involvement – could be high or moderate (Akinkunmi et al. 2018; Ogunsanmi 2015).
• Client satisfactory level of cost control and monitoring	Opportunities abound for clients to control and monitor cost and quality (Akinkunmi et al. 2018; Ogunsanmi 2015).
• Client control of building process and parties during construction	Clients are able to control construction processes (Akinkunmi et al. 2018). They are more able to control subcontractors than in the traditional procurement approach (Ogunsanmi 2015).
• Client's opportunity to monitor cost of quality	Supply of substandard material is reduced (Akinkunmi et al. 2018). Contractors are able to concentrate on labour as clients alleviate the burdens of the supply chain. Clients benefit from cost of quality, as they are able to secure their utility in material supply and monitor quality of workmanship.
• High level of flexibility	Clients are able to make changes alongside other procurement methods (Akinkunmi et al. 2018). They can vary the original design and structure of control at minimal costs (Ogunsanmi 2015).
• Opportunities for negotiation	Majority of LoPS projects were procured through negotiation. Ogunsanmi 2015 found 83 per cent of respondents used negotiated tendering in LoPS.
• Completion time is shorter	Ogunsanmi (2015) and Fagbenle (2010) found fewer LoPS projects experienced schedule overruns compared to the traditional procurement approach.
• More propensity for confrontation	Ogunsanmi (2015) found LoPS generates more claims than in traditional procurement. The authors found 64 per cent in LoPS compared to traditional procurement's 44 per cent.
• Influence on selection of trade contractors	Client can significantly influence the selection of subcontractors.

risk-sharing commitment in a way that helps them safeguard their interest rather than seeking to avoid risks and paying a heftier price for only a little outcome (Deacon and Smallwood 2016).

Trade-offs between clients' ownership of risks in LoPS and the traditional approach to risk transfer to contractors are important. The benefits that accrue to clients as they procure materials can be reinvested into how contractors make the best of project situations to maximize work quality and H&S deliverables (Wills and Hawkins 2010). This approach presents a different view to common views in normative literature where clients are portrayed as though they are impaired by the economic gains that accrue from apparent material procurement and their limited involvement in H&S costs. Rather, it suggests that there are additional benefits for clients to consider their strategic transfer of benefits in exchange for added H&S outcomes and improved quality costs. Outcomes from this are best when this starts from the prequalification stage through tender action. Client involvement at this stage presents opportunities to analyse contractors' H&S performances in the past and their current potential to achieve exceptional outcomes on new projects.

Merits of LoPS are presented in Table 10.2. Certain attributes of LoPS can be interpreted as merits and demerits. For example, opportunity to negotiate is an advantage in LoPS (see Table 10.1) but could also be a hindrance to H&S outcomes (see Table 10.3). This is because negotiation limits the participation of excluded stakeholders. Examples of these include workers, workers' families, manufacturers and subcontractors that are not nominated by clients. The exclusion of such role players may reduce their commitment to H&S and process innovation. The opportunity on the other hand is for clients to be transparent and share the perceived 'costs' and responsibilities of H&S.

Clients' involvement in the building processes is both a merit and a characteristic of LoPS. Will and Hawkins (2010:7) put this aptly: while clients are able to use contract procedures to monitor contract provisions against key performance indicators, monitoring from within the project team is more effective. LoPS presents opportunities for clients to ensure the contractors adhere to H&S requirements. They reserve the right to construct this appropriately in their own contract language. They are able to enforce their objectives during tender action,

Table 10.2 Merits of LoPS

Merit	Description
• Improved relationship	LoPS offers prospects of an improved relationship among the project team than in traditional procurement (Ogunsanmi 2015).
• Potential of higher quality and value for money	Improved quality of workmanship is likely as resource quality is ascertained (Akinkunmi et al. 2018).
• Presents opportunities	Opportunities for tradespersons to be engaged as subcontractors (Akinkunmi et al. 2018).
• Potential to save time and costs	Collaboration saves time, including during design and construction period (Akinkunmi et al. 2018). This means projects can be delivered timely (Fagbenle 2010; Ogunsanmi 2015). In addition, clients are able to save costs, including from contractors' overhead (Akinkunmi et al. 2018; Ogunsanmi 2013b).
• Satisfactory coordination and control	Clients have better control of the construction process and can achieve better satisfaction with their involvement than they usually do in traditional procurement (Ogunsanmi 2015).

Table 10.3 Demerits of LoPS

Demerit	Description
• Limited opportunities and scope	LoPS is likely to attract indigenous contractors only, mainly small contractors (Akinkunmi et al. 2018; Ogunde 2011).
• High level of client involvement	High level of involvement attracts commensurate time commitment and energy (Akinkunmi et al. 2018). Clients' diplomacy is incumbent also (Hardy 2013).
• Reduced profit for contractors	Contractors' profit is reduced (Ogunsanmi 2013b, 2015; Ogunde 2011). Incomes from material procurement and logistic are gone.
• Results in contractual disputes	Ogunde (2011) found ambiguous contracts, poor communication, distrust, and design and constructability issues in LoPS.
• Negotiated tender is mainly used	Negotiated tender offers limited tendering options (Akinkunmi et al. 2018; Ogunsanmi 2015).
• Biased risk in favour of contractors	Clients are vulnerable where work is at risk of quality issues (Hardy 2013). Client bears more of the risk (Ogunsanmi 2015).
• High level of knowledge, experience and supervision skills	Client needs strong knowledge of building construction, experience and supervision skills (Akinkunmi et al. 2018)
• Implementation issues – coordination of resources	Responsibilities can be unclear, ambiguous and overlapping (Ogunde 2011). Contractors may struggle to harness production resources with own labour resources and may fail to break even due to low overhead and profit (Ogunde 2011).

planning and control of value streams, and are able to measure contractors' capabilities and learning cultures against key risks indicators (Wills and Hawkins 2010). A typical example illustrated by Wills and Hawkins (2010) is contractors' plans and financial provisions for falsework and temporary work such as formwork and scaffolding. They argue that these could receive particular attention during tender evaluation.

In addition, Table 10.2 suggests clients' propensity to save cost and time in LoPS. Table 10.3 suggests this must come at a cost. There is 'the need for clients to have a high level of knowledge, experience and supervision skills' (see Table 10.3). Given that LoPS results in an improved relationship among the project participants, clients could take advantage of the knowledge and experience of contractors and designers and shift H&S responsibilities to them. H&S laws such as the Construction Design and Management Regulations (2015) in Britain require domestic clients to shift H&S duties to contractors and principal designers. In the same vein, it is incumbent that LoPS clients consider candidates with strong H&S backgrounds with demonstrable skills, knowledge and experience only.

A major barrier to integrating H&S into LoPS is ambiguity of contracts and the apparent lack of clarity regarding roles and responsibilities of players in LoPS.

Wills and Hawkins (2010) identify vague and generalised reference to H&S in construction contracts. They argue in favour of clear benchmarks and definitions of H&S in contracts. This would reduce or prevent disputes in LoPS (Ogunde 2011). In addition, contractors are constrained by costs and do have reduced profit. They are likely to struggle to break even. As a result, Wills and Hawkins (2010) suggest it is incumbent to include costs of H&S in the bills of quantities. This will remove the burden of H&S costs from contractors, the most naïve excuses of contractors and clients to compromise H&S (Umeokafor 2017).

The structure and characteristics of LoPS highlight a soft tendency for H&S, a propensity to favour the informal sector and a shift from verbal arrangements to standard agreements. In addition, it involves the opportunity to reach out to the informal sector and the need for appropriate regulation. Whilst Annan and colleagues (2015) put the responsibility for regulation to governments, Umeokafor (2017) has proposed a complementary alternative in self-regulation.

Integrating H&S into LoPS

Key strategies for integrating H&S into LoPS are suggested alongside the key opportunities and challenges.

Strategies

- Draw on cost-saving features of H&S and benefit transfer in favour of H&S commitment of contractors and clients. Clients have the right to know that this is to protect their interests.
- Clients and contractors must interact and negotiate as they define and revise H&S responsibilities progressively from the very early stages of LoPS.
- Clearly incorporate H&S in the invitation-to-tender stage of LoPS and in contracts. This must be clear on the roles and responsibilities of parties and relative key performance indicators for each stakeholder.
- Integrate H&S in the tender evaluation stage with more attention on contractors' design and financial capabilities for construction of falsework and temporary works such as formwork and scaffolding.
- The client, designers and contractors must develop an H&S plan for coordinating and controlling resources, including labour in the early stages of LoPS.

Key opportunities

- High level of client involvement coupled with the goal of reducing costs and high level of risk.
- Responsibilities of H&S can also be negotiated, defined and even revised in the project.
- Opportunities for clients to be well-involved in H&S including monitoring H&S during the construction process.

- Platforms for reaching out to the informal sector where H&S is difficult to address.
- Clients may witness risks in the construction process first-hand; on the grounds of morals, this may improve their attitudes towards H&S.

Key barriers

- Client involvement – unwilling clients, and clients that place H&S low or not on their priority list.
- Ambiguity in contract documents and unclear roles and responsibilities.
- Lack of relevant skills and experience of the client.
- Risk of inexperienced and incompetent clients taking on H&S roles hence compromising H&S standards.
- Client's quest to save money may result in compromising H&S.
- Lack of adequate H&S laws and regulatory system to drive the agenda.
- The risk of communication issues because of the nature of LoPS.
- Poor knowledge of H&S in terms of the project team, including clients.
- H&S will be overly dependent on client willingness and attitudes hence the need for drivers such as H&S laws that place more H&S responsibility on the client.

Implications for practice and research

The implications of the research include the imperativeness of dedicated contract standards for LoPS that make adequate provisions for H&S in construction projects, contextualised for project situations in DCs. While H&S is emerging in DCs, there is need for national and universal H&S policies that make legal provisions for the integration of H&S into procurement systems. As this study is not empirical, a survey can assess the strategies, barriers and opportunities to ascertain their workability and extent respectively. Another study may explore the impact of the characteristics of LoPS on the H&S performance of contractors. Empirical evidence on the relationship between procurement theories and project's actual H&S performance is limited, hence it needs to be looked at from different perspectives.

Conclusions

The study critically examined LoPS towards identifying the barriers, opportunities and strategies for integrating H&S into the procurement approach. Among the key features of the LoPS that also present opportunities include its cost- and time-saving attributes. The system also presents a high level of client involvement in terms of control and monitoring, an opportunity for H&S. However, the level of the client's skills and knowledge in construction and H&S and the risk that inexperienced clients or those with no knowledge of H&S may assume H&S responsibilities are challenges. Because of the nature of LoPS, it is likely to attract only private clients and domestic contractors who are known for poor H&S records. Nevertheless, the strategies include drawing on the cost-saving features of H&S and the impact of the risk to sell the benefits to the client; and negotiating, defining and revising H&S responsibilities in the early stages of LoPS. The implications for DCs include the imperativeness of establishment of standardised contracts for LoPS that make adequate provisions for H&S on projects.

References

Annan, J., Addai, E. K., and Tulashie, S. K. (2015), 'A Call for Action to Improve Occupational Health and Safety in Ghana and a Critical Look at the Existing Legal Requirement and Legislation', *Safety and Health and Work*, Vol. 6(2), pp. 146–150.

Akinkunmi, G. A. O., Aghimien, D. O., and Awodele, O. A. (2018), 'Appraising the Use of Labour-Only Procurement System for Building Construction in Nigeria', *Organisation, Technology and Management in Construction*, Vol. 10(1), pp. 1719–1726.

Deacon, C. and Smallwood, J. J. (2016), 'The Effect of the Integration of Design, Procurement, and Construction Relative to Health and Safety (H&S)'. In: Dastbaz, M. and Gorse, C. (Eds.), *International Sustainable Ecology Engineering Design for Society (SEEDS) Conference*, 14–15 September 2016, Leeds Beckett University, Leeds, UK, LSI Publishing, pp. 422–453.

Eriksson, P. and Westerberg, M. (2011), 'Effects of Cooperative Procurement Procedures on Construction Project Performance: A Conceptual Framework', *International Journal of Project Management*, Vol. 29(2), pp. 197–208.

Fagbenle, O. I. (2010), 'A Comparative Study of The Time and Cost Performance of Labour Only Subcontractors in the Construction Industry in South Western Nigeria', *Journal of Building Performance*, Vol. 1(1), pp. 94–101.

Gambatese, J. A. (2013), Prevention through Design (PtD) Project 1: Benchmarking Management Practices Related to PtD in the US and UK Final Report – Activity 2: Assess the Effects of PtD Regulations on Construction Companies in the UK. National Institute for Occupational Safety and Health (NIOSH). Retrieved on 11 July 2018 from https://designforconstructionsafety.files.wordpress.com/2017/07/niosh-ptd-in-the-uk-final-report-may-2013.pdf.

Hardy, G. (2013), The Pitfalls of Labour-Only Building Contracts, ADLS. Retrieved on 10 July 2018 from www.adls.org.nz/for-the-profession/news-and-opinion/2013/4/12/the-pitfalls-of-labour-only-building-contracts/.

Health and Safety Executive (HSE) (2018), Workplace Fatal Injuries in Great Britain 2018. Retrieved on 10 July 2018 from www.hse.gov.uk/statistics/pdf/fatalinjuries.pdf

Mahamadu, A.-M., Mahdjoubi, L., Booth, C., and Fewings, P. (2015), 'Integrated Delivery of Quality, Safety and Environment Through Road Sector Procurement: The Case of Public Sector Agencies in Ghana', *Journal of Construction in Developing Countries*, Vol. 20(1), pp. 1–24.

Ogunde, A. O. (2011), 'Conflict Resolution in Labour-Only Contracts in Nigeria'. In: Ruddock, L. and Chynoweth, P. (Eds.), *COBRA 2011: Proceedings of Royal Institute of Chartered Surveyors (RICS), Construction, and Property Conference*, Salford, The Royal Institution of Chartered Surveyors, RICS, pp. 1661–1777.

Ogunsanmi, O. E. (2013a), 'Correlates of Procurement Performance of Traditional and Labour-Only Methods in Nigeria', *Ethiopian Journal of Environmental Studies and Management*, Vol. 6(2), pp. 182–190.

Ogunsanmi, O. E. (2013b), 'Comparisons of Performance of Traditional and Labour-Only Procurement in Construction Projects in Nigeria', *Africa Management Review*, Vol. 3(2), pp. 1–15.

Ogunsanmi, O. E. (2015), 'Comparisons of Procurement Characteristics of Traditional and Labour-Only Procurements in Housing Projects in Nigeria', *Covenant Journal of Research in the Built Environment*, Vol. 3(2), pp. 53–67.

Smallwood, J. J. (1996), 'The Role of Project Managers in Occupational Health and Safety'. In: Dias, L. A. and Coble, R. J. (Eds.), *Proceedings of the First International Conference of CIB Working Commission W99. Implementation of Safety and Health on Construction Sites*, Lisbon, Portugal, Balkema, Rotterdam, pp. 203–221.

Smallwood, J. J. (1998), 'The Role of Procurement Practices in Occupational Health and Safety and the Environment'. In Fahlstedt, K. (Eds.), *Proceedings of CIB World Building Congress*, 7–12 June 1998, Gaevle, Sweden, pp. 203–213.

Tadesse, S. and Israel, D. (2016), 'Occupational Injuries among Building Construction Workers in Addis Ababa, Ethiopia', *Journal of Occupational Medicine and Toxicology*, Vol. 11(16), pp. 1–6.

Umeokafor, N. I. (2017), Realities of Construction Health and Safety Regulation in Nigeria. Doctoral thesis, University of Greenwich, United Kingdom.

Umeokafor, N. I. (2018a), 'An Investigation into Public and Private Client Attitudes, Commitment and Impact on Construction Health and Safety in Nigeria, *Engineering, Construction and Architectural Management*, Vol. 29(6), pp. 798–815.

Umeokafor, N. I. (2018b), 'Construction Health and Safety Research in Nigeria: Towards a Sustainable Future'. In: Saurin, T. A., Costa, D. B., Behm, M., and Emuze, F. (Eds.), *Proceedings of Joint CIBW99 and TG59 Conference*, 1–3 August 2018, Salvador, pp. 213–221.

Wells, J. and Hawkins, J. (2010), *Promoting Construction Health and Safety through Procurement: A Briefing Note for Developing Countries*. Engineers against Poverty. London: Institute of Civil Engineers. Retrieved on 19 August 2018 from www.engineersagainstpoverty.org/documentdownload.axd?documentresourceid=20.

11 Tertiary built environment construction health and safety education in South Africa

John Smallwood

Summary

Construction health and safety (H&S) is a multi-stakeholder issue but, historically, it has been viewed as the contractor's problem. The human cost in the form of fatalities, injuries, and disease, and the cost of accidents (COA) represents the motivation for all stakeholders to contribute to H&S. The aim of the research is to assess the status of and evolve a framework for tertiary built environment construction H&S education in South Africa. A survey was conducted among providers of tertiary built environment education to assess the status of H&S in construction education. Thereafter, built environment practitioners were surveyed to determine the extent to which they agreed with a proposed framework for tertiary built environment construction H&S education. Tertiary built environment education is inadequate in terms of H&S in construction. The degree of support for the inclusion of twenty-five aspects in thirteen disciplines' tertiary built environment programmes' construction H&S modules ranges between 66.4% (Land Surveying) and 97.4% (Construction Management). The reality is that tertiary built environment education is not empowering its graduates to contribute to H&S in construction and, therefore, compounds the problems relative to H&S. This study has provided the first comprehensive scientific framework, in terms of thirteen disciplines, and the related construction H&S modules that should be addressed.

Background to the study

A report of the Construction Industry Development Board (cidb, 2009), titled 'Construction Health & Safety in South Africa: Status & Recommendations', highlights the considerable number of accidents, fatalities, and other injuries that occur in South African construction. The report cited the high level of non-compliance with H&S legislative requirements, which is indicative of a deficiency in the management and supervision of H&S on construction sites as well as planning from the inception/conception of projects within the context of project management. The report also cited a lack of skill, experience, and knowledge amongst construction personnel to manage H&S on construction sites.

The cidb's report indicates the disabling injury incidence rate (DIIR) to be 0.98, i.e. 1 of every 100 workers experiences a disabling injury, compared with the all-industry average of 0.78 (cidb, 2009). Furthermore, the fatality rate of

25.5 per 100,000 workers does not compare favourably with international rates (cidb, 2009). The Australian construction industry fatality rate per 100,000 for the period 2009–2013 was 3.29 for all construction, and 1.76 for building construction (Safe Work Australia, 2017). The severity rate (SR), in turn, indicates the number of days lost as a result of accidents for every 1,000 hours worked. The South African construction industry SR of 1.14 is the fourth highest, after fishing, mining, and transport, the all-industry average being 0.59. Given that the average worker works 2,000 hours per year, if the SR is multiplied by 2, the average number of days lost per worker per year can be computed and shows that the construction industry lost 2.28 working days per worker.

The cidb (2009) states that the total cost of accidents (COA) could have been between 4.3% and 5.4%, based upon the value of construction work completed in South Africa. The key issue relative to the COA is that, ultimately, clients incur the COA, as the COA is included in contractors' cost structures in the form of indirect costs, as contractors do not disaggregate costs when preparing tenders. In a similar vein, workers' compensation insurance constitutes a labour overhead, and if contractors experience a poor claims ratio, their insurance premiums are increased. The COA and workers' compensation insurance thus constitute a further motivation for all project stakeholders to contribute to H&S in construction and integrate it into all stages of projects.

H&S in construction is a multi-stakeholder issue, which is underscored by the South African Construction Regulations of 2003 and revised in 2014 (Republic of South Africa, 2014), as they require a range of interventions by clients, designers, and quantity surveyors (QSs) in addition to contractors. Except for construction project managers (CPMs), 'designers' include the range of built environment practitioners (BEPs) that constitute the design team, interior designers, and landscape architects included. Furthermore, the cidb's H&S status report (2009) addresses the respective stakeholders. CPMs manage the entire project process from project initiation through to project close-out. They are in a unique position, as they effectively manage clients, design delivery, and integrate design and construction, and can thus influence H&S in construction. Designers influence H&S in construction directly through design-specific, supervisory, and administrative interventions, and indirectly through the type of procurement system adopted, pre-qualification, decision regarding project duration, adoption of partnering, and the facilitating of pre-planning. QSs also influence H&S in construction indirectly through the type of procurement system adopted, facilitating financial provision for H&S in construction, pre-qualification, decisions regarding project duration, adopting of partnering, and the facilitating of pre-planning. Contractors manage the physical process of construction, and the related activities, including a range of resources.

The cidb's (2009) H&S status report states that H&S-relevant education and training, or the lack thereof, at all levels, has a significant effect on H&S in construction. Furthermore, at the tertiary level, not all construction-related programmes in South Africa include H&S within their curricula, especially designer programmes. H&S training empowers supervisors and workers to supervise and work in a healthy and safe manner. Tertiary H&S education, in turn, creates awareness, which engenders commitment to H&S, which engenders consideration of and reference to H&S, and the necessary resourcing.

Given the aforementioned, a study was conducted to evolve a framework for tertiary built environment construction H&S education in South Africa. The study consisted of two phases. The objective of the first phase was to determine the status of H&S in construction education in the respective tertiary built environment education programmes, and the objective of the second phase was to evolve a framework for tertiary built environment construction H&S education.

Perspectives and concepts

Legislation and guidelines

The Occupational Health and Safety Act (OHSA) (Republic of South Africa, 1993) schedules a range of requirements for employers and employees. Therefore, all stakeholders, clients, CPMs, designers, and QSs, in addition to contractors, must be aware, sensitive, and educated with regard to H&S in construction, particularly given that they visit construction projects and are personally exposed to hazards and risk. Although the Construction Regulations were promulgated in 2003, designers have been liable for H&S in construction in terms of the act since 1993, as Section 10 requires designers to ensure that any 'article' is safe and without risks to health. This includes buildings and structures.

Identities of work

In South Africa, BEPs are required to register with a statutory council related to the practice of their discipline. There are six statutory councils that register seven BEPs, which include architects, construction managers (CMs), CPMs, engineers, landscape architects, property valuers, and QSs. Each of the six BEP councils (BEPCs) evolved a scope of work, which includes the interventions and deliverables for the six project stages, which is known as the 'Identity of Work' (IoW). However, a review of the respective IoWs in terms of H&S interventions determined the mean percentage involvement by the seven BEPs as follows (Deacon & Smallwood, 2016): Stage 1 – Project Initiation and Briefing (0.0%); Stage 2 – Concept and Feasibility (66.7%); Stage 3 – Design and Development (50.0%); Stage 4 – Tender Documentation and Procurement (16.0%); Stage 5 – Construction Documentation and Management (50.0%), and Stage 6 – Project Close-out (16.0%).

Importance of tertiary built environment construction H&S education

A recent study, titled 'Preventing Accidents in Construction', was conducted among Master Builders South Africa (MBSA) National H&S Competition Award winners by Smallwood (2014) to determine the importance of thirty-seven aspects in terms of preventing accidents or achieving optimum H&S. Aspects related to this chapter, accompanied by mean scores (MSs) between 1.00 and 5.00, based upon percentage responses to a 5-point scale, and their rank based upon the MSs were: H&S education (5.00; 1=); H&S training (5.00; 1=); designing for construction H&S (4.44; 1 7=); design hazard identification and risk assessments (4.22; 22=); and tertiary built environment education that includes construction H&S (4.11; 28).

A study titled 'Construction Management Health and Safety (H&S) Course Content', conducted by Smallwood (2010) among MBSA National H&S Competition Award winners or holders of a 4- or 5-star H&S grading on one or more of their projects, investigated the importance of the inclusion of construction H&S in the tertiary education programmes of nine built environment disciplines. The MSs between 1.00 and 5.00, based upon percentage responses to a five-point scale, and a rank based upon the MSs were: construction management (5.00; 1); civil engineering (4.89; 2); electrical engineering (4.89; 3); project management (4.78; 4); mechanical engineering (4.56; 5); architecture (4.50; 6); interior design (3.78; 7); landscape architecture (3.78; 8); and quantity surveying (3.33; 9).

Research method

Both studies were quantitative in nature, but respondents were requested to provide comments in general regarding tertiary built environment construction H&S education. Study 1 constituted an assessment of the extent to which tertiary built environment education addresses construction H&S, whereas Study 2 investigated practitioners' perceptions about the aspects that tertiary built environment construction H&S education should address.

Study 1

The first survey was titled 'Tertiary Built Environment Construction H&S Education Study'. Owing to the non-availability of a comprehensive contact list for tertiary built environment education departments, a snowball sample was used, i.e. contacts of the researcher were approached and were also requested to forward the survey to 'sister' departments. Data analysis included the computation of frequencies and, in the case of Likert-scale questions, a measure of central tendency in the form of a MS to enable the ranking of variables.

Study 2

The second survey was titled 'Tertiary Built Environment Construction H&S Education Framework Study'. Based upon the survey of the literature, 25 aspects were identified for possible inclusion as sub-modules of respective H&S programmes (Smallwood, 2010, 2015). Initially, nine programmes were considered in terms of the proposed framework; however, project management, and town/

urban planning were added. Furthermore, town/urban planning was included, as although the focus of the study/project may be deemed to be at the undergraduate and BTech/honours level, built environment–related coursework for master's studies should address construction H&S. However, after the launch of the survey, the author deemed it appropriate to include facilities management and land surveying, which resulted in a total of thirteen programmes. Table 11.1 presents the proposed extent to which aspects should be addressed by the H&S modules of the programmes. The bottom row presents a discipline mean percentage, and the far-right column an aspect mean percentage.

A convenience sample and, to a degree, a convenience 'snowball' sample of built environment practitioners was selected. Potential, discipline-specific respondents were requested to respond relative to their disciplines only or, if registered with more than one statutory council, to respond relative to the disciplines concerned. Furthermore, certain respondents indicated that they studied quantity surveying, for example, but now practised as a CM in the form of a managing director of a general contractor. Others practised as both an architect and an interior designer. Construction H&S agents (CHSAs), construction H&S managers (CHSMs), and construction H&S advisors such as those employed by the Master Builders Associations (MBAs) and the South African Forum of Civil Engineering Contractors (SAFCEC), were requested to respond relative to all the disciplines.

Data analysis included the computation of frequencies.

Results and discussion

Study 1

A total of 30 responses emanating from 11 institutions were included in the analysis of the data (see Table 11.2). Nine programmes are represented. It should be noted that 'Building' refers to the NDip: Building, which constitutes the undergraduate programme relative to BTech: Construction Management and BTech: Quantity Surveying. It should also be noted that, in many cases (i.e. institutions), certain programmes are not offered.

In response to the question 'Is construction H&S addressed/included in your programme/curriculum offered by your Department/School?', 64% of respondents responded in the affirmative, 32% in the negative, and 4% were unsure.

Only 28.6% of designer programmes address/include construction H&S, 64.3% do not, and 7.1% are unsure. However, 100% of construction management, property studies, and quantity surveying programmes do; 85.7% of building programmes do; and 14.3% of the related respondents did not respond.

Those respondents that responded in the affirmative were requested to indicate how it is included and addressed. Only 16.7% include H&S in construction as a separate subject, the mean hours presentation time being 23, whereas 78.6% include it as a component of a subject; and 62.5% of respondents also indicated that it is addressed on an ad hoc basis, 25% that it is not, and 12.5% were unsure.

Table 11.1 Revised proposed aspects for inclusion in construction H&S modules of tertiary built environment programmes

Aspect	Architecture	Civil engineering	Construction management	Electrical engineering	Facilities management	Interior design	Landscape architecture	Mechanical engineering	Project management	Property development	Quantity surveying	Town/urban planning	Land surveying	Aspect mean (%)
Need for construction H&S	●		●		●	●	●	●	●	●	●	●	●	100.0
OH&S Act	●	●	●	●	●	●	●	●	●	●	●	●	●	100.0
COID Act	●	●●●	●●●	●	●●●								●	23.3
Construction regulations	●	●●	●●	●	●	●	●	●	●	●	●	●	●	100.0
Other regulations		●	●	●	●●●									46.2
Occupational health	●	●●●	●●●	●●	●●●●	●●●	●●●	●	●●●●	●●●	●	●	●●	100.0
Occupational safety	●●	●●●	●●●	●	●●●	●●●	●●●	●	●●●●	●●●	●	●	●●●	100.0
Ergonomics	●●	●●	●●	●	●●	●	●	●	●●	●	●	●	●	100.0
Primary health promotion		●	●		●									30.8
Environment and construction H&S	●	●	●	●	●	●	●	●	●	●	●	●	●	100.0
Role of construction H&S in project performance	●	●	●	●	●	●	●	●	●	●	●	●	●	100.0
Economics of construction H&S	●	●	●	●	●	●	●	●	●	●	●	●	●	100.0
Role of clients in construction H&S	●	●	●	●	●	●	●	●	●	●	●	●	●	100.0
Role of project managers in construction H&S	●	●	●	●	●	●	●	●	●	●	●	●	●	100.0

(Continued)

Table 11.1 Continued

Aspect	Architecture	Civil engineering	Construction management	Electrical engineering	Facilities management	Interior design	Landscape architecture	Mechanical engineering	Project management	Property development	Quantity surveying	Town/urban planning	Land surveying	Aspect mean (%)
Role of designers in construction H&S	●	●	●	●	●	●	●	●	●	●	●	●	●	100.0
Role of quantity surveyors in construction H&S	●	●	●	●	●	●	●	●	●	●	●	●	●	100.0
Contractor H&S		●●	●●						●●					23.3
Managing subcontractor H&S		●●	●●						●●					23.3
Role of manufacturers in construction H&S	●	●	●	●	●	●	●	●	●	●	●	●		92.3
Influence of procurement on construction H&S	●	●	●	●	●	●	●	●	●	●	●			84.6
Hazard identification and risk assessment	●	●	●	●	●	●	●	●	●	●	●	●	●	100.0
H&S specifications	●●	●●	●●	●●	●●	●●	●●	●●	●●	●●	●●	●●	●●	100.0
Designer report (to clients)	●●	●●	●●	●●	●●	●●	●●	●●	●●	●●	●●	●●	●●	100.0
H&S plans	●●	●●	●●	●●	●●	●●	●●	●●	●●	●●	●●	●●	●●	100.0
H&S file	●●	●●	●●	●●	●●	●●	●●	●●	●●	●●	●●	●●	●●	100.0
Discipline Mean (%)	80.0	100.0	100.0	84.0	92.0	80.0	80.0	84.0	96.0	80.0	80.0	76.0	72.0	

Table 11.2 Respondents to Study 1

Institution	Architectural technology	Building	Civil engineering	Construction management	Electrical engineering	Interior design	Mechanical engineering	Property studies	Quantity surveying	Total
University of Technology A		●								1
University of Technology B		●								1
University of Technology C	●	●	●		●		●			5
University of Technology D		●								1
Comprehensive University A	●	●	●	●	●	●	●		●	8
Comprehensive University B			●							1
University B			●	●●	●		●		●	5
Comprehensive University B		●		●						2
University C								●		1
University D								●●	●	3
Comprehensive University C		●	●							2
Total	2	7	5	4	3	1	3	2	3	30

Furthermore, 27.3% of respondents responded that it is an 'issue in design or integrative project (included in the brief), 45.5% that it is not, and 27.3% were unsure; 22.2% responded that it is a 'criterion for assessment in design or integrative project', 44.4% that it is not, and 33.3% were unsure.

Respondents were required to indicate the intention of their departments/schools that do not address/include H&S. It is notable that 50% are unsure; 8.3% of respondents responded relative to each of 'never', one year, and three years; and 25% responded relative to two years.

Table 11.3 indicates the importance of the inclusion of construction H&S in the tertiary education programmes of nine built environment disciplines in terms of percentage responses on a scale of 1 (hardly) to 5 (very), and MSs between 1.00 and 5.00. The results show that 6/9 (66.7%) MSs were > 4.20 ≤ 5.00, which indicates the inclusion is between more than important to very important; 3/9 (33.3%) of the MSs > 3.40 ≤ 4.20, which indicates that the inclusion is between important to more than important in quantity surveying, landscape architecture, and interior design. However, it should be noted that the MS of quantity surveying is on the upper point of the range.

Respondents were requested to provide comments in general regarding tertiary built environment construction H&S education, 76.2% of which did and 23.8% did not. Selected comments recorded verbatim were:

- 'Urgently required in all built environment modules'.
- 'Indeed, it is vital for H&S to be introduced as well as highlight the impact it has in the industry at large'.
- 'H&S should be included as a separate subject in order to emphasise its importance in the construction industry'.
- 'The School of Architecture may have to look at this in future, not as a separate module, but as part of existing modules'.
- 'H&S should be taught right from 1st year'.

Table 11.3 Importance of the inclusion of construction H&S in the tertiary education programmes of nine built environment disciplines

Discipline	Unsure	Hardly 1	2	3	4	Very 5	MS	Rank
Construction management	0.0	0.0	0.0	4.8	4.8	0.0	4.86	1
Project management	0.0	0.0	0.0	5.0	15.0	0.0	4.75	2
Civil engineering	0.0	0.0	0.0	0.0	30.0	0.0	4.70	3
Electrical engineering	0.0	0.0	5.3	5.3	31.6	5.3	4.42	4
Mechanical engineering	0.0	0.0	5.3	5.3	31.6	5.3	4.42	5
Architecture	0.0	0.0	5.3	15.8	26.3	5.3	4.26	6
Quantity surveying	4.8	0.0	4.8	19.0	23.8	4.8	4.20	7
Landscape architecture	5.3	0.0	15.8	31.6	15.8	15.8	3.67	8
Interior design	0.0	5.0	20.0	30.0	15.0	20.0	3.45	9

Health and safety education in South Africa 161

- 'The BSc property development does not include any construction H&S in its curriculum other than scant reference to the relevant legislation'.
- 'H&S should be made a standard part of the curriculum'.
- 'A programme should not be accredited, if H&S is not adequately covered (or have no reference to H&S content)'.
- 'Historically, it has been addressed in the undergraduate and Honours programmes. Construction H&S is an integral part of construction management'.

Study 2

In terms of gender, 26.9% of respondents were female, 71.2% were male, and 1.9% did not indicate gender.

Respondents had worked 10.7 years on average for their current employer, and 19.9 years in construction. They were 47.3 years of age on average.

A range of qualifications was recorded, ranging from Grade 12 to PhD. Respondents were registered with a range of statutory BEPCs and were members of a range of professional built environment associations.

Table 11.4 indicates the degree of support for the inclusion of the twenty-five aspects in construction H&S modules of tertiary built environment programmes in terms of percentage responses per aspect per discipline, the mean percentage for all disciplines per aspect, and the mean percentage for all aspects per discipline. The total possible responses per discipline (No.) are recorded in the last row, i.e. the number of respondents per discipline. The highest number was thirty for project management, and the lowest was ten for land surveying.

It is notable that, in terms of the mean percentage for all disciplines per aspect, 15/25 (60%) aspects achieved percentages > 90% ≤ 100%; 20/25 (80%) aspects achieved percentages > 80% ≤ 90%; and 5/25 (20%) aspects achieved percentages < 50%. The > 90% ≤ 100% range included Need for construction H&S, Construction Regulations, Hazard identification and risk assessment, H&S specifications, Occupational safety, Environment and construction H&S, OH&S Act, Occupational health, Designer report (to clients), Role of clients in construction H&S, Ergonomics, Role of project managers in construction H&S, Role of construction H&S in project performance, Role of designers in construction H&S, and Economics of construction H&S. Then, in terms of the mean percentage for all aspects per discipline:

- 2/13 (15.4%) disciplines achieved percentages > 90% ≤ 100% – Civil engineering and construction management
- 6/13 (46.2%) disciplines achieved percentages > 80% ≤ 90.0% – Architecture, electrical engineering, facilities management, mechanical engineering, project management, and property development
- 3/13 (23.1%) disciplines achieved percentages > 70% ≤ 80% – Interior design, landscape architecture, and quantity surveying
- 2/13 (15.4%) disciplines achieved percentages > 60% ≤ 70% – Town/urban planning and land surveying

Table 11.4 Degree of support for the inclusion of aspects in tertiary built environment programmes' construction H&S modules

Aspect	Response (%)																Rank
	Architecture	Civil engineering	Construction management	Electrical engineering	Facilities management	Interior design	Landscape architecture	Mechanical engineering	Project management	Property development	Quantity surveying	Town/ urban planning	Land surveying	Aspect mean			
Need for construction H&S	100.0	96.2	100.0	100.0	95.0	95.8	100.0	100.0	100.0	100.0	100.0	95.5	100.0	98.6			1
Construction Regulations	96.6	100.0	100.0	95.5	100.0	100.0	95.5	100.0	100.0	100.0	96.0	90.9	90.0	97.3			2
Hazard identification and risk assessment	96.6	96.2	100.0	95.5	100.0	95.8	100.0	100.0	96.7	90.9	96.0	86.4	100.0	96.5			3=
H&S specifications	96.6	96.2	100.0	95.5	100.0	91.7	95.5	100.0	96.7	95.5	96.0	90.9	100.0	96.5			3=
Occupational safety	96.6	96.2	100.0	95.5	100.0	91.7	90.9	100.0	96.7	90.9	92.0	90.9	100.0	95.5			5
Environment and construction H&S	96.6	100.0	100.0	95.5	85.0	79.2	95.5	95.5	90.0	95.5	92.0	100.0	100.0	94.2			6
OH&S Act	93.1	96.2	96.6	95.5	95.0	83.3	90.9	95.5	96.7	95.5	92.0	90.9	100.0	93.9			7
Occupational health	96.6	92.3	100.0	95.5	100.0	87.5	90.9	95.5	96.7	90.9	92.0	90.9	90.0	93.7			8
Designer report (to clients)	100.0	96.2	89.7	100.0	85.0	91.7	100.0	100.0	93.3	95.5	92.0	90.9	80.0	93.4			9
Role of clients in construction H&S	96.6	92.3	100.0	90.9	95.0	83.3	86.4	95.5	96.7	100.0	88.0	86.4	90.0	92.4			10
Ergonomics	93.1	92.3	100.0	90.9	95.0	100.0	100.0	95.5	90.0	90.9	80.0	86.4	80.0	91.8			11
Role of project managers in construction H&S	93.1	92.3	100.0	90.9	90.0	87.5	86.4	95.5	96.7	95.5	92.0	81.8	90.0	91.7			12
Role of construction H&S in project performance	96.6	92.3	100.0	95.5	85.0	83.3	86.4	95.5	96.7	100.0	92.0	86.4	80.0	91.5			13
Role of designers in construction H&S	96.6	96.2	96.6	90.9	85.0	91.7	90.9	95.5	96.7	95.5	88.0	81.8	70.0	90.4			14
Economics of construction H&S	93.1	96.2	100.0	90.9	85.0	79.2	81.8	95.5	96.7	95.5	96.0	81.8	80.0	90.1			15
H&S plans	93.1	92.3	100.0	90.9	95.0	87.5	81.8	86.4	93.3	95.5	84.0	72.7	90.0	89.4			16
Role of manufacturers in construction H&S	93.1	92.3	100.0	90.9	95.0	95.8	90.9	95.5	96.7	90.9	88.0	81.8	10.0	86.2			17

Topic														Mean	No.
H&S file	82.8	88.5	96.6	90.9	100.0	83.3	81.8	86.4	93.3	90.9	76.0	68.2	80.0	86.0	18
	72.4	88.5	96.6	81.8	75.0	75.0	77.3	86.4	86.7	90.9	100.0	72.7	60.0	81.8	19
Role of quantity surveyors in construction H&S															
Influence of procurement on construction H&S	89.7	92.3	96.6	90.9	90.0	83.3	86.4	95.5	96.7	95.5	88.0	40.9	10.0	81.2	20
Other Regulations	41.4	69.2	79.3	81.8	85.0	25.0	22.7	86.4	73.3	27.3	20.0	18.2	10.0	49.2	21
Contractor H&S	27.6	84.6	100.0	31.8	25.0	25.0	13.6	36.4	96.7	27.3	36.0	13.6	10.0	40.6	22
COID Act	27.6	84.6	93.1	27.3	90.0	20.8	22.7	31.8	36.7	27.3	24.0	18.2	20.0	40.3	23
Managing subcontractor H&S	24.1	80.8	100.0	22.7	35.0	20.8	13.6	27.3	96.7	27.3	36.0	9.1	10.0	38.7	24
Primary health promotion	17.2	73.1	89.7	9.1	90.0	4.2	9.1	18.2	40.0	13.6	16.0	18.2	10.0	31.4	25
Discipline Mean	80.4	91.1	97.4	81.5	87.0	74.5	75.6	84.4	90.0	80.7	78.1	69.8	66.4	–	–
Total possible responses per discipline (No.)	29	26	29	22	20	24	22	22	30	22	25	22	10	–	–

Respondents were also requested to provide any comments in general regarding the inclusion of construction H&S in tertiary education relative to their discipline and 28.8% provided no comment; 50% provided one comment; 15.4% provided 2 comments; and 1.9% provided each of 3, 4 and 5 comments. In summary, 71.2% recorded 1 comment or more. Selected comments recorded verbatim were:

- 'Bringing awareness to H&S at an early level will help ingrain this critical element into the DNA and culture of a wide audience'.
- 'H&S is a critical component of all industry in South Africa, and in particular, to tertiary education across all built environment disciplines'.
- 'This is long overdue'.
- 'I believe that it is imperative that anyone studying towards a career in the construction industry has a good understanding and knowledge base of construction H&S'.
- 'If H&S isn't a fundamental component of all built environment qualifications the industry's H&S performance will continue to be plagued by high levels of fatalities and major accidents'.

The results underscore the findings of previous South African and international research, first, in terms of the inadequacy of tertiary built environment education in terms of construction H&S, and, second, in terms of the degree of support for the inclusion of 25 aspects in construction H&S modules of tertiary built environment programmes for 13 disciplines.

Implications for practice and research

The implications for developing countries are that construction-specific legislation exists, and that such legislation is multi-stakeholder oriented, especially with respect to designers.

However, legislation is merely the foundation upon which further actions should be based. A paradigm shift with respect to the status of H&S might be required by the respective stakeholders. This is amplified by the South African research findings, namely, that despite designers having been required to contribute to construction H&S in terms of the Construction Regulations, tertiary built environment designer programmes generally do not address construction H&S. The latter is probably attributable to a lack of awareness and knowledge, which implies that the related academics will have to attend training.

The reality is that, in general, tertiary built environment education is not empowering graduates to contribute to construction H&S. Furthermore, the IoWs of BEPCs are deficient and require amendment. Then, the education frameworks of BEPCs and professional associations/institutes for the respective disciplines might have to be amended, and accreditation panel reviews must determine whether H&S is embedded in the respective programmes.

Conclusions

The cidb (2009) report stated that H&S-relevant education and training, or the lack thereof, at all levels, has a significant effect on H&S in construction. Furthermore, at the tertiary level, not all construction-related programmes in South Africa include H&S within their curricula, especially designer programmes, despite the South African H&S legislative framework being clear with respect to the need to do so, albeit implicit.

Study 1

The degree of response to the survey might have been attributable to the non-inclusion/non-addressing of construction H&S in programmes/curricula on the part of the non-respondents.

Nearly a third of the responding departments did not address/include construction H&S in their programmes/curricula. Of those that did, the majority address/include it as a component of a subject. Therefore, it can be concluded that H&S in construction is not a pervasive 'value' in terms of programmes and that statutory BEPCs do not focus on it during accreditation visits.

The perceived importance of the inclusion of construction H&S in the tertiary education programmes of nine built environment disciplines leads to the conclusion that there is a degree of understanding and appreciation of the importance of including construction H&S, but it has not been translated into action. This is possibly attributable to a lack of knowledge relative to the subject area, especially the design disciplines.

Study 2

Given the degree of support for the inclusion of twenty-five aspects in the construction H&S modules of tertiary built environment programmes for thirteen disciplines, it can be concluded that H&S must be embedded in such programmes.

Tertiary built environment institutions, professional built environment associations/institutions and statutory BEPCs can benefit from the findings, as the framework informs which aspects should be addressed with respect to H&S in construction.

Recommendations

Statutory BEPCs should revisit their IoW. Furthermore, they and BEP associations/institutes should evolve an H&S framework for their respective tertiary built environment programmes and interrogate the extent to which H&S is embedded in such programmes when undertaking accreditation visits. In terms of the registration of persons by statutory BEPCs, the application and review process should include H&S.

Statutory BEPCs and professional associations/institutes should afford H&S status equal to or greater than other performance parameters such as cost, quality, and time. They should promote H&S and contribute to the development of their registered persons and members respectively. H&S practice notes should be evolved, and H&S continuing professional development (CPD) should be facilitated, or arranged, or presented. Newsletters, conferences and related interventions should include/address H&S.

Tertiary built environment education should establish industry liaison committees with employer associations, and request reviews of their H&S framework and module contents.

References

Construction Industry Development Board (cidb). (2009), *Construction Health & Safety in South Africa: Status & Recommendations*. Pretoria: cidb.

Deacon, C. H. & Smallwood, J. J. (2016), The effect of the integration of design, procurement, and construction relative to health and safety (H&S). In: *Proceedings International Sustainable Ecological Engineering Design for Society (SEEDS) Conference*, Leeds, United Kingdom, 14–15 September 2016.

Republic of South Africa. (1993), Occupational Health & Safety Act: No. 85 of 1993. Government Gazette, No. 14918. Pretoria.

Republic of South Africa. (2014), No. R. 84 Occupational Health and Safety Act, 1993: Construction Regulations 2014. *Government Gazette*, No. 37305. Pretoria.

Safe Work Australia. (2017), Key work health and safety statistics Australia 2017. Canberra: Safe Work Australia.

Smallwood, J. J. (2010), Construction management health and safety (H&S) course content: Towards the optimum. In: P. Barrett, D. Amaratunga, R. Haigh, K. Keraminiyage & C. Pathirage (Eds.), *Proceedings of the CIB World Building Congress, Salford Quays*, 14–17 May, 2010. [Online]. Available at: www.disaster-resilience.net/cib2010/files/papers/908.pdf.

Smallwood, J. J. (2014), Preventing accidents in construction. Unpublished research findings.

Smallwood, J. J. (2015), The need for the inclusion of construction health and safety (H&S) in architectural education to assure healthier and safer construction. In: *Proceedings of CIB W099 Workers and Society through Inherently Safe(r) Construction*, Belfast, Northern Ireland, United Kingdom, 10–11 September, pp. 169–179.

Part III
Behavioural and cultural issues in safety management

12 Safety communication and suggestion scheme in construction

Victor N. Okorie and Fidelis Emuze

Summary

Project actors do not dispute the importance of information sharing in a project environment. In construction, the collection and dissemination of information on risks and hazards are critical to the realisation of zero harm on a project. Two-way communication between management and workers on safety issues creates a work environment characterised by trust and harmony. Workers are involved in workplace safety management, as they are always on the front line of site operations. Safety suggestions made by workers with feedback from management promotes a positive safety culture. When workers are actively involved in the workplace safety decision-making process, there are tangible benefits such as cost savings, higher productivity and reduction in site accidents and incidents. In addition, there could also be intangible benefits like higher levels of morale and trust among workers and management. This chapter examines the relevance of workers' involvement and participation in construction safety management through a communication and suggestion scheme. A proactive safety communication and suggestion scheme is effective in promoting a positive safety culture on the construction site.

Introduction

The literature shows that construction firms tend to have a reactive approach to safety, as they only take actions after accidents or incidents as opposed to being proactive with communication and suggestion schemes (Hughes and Ferrett, 2010). Phoya (2017) emphasised that in an organisation where workers can freely communicate about safety matters, there will be a reduction in site fatalities and injuries. In effect, workers' participation in the safety decision-making process creates lifesaving trust between them and management (Reese, 2016). Workers find their jobs more meaningful and exciting when they are involved in a safety decision-making process that concerns their welfare and well-being (Vecchio-Sadus, 2007). Safety information is critical to project performance. As a result, the Construction Design and Management (CDM) Regulations 2015 in the United Kingdom (UK) place specific duties and responsibilities on key project actors to provide the workforce necessary information regarding health and safety (H&S) of projects and worksites (HSE, 2015). Safety communication thus provides workers with information about potential risks and their control.

Improving construction safety performance is preventing accidents through communication and other interventions (Health and Safety Executive [HSE], 2010). Toolbox talks, site meetings, induction training, safety committee, notice boards, signs, symbols and bulletins help communicate safety issues to workers. With a communication and suggestion scheme, management and workers become aware of risks and hazards on worksites. Additionally, communicating safety to workers brings about attitude and behaviour change that is supported by increased risk perceptions.

Communication is a veritable tool in safety management. As an illustration, it is through safety communication that workers are informed and educated about the risk and hazards inherent at worksites, apart from the opportunity to learn safe work practices and the proper use of personal protective equipment (PPE). It also brings about attitude and behaviour change among workers (Reese, 2016).

Safety communication and a suggestion is an influencing factor in construction (Hughes and Ferrett, 2010), as most safety issues emanated from either lack of information and communication, or inadequate information and communication (Phoya, 2017). Two-way communication between management and workers on safety issues are critical for effective safety management. A proactive safety management system such as a safety suggestion scheme allows workers to contribute to the decision-making process of an organisation. Worker involvement in the safety decision-making process has been recognised as an essential step for workplace H&S culture (Krause, 2003; Okorie, 2016). In line with this, Phoya (2017) states that safety suggestions made by workers have the potential to contribute positively towards workers' safety behaviour, as they have direct experience of site conditions and are often the first to identify the potential problems.

Sharing and gathering of safety information on potential sites hazards and their control could be beneficial to both management and workers. Unfortunately, in sub-Saharan Africa countries, the seamless sharing of information is hindered by the low literacy levels among workers in the region (Okorie et al., 2015). As a result of the low education level among these categories of site workers, effective implementation of a suggestion box suffers in this part of the world. Also, there is likely to be a lack of confidence among rural and unskilled site workers to speak about workplace safety issues (Okorie et al., 2015). This observation resonates with research conducted in Australia by Vecchio-Sadus (2007, as cited in Phoya, 2017). The research on safety culture through effective communication identified illiterate and unskilled workforce, culture and language barriers, nature of engagement and a transient workforce as hindrances to information sharing.

A suggestion scheme as an information gathering tool can be beneficial to workers and management if it is well implemented (Chinwuba and Chika, 2013). According to Phoya (2017), a safety suggestion scheme is unlike the safety committee or safety representative that acts as a communication link between worker and management. A safety suggestion scheme has the potential to act as an essential mechanism that gives a worker the right(s) to communicate directly with management by coming up with ideas, suggestions, complaints or recommendations on how to improve construction safety (Okorie, 2016).

Based on the foregoing, this chapter presents a safety communication and suggestion scheme as a way of highlighting its contributions to the improvement of safety performance in construction, particularly in the developing countries. Before the schema is presented, the next section of the chapter provides an overview of the theoretical perspectives on the discourse.

Perspectives on safety communication and suggestions

Safety communication

The word *communication* is from the Latin word *communicare* or *communico*, which means sharing information. Information and ideas are shared among individuals and organisations for mutual benefit. Communication connotes getting information for analysis and assimilation (Chinwuba and Chika, 2013). In this chapter, communication is addressed concerning workplace safety information sharing between the management and the workforce on a construction site. Safety communication is a process by which information about hazards is shared between management and workers who may be exposed to hazards on a worksite (Vecchio-Sadus, 2007; HSE, 2010). In order words, it is a social process by which people are informed about risks and hazards associated with their work environments; materials they come in contact with in the line of duty; and what is expected to be done to enhance personal safety, health and well-being.

Safety communication refers to information sharing between management and the workforce regarding workplace safety. It requires that both management and workers regularly communicate safety policies and values. Workers need to be informed of risks and hazards as it patterns to their work environments. Safety communication, according to Chinwuba and Chika (2013), plays vital roles in construction safety management, though it has remained one of the least applied and least known variables for effective safety management in the construction industry, particularly in developing countries. Safety communication is regarded as an indispensable tool in informing workers of risks and hazards associated with the work environments and materials. Safety communication to the workforce has been linked to positive safety work behaviour. A well-informed and educated worker will take care of himself and others.

Safety communication plays a vital role in nurturing a workplace safety culture (International Labour Organisation [ILO], 2011). While the importance of communicating is not in dispute, some factors tend to make it ineffective. Such factors include lack of knowledge of risk factors, lack of attention to details by designers, unclear goals, selective listening, lack of empathy, self-image, status prejudice, accents/language of the safety communicators and differing perception (Vecchio-Sadus, 2007, as cited in Phoya, 2017). Additional factors include power conflicts, mistrust between sender and receiver, poor working environment and disturbance such as noises, and language barriers as factors impacting on risk communication in the construction industry (HSE, 2010). However, the channel used in sharing information determines the influence of these obstacles.

Channels are physical means through which languages and signals are shared (HSE, 2010).

In a typical construction organisation, there are various safety communication channels through which information is shared (HSE, 2010). Safety communication channels are classified into written, electronic, and verbal or oral. An example of the written safety communication channel can be in the form of written correspondence such as letters and posters, safety manuals and books. Electronic safety communication channels can take the form of email and fax, while verbal communication channels occur through site meetings, toolbox talks, safety induction training, staff safety meetings, safety briefing sessions and union meetings.

For instance, notice boards on headquarter and sites offices are mainly used to transmit information of a short term such as general safety performance, hazards refresher information, accidents and incidents report summaries, procedure updates, minutes of safety review meetings and safety inspection reports. A written circular/brochure that informs workers about the risks and hazards associated with their worksites and materials is essential to improve safety performance. Newspapers and safety newsletters also play the significant role of written safety communication channels on risks and hazards in more detailed for both internal and external use. Safety signs and symbols are also examples of written safety communication that provide safety information to workers and the public. According to the ILO (2011), employers are required to provide safety signs and symbols in the workplace to inform workers of the presence of such risks and hazards. Signs need to be in strategic locations so that workers and the public can see and quickly read it.

The electronic safety communication channel is increasingly becoming popular with the advance in technology. For instance, there are web-based company sites where management shares information regarding company's policies and values and as well as web-based opinion pools that can be used to gather information from workers and the general public on how to improve services. However, with the nature of the construction industry and its illiteracy level, electronic safety communication channels may not be solely used to communicate safety to the workforce.

The verbal safety communication channel takes the form of site safety meetings, toolbox talks, safety training, safety inductions, safety committee meetings and safety representatives. Safety officers, site managers, resource persons and supervisors verbally communicate safety to workers on a daily, weekly or monthly basis. With verbal safety communication, the communicator has physical contact with workers. The communicator can employ different languages and safety symbols to get the messages across with immediate feedback from the audience, unlike safety bulletins or safety newsletters. Verbal safety communication channels have been found to be more effective due to the nature of the construction industry, where the majority of construction workers have a low level of education (Vazquez and Stalnaker, 2004; Okorie and Smallwood, 2012).

In sum, small and large construction companies have identified effective safety communication as a necessity for reducing construction fatalities and injuries (Vecchio-Sadus, 2007; HSE, 2010).

Suggestion scheme

The concept of a suggestion scheme was recorded in 1770 when a British naval officer asked his soldiers to express their ideas freely without fear of punishment (Koch, 2017, as cited in Ostrowski, 2017). John Patterson in 1880, a chief executive officer of an organisation in the United States of America, carried out the first methodical implementation of this scheme. He discovered that employees who were always on the front line of sites activities had valuable ideas to share with management. However, the prevailing management structure prevented the sharing of ideas due to hierarchy. However, the scheme became popular after the Second World War (Chinwuba and Chika, 2013).

Through the application of suggestion schemes, contractors can improve their profits, reduce costs resulting from site accidents and incidents, enhance employee's morale, improve workers' welfares and well-being, facilitate innovation and improve responsiveness to the public (Phoya, 2017). Workers are the ones to identify problems first, as they are always on the front line of sites activities; therefore, construction companies should recognise safety suggestions and recommendations made by them. Thus, the suggestion box gives workers the freedom to communicate directly to management (HSE, 2010).

The anonymity and cost-effectiveness of a suggestion scheme underscores its importance as a veritable tool to improve construction safety (Cesarini et al., 2013), especially in developing countries where financial provisions for safety is minimal. The scheme encourages frankness and open feedback, thereby providing opportunities for obtaining specific safety issues on worksites leading to a reduction of site accidents and incidents. Safety suggestions put in or made by construction site workers give more information on site hazards and risks. Chinwuba and Chika (2013) posit that the suggestion scheme could improve workers' welfares because detailed information is received from workers as people who are physically in contact with risks or hazards. When such information is investigated and feedback communicated back to workers, they will be motivated. Further, the risks identified by workers and made known to management could receive prompt attention. Thus, suggestion schemes prevent serious accidents on worksites. Phoya (2017) contends that early detection and prevention of site accidents and incidents through information sharing in the form of suggestion schemes promotes positive safety that increases site productivity.

The use of a suggestion scheme on site is not without its challenges. Okorie and colleagues (2015) observe that the high level of illiteracy among rural migrant and unskilled site workers pose severe challenges to safety management in Nigeria. The illiteracy levels of construction workers in developing countries are not peculiar to Nigeria. Therefore, implementation of safety suggestion schemes is difficult. For example, an illiterate site worker who cannot read or write cannot make use of this suggestion scheme. The literature says that not only does a low level of education among construction workers contribute to the high rate of accidents, but it also increases non-compliance to simple H&S rules and procedures. For example, Vazquez and Stalnaker (2006) maintain that low education level has a negative impact on construction safety performance. Effective utilisation of a suggestion scheme in construction requires that a worker should be able to write suggestions, complaints and recommendations regarding worksite conditions.

An illustrated safety communication and suggestion scheme

Construction by its nature is labour intensive, and therefore, site workers are the most valuable asset, of which their health, safety and wellbeing (HSW) should a priority to management. Thus, educating and informing workers on risks and the hazardous nature of materials on sites is essential. For instance, in a typical construction organisation, there are various methods through which management communicates safety to the workforce: notice boards, newsletters, bulletins, toolbox talks, safety induction, safety training, union safety meetings, safety committees, signs and symbols, and web-based and opinion-based pools. However, there are some challenges to effective communication of safety to workers. Okorie and colleagues (2015) noted that the majority of the workforce is unskilled and illiterate rural migrants in Nigeria, for example. In such situations, verbal communication is more effective. According to Phoya (2017), verbal safety communication that takes place in a typical construction site has proved to be an active channel of communication because of the interactive nature.

The communicator (manager or supervisor) has face-to-face communication with workers, and can use or employ different languages. The use of different languages by the communicator brings the message home. The receivers readily understand and ask questions where necessary. Workers are always on the front line of site operations, directly affected by site activities and the ones closest to the problems and can find problems first. Suggestion schemes give workers the privilege to make safety suggestions, complaints and recommendations directly to management. Workers initiate ideas and suggestions and put them forward for investigation and implementation to management. Management, in turn, investigates safety suggestions and complaints from workers through a safety committee, safety representative or safety union that reports to the management for implementation of usable ones, as illustrated with the link between safety suggestion scheme and safety communication. When workers are encouraged or empowered to make suggestions regarding their workplace safety, workers become more committed to organisational safety values and culture.

Figure 12.1 depicts an issue resolution procedure that is usable at a construction project in developing countries. The figure shows that a worker is at liberty to raise HSW issues based on an observation. The worker raises the issue, then submits it for resolution through a suggestion box. Depending on the organisational structure of the project, site supervisors or HSW personnel get the notice of

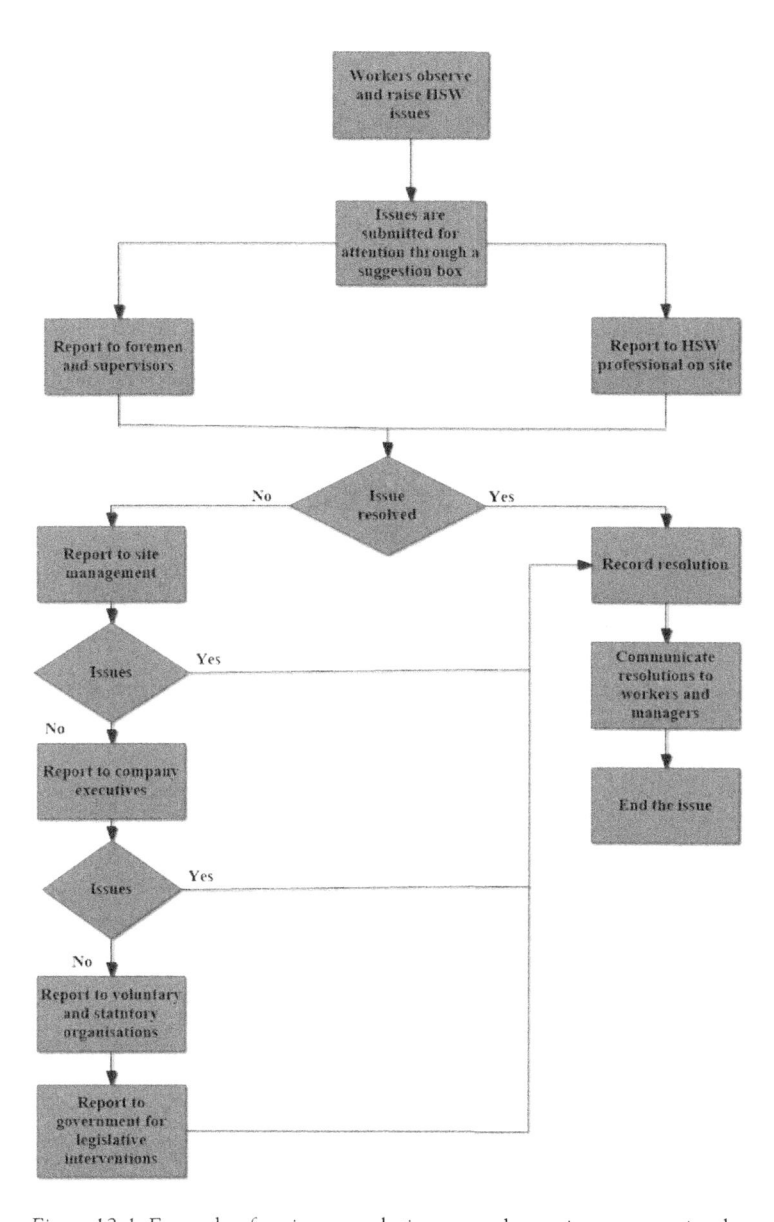

Figure 12.1 Example of an issue resolution procedure using a suggestion box as a tool.

the matter. When the issue is resolved at the supervisory level, it is documented before the information is shared throughout the site. However, if the issue is not resolved at the supervisory level, the matter is brought to the attention of site management. Figure 12.1 thus shows that the resolution of HSW matters could extend beyond the site and contractors to industry organisations and relevant

[Figure: Positive Safety Culture diagram showing: Management commitment, Goals for frontline management, Employee involvement, Responsive to change, Effective communication, Mindfulness and attention to detail, Problem solving without looking for faults, Trust and safety-first led decisions — all contributing to "Value placed on safety in a workplace"]

Figure 12.2 Characteristics of a positive safety culture.
(Adapted from Reese, C. D., 2016, *Occupational Health and Safety Management: A Practical Approach*, CRC Press, Boca Raton, FL, pp. 395–396).

governmental agencies. It is however important to note that the procedure may unfold with ease in a company with a positive safety culture.

The implementation of a suggestion scheme on project sites helps to strengthen a positive safety culture. A positive safety culture is doing the right thing at the right time to address either routine or emergency operations (Reese, 2016). A contractor with a positive safety culture exhibits the features in Figure 12.2 and is one that prioritises all safety-related matters. Such a contractor shows commitment to safety by explaining safety expectations to front-line management while involving all categories of employees in safety management through effective communication (Reese, 2016).

Conclusions

The complexity of construction increases in tandem with project size and type. A high number of illiterate and unskilled workers compounds project complexity. These categories of workers form the majority of site workers, particularly in developing countries. As a result, the deployment of safety tools such as a suggestion scheme is complicated. However, not only does the low level of education among construction site workers in developing countries contribute to non-compliance to H&S rules and regulations, but could also create lack of confidence among workers to speak about site H&S issues.

Safety communication and suggestion schemes form proactive measures for sharing ideas and information on the risks and hazards associated with construction activities. When workers are aware of existing risks, there is a tendency for improved safety behaviour and attitudes. Communicating safety to workers increases their risk perceptions in favour of reduced workplace incidents. Safety suggestions made by workers alert management on the existence of hazards at worksites, apart from their participation in the safety decision-making process, a situation that gives them a sense of belonging.

When organisations with a positive safety culture are asked if workers are comfortable voicing their concerns or ideas on safety issues to management, the response is affirmative. A safety suggestion box is a tool for hearing the voice of workers in the workplace on matters that concerns them, especially in relation to safety. In other words, suggestion schemes could create trust and harmony between management and workers. This chapter postulates that safety communication and suggestion schemes are a useful tool for improving construction safety performance in developing nations, particularly in sub-Sahara African countries where cost hinders the use of high-end technology to enhance HSW.

References

Cesarini, G., Hall, G. & Kupiec, M. (2013) Building a Proactive Safety Culture in the Construction Industry, ACE Group, Philadelphia, PA.

Chinwuba, M. S. & Chika, O. J. (2013) Effective management of corporate organisation through the use of suggestion box: An empirical insight from banking industry. *International Journal of Business and Management Invention*, Vol. 2(51), 19–26.

Health and Safety Executive (HSE) (2010) Improving Health and Safety: An Analysis of Health Risk Communication in the 21st Century, HSE, London, UK.

Health and Safety Executive (HSE) (2015) *Managing Health and Safety in Construction, Guidance Regulations*. www.hse.gov.uk/pUbns/priced/l153.pdf.

Hughes, P. & Ferrett, E. D. (2010) *Introduction to Health and Safety in Construction*, Butterworth-Heinemann, Oxford, UK.

International Labour Organisation (ILO) (2011), ILO Standard on Occupational Safety and Health – Promoting Safe and Healthy Working Environment, ILO, Geneva.

Krause, T. R. (2003) A behaviour-based safety approach to accidents investigation. *Professional Safety*, Vol. 45(12), 342–358.

Okorie, V. N. (2016) Suggestion box as tool for improving construction site health and safety performance in Nigeria, *PM World Journal*, Vol. 5(9), 1–12.

Okorie, V. N., Okoile, K. C. & Ukah, G. C. (2015) Impact of illiterate rural migrant workers on the effectiveness of safety induction in Lagos State Nigeria. *International Journal of Civil Engineering, Construction and Estate Management*, Vol. 3(4), 31–40.

Okorie, V. N. & Smallwood, J. J. (2012) The impact of poor communication skills on construction site health and safety performance in South Africa. In *Proceedings: CIB W099 International Conference on Modelling and Building Health and Safety* (pp. 589–598), Singapore.

Ostrowski, D. (2017). Assessment of employee engagement in the implementation of an employee suggestion program in company X – Research results. *Economic and Environmental Studies*, Vol. 17(4), 985–1002.

Phoya, S. (2017) The practice of communication of health and safety risk information at construction sites in Tanzania. *International Journal of Engineering Trends and Technology*, Vol. 7, 384–393.

Reese, C. D. (2016) *Occupational Health and Safety Management: A Practical Approach*, CRC Press, Boca Raton, FL.

Vazquez, R.F. & Stalnaker, C.K. (2006) Latino workers in the construction industry: Overcoming the Language barrier improves safety, *Professional Safety*, Vol. 24(2), 121–134.

Vecchio-Sadus, A. M. (2007) Enhancing safety culture through effective communication. *Safety Science Monitor*, Vol. 11(3), 12–24, Article 2.

13 Making sense of safety in systemic and cultural contexts

Andrea Y. Jia, Steve Rowlinson, Martin Loosemore, Mengnan Xu, Baizhan Li, Marina Ciccarelli and Heap-Yih Chong

Summary

This chapter discusses how the socially constructed meaning of 'safety' in a developing society differs from that in a developed society. We take a grounded theory approach to compare the contents and consequences of sensemaking on safety between Chongqing in mainland China and Hong Kong. We find that in the context of Chongqing, safety means workers' self-interest and employers' benevolence; while in the context of Hong Kong, both employers and workers make sense of safety as compliance to rules. The results show that in a developed society where regulations and management systems are well established, institutions are mediating the sensemaking between safety and the actors. While in a developing society where formal rules and systems are underdeveloped or absent, safety is processed in a more personal approach as reciprocity of benevolence between the workers and the employers. The finding implies that in the systemic and cultural context of a developing country, rules developed organically bottom-up will tackle safety issues better than a top-down coercive approach.

Introduction

Worker engagement as a key intervention for site safety improvement is receiving more in-depth investigation in recent years (Maloney et al., 2007; Lawani et al., 2017, 2018). This recent development reminds us that workers are motivated differently from their managers; and stakeholders of different interests and identities have different schema of what safety means (Love et al., 2011). Weick (1979, 1988, 1993, 1995, 2001a, 2012), for example, asserted that people act upon and respond to a self- or socially constructed 'script' which constitutes a chain of actions and consequences. This cognitive script works as a roadmap in people's mind to guide their reasoning of actions in response to emerging situations, as reality develops according to the personal roadmap, to arrive at a destiny, in this case, safety. The process of constructing these scripts is widely referred to as 'sensemaking', whereas an accident is a result of interrupted sensemaking (i.e., confused mental roadmap) and, consequentially, sightless decisions leading to an unsafe destiny.

However, Weick's focus on individual cognitive process and the high reliability organisations left a few important gaps in the safety body of knowledge.

Leveson (2011), for example, from a systems thinking perspective, questioned his assumption of equating safety for reliability of the production procedure and his postulation that there is only one correct mental model to be enacted. Furthermore, the solution for safety derived from Weick's sensemaking model favours an underspecified organisational form to stretch collective mindfulness and therefore organisational resilience (Weick et al., 1999). This solution concentrates on the process of sensemaking but ignores the content of sensemaking, e.g., the constructed meaning itself, and the fact that actors of different contexts can make diverse meanings out of the same situation.

From a systems thinking perspective, the aim of this study is to explore the constructed meaning of safety as a result of sensemaking in the context of a developing society in comparison to a developed society, using climatic heat risk as a central event of analysis. A comparative study was undertaken between two socio-political systems of one country: Chongqing Municipality in mainland China and Hong Kong Special Administrative Region, the former to represent a developing society, the latter a developed society. Hong Kong was a British colony from 1842 until 1997 when its sovereignty was returned to China with a prospect of 50 years' continuation of the same system. Two research questions guided the inquiry: What does safety mean to construction workers and managers in these two societies? How does the sensemaking lead to interpretations of individual and organisational safety behaviour?

Systems thinking

Systems thinking takes accident investigation from individual mistakes to the underlying structures, from discrete components to their interconnections, and from static snapshots to the changing process (Senge, 2006). Viewed through a system lens, the distinction between individual accident and organisational accident is blurred, as all accidents are eventually organisational (Reason, 1997; Goh et al., 2012). However, two fundamental questions need to be clarified before we take it further: what is 'the system' and what is the nature of 'the connections'?

Two assumptions of 'the system' are underpinning the existing body of literature: a socio-technical system and a socio-political system. The former focuses on a mechanical system around which work activities are organised, for example, a book sorting system based on which a library operates (Rasmussen et al., 1994). An essential characteristic of this system is a technical core structure, which largely simplifies the otherwise confounding connections and interactions into a manageable few that can be responded as deterministic causal relationships. Safety is seen as an attribute of the socio-technical system, while accident is a result of drifting boundaries down the decision hierarchy. The drifting effect happens when local operators at multi-levels make discretionary decisions utilising their legitimate safe operation margins, although all are locally safe decisions, as the safe margins build up across levels, the system as a result drifts towards an accident. In this perspective, the solution for accident prevention is to continuously engineer the system toward perfection through refined learning about causes, consequences and top-down controls (Leveson, 2011). This has been proved effective in preventing major accidents in the high reliability organisations and is largely taken for granted as the default system thinking approach to safety.

Can this frame of reference effectively address the 'traditional, and often seemingly simple and trivial occupational accidents' (Hovden et al., 2010) in the organisational context of a construction project that is based on human-to-human interactions without the mediation of a central technical system? Many have done their applied research without questioning its applicability. Yet in a systemic context without an anchoring mechanical system, the socio-technical approach tends to produce simplistic solutions that are either ineffective or unrealistic.

Recognising the autonomy of human agents, the socio-political assumption of 'the system' embraces uncertainty and complexity in the system, and interdependency as the nature of the connections. This leads to systemic models of construction accident causalities which do not attempt to define deterministic causal relationships but rather act more as a guide to stakeholders to better understand how their actions might affect workers' personal safety on site, a consequence seemingly rather distant from their immediate influence (e.g. Gibb et al., 2006; Hale et al., 2012; Rowlinson and Jia, 2015).

Weick (1976) conceptualised the 'loosely coupled system' to provide an alternative to the taken-for-granted tight-coupling nature of the socio-technical systems. A loosely coupled system is characterised by components that respond to each other but with independent identities. The interactions are characterised as sudden (rather than continuous); occasional (rather than constant); negligible (rather than significant); indirect (rather than direct); and eventual (rather than immediate) (Weick, 1982; Orton and Weick, 1990). Compared against these criteria, the organisation of a construction project is typically a loosely coupled system (Doree and Holmen, 2004; Dubois and Gadde, 2002), where an individual malfunction is unlikely to set off a chain effect to cause a catastrophic accident. This is because components 'seal off', thus limit the spread of the consequence of a local error (Orton and Weick, 1990). However, these characteristics also imply that a top-down initiative is unlikely to work effectively because its implementation is subject to the interpretations of actors of the local sub-systems that are somewhat isolated from each other (Weick, 2001b, 2012). Thus although a construction project organisation seems to be hopeless in controlling the countless local hazards in daily practice, the lost-of-control rarely develops into catastrophic accidents due to the organic adaptations at different levels of sub-systems.

Sensemaking

Sensemaking is the process of constructing meaning out of a situation. In an organisational context it is a collaborative process of creating shared awareness through negotiating and resolving multiple stakeholders' perspectives and interests around ambiguous situations (Weick, 1993, 1995, 2001a, 2012). The 'sense' is a personal gestalt of an event or action, according to which people interpret and respond to the changing situation. Two essential questions for sensemaking are 'What is going on here?' (situational frames) and 'Who am I?' (identity) (Goffman, 1974; Weber and Glynn, 2006). These two questions are interdependent and condition each other. Based on this, we can expect managers and workers make different senses of the same situation to reflect their respective identities in the project systemic context, which is partially defined by the 'stakes' they have in a project organisation (Fellows and Liu, 2016). Sensemaking shapes people's minds through identities (people's identity shapes what they enact and how they interpret events), preferences in social relations (as people interact with others and build narrative accounts, it helps them understand what they see), retrospection (retrospection in time affects what people notice), cues (people select what information is relevant and what explanations are acceptable as points of reference for linking ideas to broader networks of meaning), updating (sensemaking is an ongoing process that emerges over time), criteria for plausible stories (people prefer information that is plausible rather than accurate) and coping actions (people learn from their environments and adjust their accounts of the world) (Weick, 1995; Weick et al., 2005).

Whilst the concept of sensemaking was predominantly utilised to predict how system safety could be improved through building up mindfulness among the individual actors, Weber and Glynn (2006) took the concept to a system level, affirming that institutions (rules of the system) prime, edit and trigger sensemaking. Institutional interventions, through introducing new rules, regulations or guidelines, change responsibilities and roles, thus identities, and interrupt the existing way of doing things (Hale and Borys, 2013). The interruption activates a process of sensemaking among the impacted personnel as they try to figure out their new identities in the changing systemic context, based on which situational frames are updated. Systems at the society level influence the organisational level systems through internalisation of social cultures within the individual actors (Fellows and Liu, 2016), the autonomous agents who are interpreting and enacting the meanings of rules and responding accordingly. Individual actors make sense of the rules through the cultural logics embedded in state, market, family, profession, community and corporation in which they are situated (Thornton et al., 2013). Therefore the social cultural context of a society is regarded as an important condition shaping the individual identities (Hong et al., 2001), and consequently, the result of sensemaking (Ivanova-Gongn and Törnroos, 2017).

Research approach

The research inquiry was conducted through comparative case studies on the practices of managing heat stress on a construction site. An ethnographic approach to data collection and a grounded theory approach to analysis were adopted to enable us to gain an insider's view, triangulate sources of information and derive valid conclusions (Lincoln and Guba, 1985; Corbin and Strauss, 2008). Chongqing in mainland China and Hong Kong were selected for a comparative case study on the sensemaking of safety as typical contexts of a developing society and a developed society respectively. The Hong Kong workers were situated in a post-British colonial, individualistic, pragmatic culture with a well-developed formal institutional environment; while their Chongqing counterparts were situated in a post-communism, Confucianism culture with underdeveloped formal institutions in construction safety.

Data was collected from construction projects in Hong Kong and Chongqing on workers and managers' perceptions of institutional interventions on the safety risk of heat stress, i.e., the dissemination and implementation of heat stress guidelines on site. Data collection protocols were developed to ensure trust building between researchers and the participants on site and researchers' immersion in the fieldwork (see Jia et al., 2016, 2017 for more details). The collected data included environmental heat stress, workers' personal health data, observations of work activities and routines, and informal interviews with workers and managers on site. The sample from the Hong Kong context comprised 246 workers and 92 managers from 26 construction sites, collected from 2011 to 2012. The Chongqing sample included 6 workers and 9 managerial personnel collected in 2013. The data was regarded adequate as theoretical saturation was reached. Multiple sources of data were first triangulated for verification. Invalid information was dropped from the database. The verified data were analysed with a grounded theory approach, going through open coding, axial coding, categorising and reflective memo writing (Glaser and Strauss, 1967). The axial coding was guided by the themes of constructed meaning, systemic context and social cultural context. The contradiction between safety and production have been long embedded in the practices of construction projects. Therefore the constructed meanings of 'safety' were analysed against the meanings of 'production'. Individual narratives were treated as composite narratives (Vaara et al., 2016) to capture the collective meanings of the interest group they identified with.

Findings

The meanings of safety and production

The constructed meanings in the two social contexts are illustrated in two matrices as shown in Figures 13.1 and 13.2.

In the Hong Kong context, the meaning of production is constructed as the core interest of both employers and workers; from it employers are looking for profit, and workers, wages. In terms of safety, for employers, safety means compliance to external institutions; for workers, safety means compliance to organisational safety rules. In the Chongqing context, there is not a mediating formal institution to represent safety. Safety is understood as workers' self-interest and organisations' benevolence, while production is constructed as employer's interest and workers' self-sacrifice to 'the whole'. Safety and production are done through reciprocity of favour between the two parties. The consequences of the sensemaking are further illustrated in three scenarios.

184 *Jia et al.*

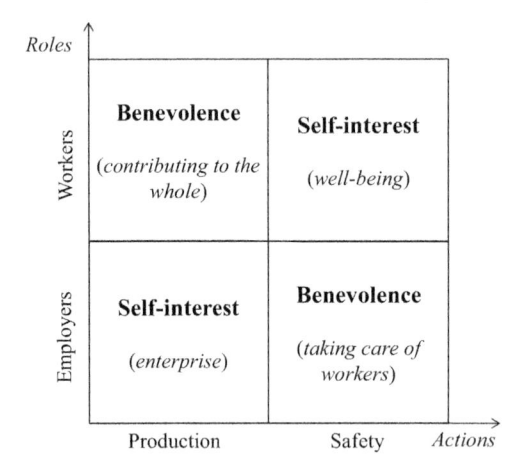

Figure 13.1 Sensemaking in a developing society (Chongqing).

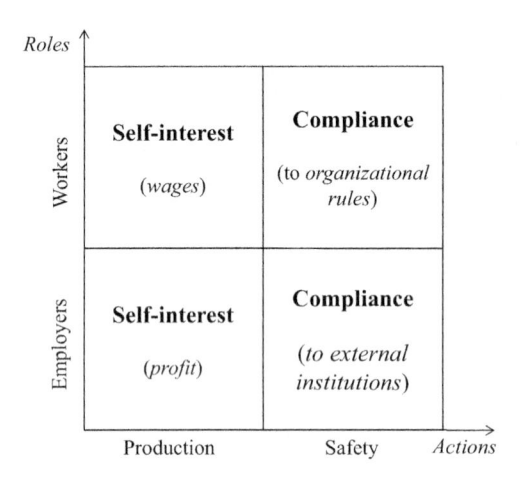

Figure 13.2 Sensemaking in a developed society (Hong Kong).

Scenario 1: 'Not at all!'

Cynicism is a manifestation of employee disengagement that exists in all workplaces to a more or less extent. The reasoning that justifies cynicism reveals how workers make sense of what is going on in the specific work environment. In this study, the patterns of workers' cynicism were captured in the analysis of whether and how workers were informed of the relevant heat stress management guidelines. The results showed that the Hong Kong workers and managers received the information mainly through the formal system, including training programs, supervisors' instructions, or Internet research (as part of the manager's job).

While in Chongqing, formal training was a source of information for managers but not for workers, because there was no training program for workers.

In spite of this, the Chongqing workers were better informed than their managers through personal networks and public media. A major concern among workers was the monetary allowance, as a worker commented, 'The newspaper said we should be paid "hot weather allowance" when the daily highest temperature is over 37°C, but we've never received any!' When asked if they had relevant training, another worker said, 'Not at all! They'd rather keep us ignorant. They are thinking how to put our money into their own pocket!' Here, a contradiction of interests between the workers and the employers was observed. Safety was interpreted as a sacrifice of profit on the employers and self-interest of the workers. The processing of safety is through a personal reciprocity, as depicted by a site manager's statement of how an organisation works: '[A] leader should be an example to his followers; a boss should be an example to his employees. If a manager respects and cares about his workers, workers will follow his example (of benevolence) to work hard for the company.'

In the Hong Kong context, safety programmes were pragmatically practiced as well-defined tasks in addition to the production tasks. The accountability was to fulfil external institutional requirements on health and safety management. Consequently, workers perceived it an extra workload for entertaining the safety programmes on top of their production workload. This was illustrated in one worker's complaint, 'The project is already behind schedule; (because) everybody is exhausted by safety!' 'But don't you appreciate that managers are initiating it for your safety and health?' 'Not at all! They are doing it for their own rice bowl!' In this context, the worker perceived that safety programmes were an extra workload on top of the existing job obligation, while managers were practising the programmes to fulfil their own job performance criteria. To both workers and managers, safety initiatives were a job requirement and compliance to rules.

In both contexts, the cynical workers denied trust of their managers with a 'Not at all!', which however had different underlying meanings. To the Chongqing worker, the cynicism was against the perception that employers withheld from doing safety, while to the Hong Kong worker, the cynicism was against extra safety workload as an institutional obligation which did not fit for purpose.

Scenario 2: Patterns of hazards associated with the use of PPE

What was the most effective practice for achieving safety in heat? To this question, the Hong Kong workers identified the enactment of formal rules, while their Chongqing counterparts practiced self-adaptation. An experienced steel worker in Hong Kong suggested that a compulsory extra break in the morning should be set up as standard practice on all construction sites in summer. Unless the break is compulsory and made a standard, workers will not have the effective protection from heat hazard. While in Chongqing, the working time in summer was negotiated between the supervisors and managers by daily weather conditions.

The consequences of the two different mechanisms were manifested in the fact that personal protective equipment (PPE) was perceived as a heat hazard by the Hong Kong workers, but not by their Chongqing counterparts. Regulations in Hong Kong require compulsory wearing of safety helmet, reflective vest and safety boots on construction sites. So far, all of these items are made of materials of low water vapour permeability, which multiplies heat stress in hot and humid weather (see Jia et al., 2015, for details of on-site records). Thus in Hong Kong, when the rigidly enforced safety rules demand wearing of the heat-aggravating PPE, workers were protected from one risk (hit by falling objects) but were exposed to another (heat stress). By contrast, mainland China only made safety helmets compulsory on construction sites. Researchers observed on the Chongqing sites that workers frequently took off their safety helmets to ventilate their heads, while supervisors sympathetically ignored workers' non-compliance actions. Here, supervisors were managing by personal judgment of the situations, while workers were practicing self-management. The underspecified formal system somewhat corresponds to Weick and colleagues's (2005) ideal of underspecified organisational forms as a springboard for collective mindfulness. However, this is subject to further development, as workers are thus exposed to the other common risks on site.

Scenario 3: Patterns of the use of local cost-effective interventions

A third scenario that illustrates the contrasting sensemaking is on the preservation or elimination of local heat-dissipative drinks. In the Chongqing context, workers identified a Chinese traditional medicine, ageratum liquid (藿香正气液), as the most effective measure for tackling heat stress. Ageratum liquid is widely used in Southwest China as a common preventive medicine and is known to have the effect of improving human body's immune system and calming the vomiting digestive system (Chinese Pharmacopoeia Commission, 2010). Most contractors provided it to the workers in summer. Similarly, in Hong Kong, the traditional Chinese herbal tea (Leung Cha) is widely used among the Chinese population to assist in cooling the body in summer time. Medical advice suggests it helps to improve kidney function and thus facilitates efficient sweating for heat dissipation. Thus herbal tea used to be provided on many construction sites in Hong Kong for coping with heat. However, at the time of the fieldwork, researchers found the provision of herbal tea had been gradually eliminated from construction sites through the formalisation of a site management system. In replacement, bottled drinks sold through vending machines became dominant on site. This substitution multiplied the cost of fluid intake in summer and shifted the cost from the organisations to the workers. Moreover, the high sugar ingredient in most bottled drinks introduced secondary health risks such as diabetes to the workforce. This contrast indicates that bottom-up methods of risk mitigation, organically grown out of local lifestyle, climate and demography, can be cost efficient and holistically effective. While formalised procedures and rules without learning from the local context may result in solutions that worsen the situation.

Implications

Findings of this study suggest that in a developed society where formal regulations are well established, both employers and workers made sense of safety as compliance to rules. While in a developing society, a systemic context lacking safety management infrastructures, safety means workers' self-interest and employers' benevolence; safety and production are processed as reciprocity of favour between workers and employers. It should be noted that the systemic context of Chongqing in this study was similar to that of Hong Kong in the 1990s, in which a safety management infrastructure did not exist and accident rates were high (Lingard and Rowlinson, 1998). In the absence of an established formal system, individual initiations were ad hoc reactions to immediate risks. Complexity and chaos theorists argue that self-organizing systems tend to be highly efficient (compared to centrally managed systems) in finding a natural equilibrium (Koubatis and Schonberger, 2005). However, in doing so they tend to settle at a 'critical edge' where a small change in the system can lead to catastrophic changes in the overall system, as manifested in the occurrence of the global financial crises. This is because in a truly complex system, each component will tend to act only locally, being unaware of its impact on the wider system. Falling short of coordination and proactive planning, the sites were unable to cope with risks beyond individual workers' control. However, in retrospect, the pathway of Hong Kong's development suggests that whilst a reliable system is essential for construction site safety management, over-reliance on formalisation and a tangible system can generate pragmatic compliance among the actors, which decouples safety from its original meaning.

In the European context, Blazsin and Guldenmund (2015) explored the safety discourses among three levels of field staff in a natural gas company. They found that safety meant self-interest to the workers, and rule compliance to the supervisors. Acting upon the constructed meanings, both tried to cope with the rules pragmatically: workers skipped some unrealistic operation procedures to achieve personal safety; supervisors decoupled safety rules to achieve efficiency in production. The results are compatible with the meaning matrices found in our study. Two clues stand to explain the difference: first, essential individual identities are different between the European and Chinese cultural contexts and, second, the difference between tightly and loosely coupled systems that feature the oil and gas and construction industries respectively. These merit further cross-industrial and cross-cultural research.

Conclusions

This study compared the socially constructed meanings of construction safety as outcomes of sensemaking in the systemic and cultural contexts of a developed society (Hong Kong) and a developing society (Chongqing in mainland China). Findings of this study illustrate stakeholders at different locations of the system have their own schema of what, why and how safety should occur. Individuals as agents of social culture interpret, define and alter the system, responding differently to safety and production goals. In a developed society where formal institutions are well established, safety means compliance to rules to both managers and workers. In a developing society, without the mediation of a formal institutional system, the meaning of safety is constructed as reciprocity of benevolence between employers and workers. The results suggest the making of effective safety rules needs to be informed by the knowledge of stakeholders' sensemaking in their specific sociocultural contexts. Rules and interventions organically developed bottom-up will tackle safety issues more effectively than imposed top-down.

Acknowledgments

The authors would like to thank Mark Sharp, Gigi Tsang, Tas Yong Koh, Martin Tuuli, Yin Xianting, Sam Hui, Zhang Yu, Shen Yuzhong, Chi Wah Lau, Glen Joe, Michael Tse, Lu Wei, Ma Shichao, Ju Chuangjing, Zhang Dan, Liang Yanhong, Li Jingkai, Bi Mo, Zhou Zhipeng, Li Xiaoyang, Betty Chiu, Goh Ching Siew, Chen Nazhen, Liu Meng, Xincheng and Zheng Qiang for their contribution to the research. The research was financially supported by the Hong Kong Construction Industry Council; Seed Funding Grant for Applied Research (Project No. 201109160027); Leung Kau Kui Research and Teaching Endowment Fund (Project No. 201011165012) at the University of Hong Kong; Hong Kong Research Grant Council General Research Fund (Project No. 17206514); Key Laboratory of Eco-environment in Three Gorges Reservoir Region under Ministry of Education, Chongqing University, China; and Eminent Visiting Fellowship and Seed Funding for Future Practice Research from School of Built Environment at Curtin University, Australia.

References

Blazsin, H. & Guldenmund, F. (2015), 'The social construction of safety: comparing three realities', *Safety Science*, 71, pp. 16–27.

Chinese Pharmacopoeia Commission. (2010), *Chinese Pharmacopoeia*. Beijing: China Medical Science Press.

Corbin, J. & Strauss, A. (2008), *Basics of Qualitative Research: Techniques and Procedures for Developing Grounded Theory* (3rd Ed.). Thousand Oaks: Sage.

Doree, A. G. & Holmen, E. (2004), 'Achieving the unlikely: innovating in the loosely coupled construction system', *Construction Management and Economics*, 22, pp. 827–838.

Doubois, A. & Gadde, L. (2002), 'The construction industry as a loosely coupled system: implications for productivity and innovation', *Construction Management and Economics*, 20, pp. 621–631.

Fellows, R. & Liu, A. (2016), 'Sensemaking in the cross-cultural contexts of projects', *International Journal of Project Management*, 34, pp. 246–257.

Gibb, A., Haslam, R., Gyi, D., Hide, S. & Duff, R. (2006), 'What causes accidents?' *Civil Engineering*, 159, pp. 47–50.

Glaser, B. G. & Strauss, A. L. (1967), *The Discovery of Grounded Theory: Strategies for Qualitative Research*. London: Weidenfeld & Nicolson.

Goffman, E. (1974), *Frame Analysis: An Essay on the Organization of Experience*. Cambridge, MA: Harvard University Press.
Goh, Y. M., Love, P. E. D., Brown, H. & Spickett, J. (2012), 'Organizational accidents: a systemic model of production versus protection', *Journal of Management Studies*, 49, doi: 10.1111/j.1467-6486.2010.00959.x.
Hale, A. R. & Borys, D. (2013), 'Working to rule, or working safely? Part I: a state of the art review', *Safety Science*, 55, pp. 207–221.
Hale, A. R., Walker, D., Walters, N. & Bolt, H. (2012), 'Developing the understanding of underlying causes of construction fatal accidents', *Safety Science*, 50, pp. 2020–2027.
Hong, Y., Ip, G., Chiu, C., Morris, M. W. & Menon, T. (2001), 'Cultural identity and dynamic construction of the self: collective duties and individual rights in Chinese and American cultures', *Social Cognition*, 19, pp. 251–268.
Hovden, J., Albrechtsen, E. & Herrera, I. A. (2010), 'Is there a need for new theories, models and approaches to occupational accident prevention?' *Safety Science*, 48, pp. 950–956.
Ivanova-Gongn, M. & Törnroos, J. K. (2017), 'Understanding cultural sensemaking of business interaction: a research model', *Scandinavian Journal of Management*, 33, pp. 102–112.
Jia, A. Y., Rowlinson, S. & Ciccarelli, M. (2016), 'Climatic and psychosocial risks of heat illness incidents on construction site', *Applied Ergonomics*, 53, pp. 25–35.
Jia, A. Y., Rowlinson, S., Loosemore, M., Xu, M., Li, B. & Gibb, A. (2017), 'Institutions and institutional logics in construction safety management: the case of climatic heat stress', *Construction Management and Economics*, 35, pp. 338–367.
Koubatis, A. & Schonberger, J. Y. (2005), Risk management of complex critical systems, *International Journal of Infrastructures*, 1(2/3), pp. 195–215.
Lawani, K., Hare, B. & Cameron, I. (2017), 'Developing a worker engagement maturity model for improving occupational safety and health (OSH) in construction', *Journal of Construction Project Management and Innovation*, 7, pp. 2116–2126.
Lawani, K., Hare, B. & Cameron, I. (2018), 'Empowerment as a construct of worker engagement and wellbeing'. In: *Proceedings of the Joint CIB W099 and TG 59 Conference*, 1–3 August 2018, Salvador, Brazil.
Leveson, N. G. (2011), 'Applying systems thinking to analyse and learn from events', *Safety Science*, 49, pp. 55–64.
Lincoln, Y. S. & Guba, E. G. (1985), *Naturalistic Inquiry*. Beverly Hills: Sage.
Lingard, H. & Rowlinson, S. (1998), 'Behaviour-based safety management in Hong Kong's construction industry: the results of a field study', *Construction Management and Economics*, 16, pp. 481–488.
Love, P. E. D., Smith, J. & Han, S. (2011), 'Sensemaking of rework causation in offshore structures: people, organisation and project'. In: *COBRA 2011: RICS Construction and Property Conference*, September 2011. University of Salford, United Kingdom.
Maloney, W. F., Cameron, I. & Hare, B. (2007), 'Tradesmen involvement in health and safety', *Journal of Construction Engineering and Management*, 133(4), pp. 297.
Orton, J. D. & Weick, K. E. (1990), 'Loosely coupled systems: a reconceptualization', *Academy of Management Review*, 15, pp. 203–223.
Rasmussen, J., Pejtersen, A. M. & Goodstein, L. P. (1994), *Cognitive Systems Engineering*. New York: John Wiley & Sons.
Reason, J. (1997), *Managing the Risks of Organizational Accidents*. Surrey: Ashgate.
Rowlinson, S. & Jia, A. Y. (2015), 'Construction accident causality: an institutional analysis of heat illness incidents on site', *Safety Science*, 78, pp. 179–189.

Senge, P. M. (2006), *The Fifth Discipline: The Art and Practice of the Learning Organization.* New York: Doubleday/Currency.

Thornton, P. H., Ocasio, W. & Lounsbury, M. (2013), *The Institutional Logics Perspective: A New Approach to Culture, Structure and Process.* Oxford: Oxford University Press.

Vaara, E., Sonenshein, S. & Boje, D. (2016), 'Narratives as sources of stability and change in organizations: approaches and directions for future research', *The Academy of Management Annals*, 10, pp. 495–560.

Weber, K. & Glynn, M. A. (2006), 'Making sense in institutions: context, thought and action in Karl Weick's theory', *Organizational Studies*, 27, pp. 1639–1660.

Weick, K. E. (1976), 'Educational organizations as loosely coupled systems', *Administrative Science Quarterly*, 21, pp. 1–19.

Weick, K. E. (1979), *The Social Psychology of Organizing.* New York: Random House.

Weick, K. E. (1982), 'Management of organizational change among loosely coupled elements'. In: Goodman, P. S. & Associates (eds.) *Changes in Organizations.* San Francisco: Jossey-Bass.

Weick, K. E. (1988), 'Enacted sensemaking in crisis situations', *Journal of Management Studies*, 25, pp. 305–317.

Weick, K. E. (1993), 'The collapse of sensemaking in organizations: the Mann Gulch disaster', *Administrative Science Quarterly*, 38, pp. 628–652.

Weick, K. E. (1995), *Sensemaking of Organizations.* Thousand Oaks, CA: Sage.

Weick, K. E. (2001a), *Making Sense of the Organization.* Oxford: Blackwell.

Weick, K. E. (2001b), 'Management of organizational change among loosely coupled elements'. In: *Making Sense of the Organization.* Oxford: Blackwell.

Weick, K. E. (2012), *Making Sense of the Organization*, volume 2 – *The Impermanent Organization.* Chichester: Wiley.

Weick, K. E., Sutcliffe, K. M. & Obstfeld, D. (1999), 'Organising for high reliability: processes of collective mindfulness'. In: Sutton, R. S. & Staw, B. M. (eds.) *Research in Organizational Behaviour.* Stamford, CT: Jai Press.

Weick, K. E., Sutcliffe, K. M. & Obstfeld, D. (2005), 'Organizing and the process of sensemaking', *Organization Science*, 16, pp. 409–421.

14 Managing construction safety in developing countries

From the viewpoint of Chinese international contractors

Ran Gao and Albert P. C. Chan

Summary

Chinese contractors are getting more important in the international construction market. They endeavour to improve their management level and acquire the competitive advantage gradually. Safety management is an important topic in construction. This study aims to understand the safety management practices and identify the problems encountered by Chinese international contractors (CICs) in their overseas projects. Six international projects from three CICs' major regional markets – Southeast Asia, the Middle East and Africa – were investigated through structured interviews. The study results in the identification of three problems: environmental- and societal-level problems, i.e. harsh climate, low social-economic development level, and culture differences; organizational-level problems, i.e. immature short-term corporate strategy and lack of contracting experiences, tactical management problems, and inadequate and changeable supply of human resources; and personal-level problems, i.e. communication problems and mental fatigue of Chinese workers. Safety issues of the CICs' overseas projects should carry weight and thus be improved in the future.

Introduction

Globalization has advanced the international construction market, resulting in increasingly advanced technology, rapid transportation, and convenient communication. As a response to this trend, an increasing number of construction companies have become involved in the international construction market to secure more opportunities to guarantee survival and seek continuous development (Zilke and Taylor, 2014). Chinese contractors are no exception. These contractors are encouraged further by the Chinese government's "going out" strategy to compete in overseas markets. Governmental support has significantly promoted the development of Chinese contractors in the international construction market (Zhao and Shen, 2008). According to the *China Statistical Yearbook 2015*, the value of turnover fulfilled in international projects undertaken by CICs in 2014 was 45.5% in Asia and 37.2% in Africa, forming two of their most conspicuous major international markets, which mainly consist of developing and undeveloped countries (National Bureau of Statistics, 2015).

With the process of industrialization, safety is increasingly considered to be a priority by society (Mahalingam and Levitt, 2007). The construction industry is considered one of the most dangerous industries in the world because of its special characteristics in the production process, and thus safety can be a rather complicated issue (Swuste et al., 2012). Construction safety management has been considered as unsatisfactory in Chinese domestic projects with many inadequacies such as poor safety awareness of both top management and project managers, lack of training, reckless operations, and reluctance to input resources to safety (Tam et al., 2004). In comparison with domestic projects, international projects involve participants from various countries and regions, leading to complexities in national, organizational, and individual perspectives that may have adverse effects on construction safety management leading to unsatisfactory safety performance (Gao et al., 2016). Moreover, CICs face additional challenges and difficulties in their international projects in developing countries (Gao et al., 2018).

CICs have special characteristics in managing construction safety in their overseas projects. A good example of these specialties is the gradual localization processes of the workforce. In the 1990s, Chinese state-owned construction companies tended to export workers as part of the construction package. The workers received proper training and adequate skills before they were sent to overseas construction sites (Zhao and Shen, 2008). With the gradually growing trend of labour price in the Chinese domestic market, working in overseas projects, especially those in undeveloped countries, has already become unattractive to Chinese construction workers as of 20 years ago. Moreover, as CICs' major markets are in developing and undeveloped countries, easy access to cheap local workers has also accelerated the processes of CICs' labour localization. An overwhelming number of unskilled local workers have been engaged in CICs' overseas projects, resulting in various difficulties and barriers in managing construction safety effectively. Other characteristics such as management style, technology, and contracting philosophy could also affect the safety performance of CICs (Kjellén, 2012).

The increasingly important role of CICs in the international construction market, the significance of safety management in construction projects, and the unique characteristics of CICs' overseas projects all highlight the importance of studying construction safety management in CICs' overseas projects. However, the amount of literature available in this area is quite limited. Thus, a comprehensive study related to construction safety management from CICs' perspective in a multi-cultural and overseas environment is necessary. This study seeks to bridge some of the gaps in this area. The focus of this study is to develop an understanding of the safety management practices and to identify the problems encountered by CICs in their overseas projects. Other international contractors, especially those from developing countries, could also benefit from this study by understanding the similar situations they might meet in their own overseas projects.

Construction safety management in multi-cultural context

As CICs' overseas projects involve participants from various countries who share different cultures and values, implementing construction safety management actually involves managing construction safety within a complicated multi-cultural context. Construction safety management within a multi-cultural context encounters specific problems compared to that in a mono-cultural context. In a study conducted by Schubert and Dijkstra (2009), half of their interviewed experts could perceive higher risks among works involving foreign contractors and employees. They also identified five common problematic areas when people worked together with foreign personnel and contractors. These areas include communication, education level, cultural differences, cooperation, and specific employment situations.

Kjellén (2012) evaluated two different safety management approaches in two different international construction projects located in India and the Philippines. The author indicated that the proactive approach had the ability to present satisfactory safety outcomes at reasonable costs of execution, whereas the reactive approach tended to result in poor safety outcomes. Kjellén (2012) also identified eight factors that influenced safety performance in international projects, which are composed of three external factors (national regulations and authority handling, non-governmental safety standards, and national economic wealth and culture) as well as five internal factors (nature of the project and site layout, management systems and practices, contracting philosophy, technology standards, and human resources).

Human resources factors are also crucial in influencing safety management, and the risks of these factors contribute significantly to unmanageable construction safety in international projects, leading to worse practices of safety management in international construction projects than in local projects. The reasons are embodied in the following three points. First, language and communication is a problem (Loosemore and Lee, 2002). In most cases, the transfer of information between supervisors and workers or among workers from different countries and speaking various languages is challenging and ineffective. Jaselskis and colleagues (2008) stated that the main problem leading to higher accident rates by Hispanic workers on the jobsite is bad safety communication because of the language barrier. Second, project participants coming from diverse countries may hold different opinions of safety management. For example, practices that international contractors from developed countries may consider as unsafe or defective could be deemed safe enough by subcontractors from developing countries. This contrast and opposition causes conflicts and time losses in rectifying work (Mahalingam and Levitt, 2007). Third, a high turnover rate hinders the effectiveness of safety management in international projects. With a relatively frequent turnover rate of the management staff and front-line workers in different phases of the international construction projects, those who have improved their safety performance during the previous process may be substituted by new members, and the overall safety performance will inevitably deteriorate without continuously effective management (Zhang and Fang, 2013).

The research direction that focuses on the relationship between national culture and safety has received increasing scholarly interest in recent years. These studies were basically developed from Hofstede's cultural dimension theory (Hofstede, 1991), which established a systematic framework to assess national cultures from five cultural dimensions: large vs. small power distance, individualism vs. collectivism, strong vs. weak uncertainty avoidance, masculinity vs. femininity, and short-term vs. long-term orientation. Starren and colleagues (2013) examined the possible effects of national culture on the major antecedents of safety behaviour such as safety knowledge, safety motivation, and safety climate. He indicated that safety knowledge could be improved gradually with proper evidence, pictograms, and training. However, the effects of these instrumentalities could change from one person to another because of the various cultural backgrounds of project participants who received the instruments. Mearns and Yule (2009) examined the relationships among Hofstede's cultural dimensions, safety climate, and risk-taking behaviours at work through a study of six national workforce groups, and found that only masculinity and power distance could significantly predict risk-taking behaviours. Mohamed and colleagues (2009) verified that collectivism, femininity, and higher uncertainty avoidance could enhance workers' safety awareness and beliefs, thereby leading to safer jobsite behaviours.

Case study

Asia and Africa are the two most conspicuous major international markets of CICs. Table 14.1 shows CICs' fulfilled turnover value of contracted projects based on continents and regions in 2014. Specifically, Southeast Asia, the Middle East, and Africa are the 3 regions that separately account for over 10% of the total

Table 14.1 CICs' value of turnover fulfilled for contracted projects in particular continents and regions in 2014

Continent/ Region	Value of turnover fulfilled (billion USD)	Percentage
Asia	64.84	45.5%
Southeast Asia	22.41	15.7%
China – Hong Kong, Macao and Taiwan	4.61	3.2%
The Middle East (w/o Cyprus and Egypt)	19.54	13.7%
Asia other than above	18.27	12.8%
Africa	52.97	37.2%
Europe	7.15	5.0%
Latin America	13.18	9.3%
North America	2.02	1.4%
Oceanic and Pacific Islands	2.25	1.6%
Others	0.00	0.0%
Total	142.41	100.00%

Source: National Bureau of Statistics, 2015, *China Statistical Yearbook 2015*, Beijing: China Statistics Press.

value of turnovers fulfilled by CICs, and they accumulate almost two-thirds of the total value in that year. Thus, the international projects in these three regions are relatively typical among all projects conducted by CICs.

Six international projects from the three aforementioned regions were finally chosen to be the research sample. Table 14.2 presents the basic information of these sample projects. Among the six projects, five are building projects, including satellite city development, residential building, commercial building, conference centre, and university. Five out of the six selected projects are Engineering-Procurement-Construction projects, which are the most commonly contracted type of CICs' international projects nowadays. The contract values of the selected projects vary from USD 121 million to USD 3.5 billion. In the six projects, the durations of the completed projects are all deferred without exception, implying a high uncertainty in international contracting. Another question that should be considered is whether the contractors are listed in the *Engineering News-Record* (*ENR*) top 225 or top 250 international contractors (Reina and Tulacz, 2016). Actually, although some contractors had a major domestic market share, they entered the international construction market relatively late. Project C and Project D in our study were both the first projects of contractors involved in particular regional markets, and they might lack contracting and management experiences to some extent. The six selected projects have good representativeness among all the overseas projects of CICs.

Pragmatism emphasizes research problems and suggests the application of all available approaches to understand and solve the problem (Creswell, 2013). As the current study focuses on safety management practices of overseas projects of CICs, a pragmatic worldview was applied in the research. The qualitative research method in the form of the structured interview was employed to gather information from all projects in the sample. The research interview is broadly

Table 14.2 Basic information of selected projects

No.	Location	Project type	Contract type	Contract value (USD 100 million)	Duration	Whether **ENR** Top 225/250
A	Africa	Satellite city developing	EPC	35.00	60 months	Yes
B	Middle East	Residential and commercial building	DB	4.40	74 months	Yes
C	Southeast Asia	Thermal power station	EPC	15.00	4 years	No
D	Middle East	University	EPC	6.12	5 years	No
E	Africa	Conference centre	EPC	8.00	42 months	Yes
F	Southeast Asia	Commercial building	EPC	1.21	43 months	Yes

Notes: EPC, engineering, procurement, and construction; DB, design–build.

employed in construction management research (Fellows and Liu, 2009). The structured interview utilizes a prepared interview list that specifies the description and sequence of questions thereby in favour of comparing participant responses (Shan et al., 2016).

A project manager or safety manager from a particular project was invited to participate in the structured interview conducted between May and June 2014. A list of the interview arrangement was sent to each participant separately before the interview was conducted. The participants were also informed that the responses would remain confidential and their participation was voluntary. We enquired about the project background information and asked several questions concentrating on safety management issues in the interviewee's project, such as "what are the major safety and health problems" and "please introduce the safety management measures and difficulties to implement them". As most of the interviewees were working abroad when the interviews were conducted, only interviews for Project C and Project D were conducted face-to-face when the interviewees had vacations in China, whereas the other four involved video interviews. The interviewer did not set a time limitation, and the interview continued until the interviewee exhausted conversation topics. Each interview took about 45 to 90 minutes. The interviews were recorded with prior consent from the interviewees, and all records were transcribed in English although the interviews were conducted in Chinese. The transcription was coded using a three-step constant comparative method in NVivo 10. The constant comparison procedure is composed of the following steps: (1) comparison within a single interview or project; (2) comparison between interviews or projects within the same group or region; and (3) comparison of interviews or projects from different groups or regions.

Results and discussion

The idea of distinguishing identified problems into different levels is derived from Spangenberg et al. (2003). Three levels of problems described as environmental and societal level, organizational level, and personal level are presented accordingly.

Environmental- and societal-level problems

1. Harsh climate. The climate in the countries of some overseas projects is quite different from that in China, and sometimes could be very harsh and extreme, which is adverse to safety management. For example, many local workers do not follow the rules to wear personal protective equipment (PPE) because of the local climate and lifestyle. A safety manager in Project C remarked:

   > As the weather here is quite hot, many local workers prefer to wear shorts and T-shirts on site, and many of them even do not wear shoes at all. We have exerted considerable efforts to change this problem,

Safety management by Chinese international contractors 197

and we even give them free shoes. However, many local workers are still barefoot on site because they find it more comfortable to not wear shoes. Meanwhile, in order to work conveniently, many local workers also do not wear other PPEs.

2. Low socio-economic development level. As CICs' overseas projects are located mainly in undeveloped and developing countries, relatively low socio-economic levels have been developed. Both the prices of local labour and compensation standards for accidents are low. For example, a fatal accident compensation amount is only equivalent to one year's wage in the country of Project F, whereas that in China is 20 times of a city's annual wage. The inadequate valuation of life leads to low safety awareness and poor safety performance. A good explanation to the relationship between low socio-economic development and poor safety performance is Maslow' hierarchy of needs, stating that workers only pursue safety or security needs after they have fulfilled the basic physiological needs (Mearns and Yule, 2009). Low socio-economic development also hinders the growth of the entire construction industry, leading to backward technology and uneven quality of local subcontractors. Not surprisingly, health care facilities in these locations are also far from perfect. As the project manager of Project E stated:

> The most important safety problem here is emergency response. Although we have on site first-aid equipment, if the injury is heavier, the injured worker could not receive timely and effective medical care due to local backward medical facilities.

In addition, the legislation on occupational safety and health in the aforementioned countries tends to be limited. With this limitation, governmental sectors implement inadequate supervision on construction sites, and local owners and subcontractors pay little attention to construction safety. The problems of construction safety could only be tackled with efforts from the government, clients, designers, contractors, subcontractors, and workers together (Debrah and Ofori, 2001).

3. Cultural differences. Although the degree to which national culture affects construction safety is pending further discussion (Mearns and Yule, 2009; Starren et al., 2013), some customs derived from various cultures would definitely affect safety performance. For example, as Project B, D, and E all involve Muslim workers, the dietary deficiency during Ramadan is considered a significant hazard in their projects. The interviewee from Project D described a falling accident arising from this problem.

Organizational-level problems

1. Immature short-term corporate strategy and lack of contracting experiences. Some CICs focus their main businesses in the domestic market rather than the overseas market, and overseas markets are deemed to be potentially risky

instead of profitable to them. A major difference between international contractors and typical multi-national enterprises is the time range for exploring overseas. Typical multi-national enterprises tend to look towards the future and involve themselves in a host market for decades, whereas contractors could gain their returns in a relatively shorter period. The short-term corporate strategy could lead to poor project performance. Investment in advanced plants and facilities in a particular overseas branch is prudent with only small scale in the beginning because of an unstable and unsustainable market expectation, and this kind of savings may be adverse to effective safety management.

Projects C and D were both the first projects of the contractors in the particular regional markets, although they have occupied a good domestic market share. The lack of contracting experiences in an overseas market could make them unprepared in safety management. As the safety manager in Project C described:

> We once required all subcontractors to conduct hazard identification similar to what we do in China, but local ones do not carry them out, and even failed to understand why it is a necessary step. Finally, we omitted this requirement as it was relatively less important than some other requirements, and only insisted on the most crucial ones.

Some CICs do not pay significant attention to self-management on construction safety even when they enter contracts with domestic projects. As local governments, owners, subcontractors, and workers are all in the habit of paying limited attention to construction safety, the safety performance in these international projects could be extremely unsatisfactory.

2. Tactical management problems. The biggest problem for CICs is the lack of sufficient management capacity, although they are strong enough to gain contracts in the international market (Lu et al., 2009). Compared to their domestic management capacity, the overseas management capacity of CICs is weak. Several obvious management problems exist in managing safety as well. Table 14.3 summarizes the major management problems related to safety from four levels.

First, management problems related to construction safety at the headquarters level include a lack of positive safety culture among corporate levels and failure to publish unified safety guidelines to overseas projects. Second, senior management–level problems include a lack of management commitment to safety, inadequate safety input, and unclear prescription of the responsibility for safety. CICs usually win their overseas projects with relatively low bidding prices, and projects with competitive biddings tend to have higher accident rates (Hinze and Raboud, 1988). Third, site management level problems include conflicting and confusing safety standards, low priority given to safety than productivity, lack of effective supervision, inadequate safety training, ineffective safety notices, and late emergency response. Fourth, Chinese foremen sometimes behave adversely toward

Table 14.3 Summary of tactical management problems

Level	Problem
Headquarters level	Lack of safety guidelines from headquarters level
	Lack of "safety first" culture
Senior management level	Lack of management commitment to safety
	Inadequate resources/input given to safety
	Responsibility is not clearly prescribed
Site management level	Conflicted and confused safety standards
	Low priority given to safety than productivity
	Lack of effective supervision
	Inadequate safety training
	Ineffective safety notices
	Emergency response is not timely
Foreman level	Chinese foremen are averse to high safety standards

safety management mainly because some jobs are based on piece rates, and meeting safety requirements could drop the rates of production. The safety manager of Project B complained:

> Compared with the safety communication barriers between safety staff and workers due to language barriers, the barriers between safety staff and Chinese foremen are even more serious, mainly because of the weak safety consciousness of the foremen. Many foremen are averse to high safety standards themselves, not to mention implement them to the workers.

3. Inadequate and changeable supply of human resources. An unsteady human resource provision is common in construction projects. For CICs' overseas projects, inadequate professional safety personnel and high turnover rate of staff and workers can influence safety management. First, both Chinese and foreign safety officers have been employed among the overseas projects investigated in this study, and each of them have their pros and cons. A foreign safety officer is more proficient in local safety rules and regulations and usually implements them strictly. However, Chinese workers may think the officer to be inflexible because of culture differences. If Chinese workers account for the majority in an overseas project, then employing Chinese safety officers is a better option instead of employing foreign ones to avert cultural conflicts. Employing Chinese professional safety personnel who understand the local system and regulations well would kill two birds with one stone. Second, the high turnover rate of front-line workers is considered harmful to safety management. Workers who work repeatedly for one contactor tend to have better safety awareness (Chen et al., 2012), whereas workers on a short tenure of employment are vulnerable to accidents (Debrah and Ofori, 2001). During the progress of project localization, many non-Chinese workers are substituted for Chinese ones, leading to a relatively high turnover rate and an adverse effect to safety management.

Personal-level problems

1. Safety communication problems. Communication barriers, particularly language barriers, pose serious challenges to the management of construction safety in CICs' overseas projects. English, as an international language, is commonly used for communication in international projects. Many Chinese employees of CICs, especially the ones working on site, including the foremen and front-line workers, have inadequate English skills (Zhao and Shen, 2008). English is often confined to the managerial level, as stated by the safety manager in Project C:

    > It is very important to reduce communication range, that is to say, we mainly conduct safety communication with local management staff that has adequate English language ability, and then they transmit safety information to the front-line workers. There is a principle used to increase safety communication between local people and to reduce that between local people and Chinese staff.

    All of the interviewees mentioned using translators or interpreters on the construction sites. However, onsite translation could cause ineffective verbal communication because of wrong interpretations or translations. Language barriers are also difficult to tackle in cases of emergency, which transform avoidable situations into accident cases. Written communication is also deficient in workers. The safety notices, posters, and signage were not always translated into Chinese, English, and local languages, and some of the workers cannot understand them well. In Projects E and F, with long-term development and localization plans in particular countries, local language courses are also provided to the Chinese managerial staff to assist them in communicating smoothly with local front-line workers.

2. Mental fatigue of Chinese workers. Fatigue covers physical, mental, and emotional aspects (Chan, 2011). Hinze's theory of distraction and Suraji and colleagues's (2001) constraint-response theory both indicate that one's mental state and emotional distress, especially miserable factors, could distract workers from the hazards, thereby increasing the likelihood of accidents (Hinze, 1996; Suraji et al., 2001). A majority of Chinese workers shoulder the responsibility of supporting their families and thus prefer to work overtime to earn more money. In addition, their boring life overseas aggravates the workers' exhausted spirits, which affect their safety performance. This phenomenon is more common in the projects in Africa and the Middle East than those in Southeast Asia, possibly because the lifestyle in Southeast Asian countries is kind of similar to the Chinese lifestyle. As described by the interviewee from Project A:

    > During the first two years of the project, no fatality accident happened among Chinese workers. However, in the following two years, there were four fatalities. One of them was a Chinese worker who rested just below the tower crane basket full of blocks and was killed when

Table 14.4 Summary of identified three problem levels

Level	Problem
Environmental- and societal-level problems	Harsh climate
	Low socio-economic development level
	Cultural differences
Organizational-level problems	Immature short-term corporate strategy and lack of contracting experiences
	Tactical management problems
	Inadequate and changeable supply of human resources
Personal-level problems	Safety communication problems
	Mental fatigue of Chinese workers

the blocks slipped off the basket. Actually, everybody knows it is very dangerous to stay below the basket. But why did he stay there? It was because of mental fatigue. The long-term boring life out of their home country makes workers minds numb.

In addition, the safety manager in Project B stated:

Due to unfamiliar environment, harsh climate, distasteful food supply, bad accommodation conditions and lack of entertainment, many Chinese workers suffer severe depression and mental fatigue. These bad emotional and mental conditions enlarge safety risks. One time in my project, there were over ten Chinese workers called in sick in the same day. What's worse, some sick workers forced themselves to work to earn daily wages, which highly threatened safety management on site.

Table 14.4 presents a summary of the identified three problem levels.

Conclusions

This study has successfully identified three problem levels of managing construction safety in CICs' overseas projects by case studies of six samples from three conspicuous major international markets – Southeast Asia, the Middle East, and Africa. Findings point out some unique safety management problems in the international setting, but at the same time they have certain likenesses to common safety problems of general construction projects. These problems have been discussed from three levels including environmental and societal, organizational, and personal. A harsh and extreme climate, low socio-economic development level of the host countries, and cultural differences between the Chinese employee and local employee are three important environmental- and societal-level problems. Problems of this level are objective and unavoidable. CICs should become ready to encounter and strive to resolve these problems, and conduct effective safety management without neglecting local culture.

Besides these macro level issues, some organizational-level problems have also been identified. An immature short-term corporate strategy and lack of contracting experiences, tactical management problems, and an inadequate and changing supply of human resources are adverse to construction safety. Problems of this level could be gradually improved with the contracting experiences accumulation and standard management implementation. CICs should take their corporate social responsibility and build up corporate reputation. In addition, two personal-level problems – ineffective safety communication and mental fatigue of Chinese workers – are identified. Both physiological and psychological health of front-line workers should be paid more attention. This study has revealed several major problems of managing construction safety in CICs' overseas projects from a qualitative perspective. In the future, more research could be done on investigating influencing factors of construction safety in international context from a quantitative view. It is hoped that this study could inspire a new research interest on construction safety in international context; and safety issues of the CICs' overseas projects could carry weight and thus be improved in the future. In this research, structured interviews were conducted to understand the opinions from industrial practitioners. The research results rely on the experience and knowledge of these practitioners and experts. The subjectivity of the evaluation could not be completely eliminated. The influence of this limitation could be further reduced by conducting the research with a larger sample size.

Acknowledgement

This study has been supported by the Hong Kong PhD Fellowship Scheme established by the Research Grants Council in Hong Kong. The authors gratefully acknowledge the Department of Building and Real Estate at the Hong Kong Polytechnic University for providing technical support to conduct this research. The contributions of all construction industry practitioners involved in the study are also highly appreciated.

References

Chan, M. (2011). Fatigue: The most critical accident risk in oil and gas construction. *Construction Management and Economics*, 29(4), 341–353.

Chen, Q., Jin, R., & Soboyejo, A. (2012). Understanding a contractor's regional variations in safety performance. *Journal of Construction Engineering and Management*, 139(6), 641–653.

Creswell, J. W. (2013). *Research Design: Qualitative, Quantitative, and Mixed Methods Approaches*. Sage Publications.

Debrah, Y. A., & Ofori, G. (2001). Subcontracting, foreign workers and job safety in the Singapore construction industry. *Asia Pacific Business Review*, 8(1), 145–166.

Fellows, R. F., & Liu, A. M. (2009). *Research Methods for Construction*. John Wiley & Sons.

Gao, R., Chan, A., Lyu, S., Zahoor, H., & Utama, W. (2018). Investigating the difficulties of implementing safety practices in international construction projects. *Safety Science*, 108, 39–47.

Gao, R., Chan, A., Utama, W., & Zahoor, H. (2016). Workers' perceptions of safety climate in international construction projects: Effects of nationality, religious belief, and employment mode. *Journal of Construction Engineering & Management*, 143(4), 04016117.

Hinze, J. (1996). The distractions theory of accident causation, in implementation of safety and health on construction sites. In Proceedings of the 1st International Conference CIB Working Commission W099 - Safety and Health on Construction Sites CIB Report, 112–121.

Hinze, J., & Raboud, P. (1988). Safety on large building construction projects. *Journal of Construction Engineering and Management*, 114(2), 286–293.

Hofstede, G. (1991). *Cultures and Organizations: Software of the Mind*. New York: McGraw-Hill, 1997, 23–47.

Jaselskis, E. J., Strong, K. C., Aveiga, F., Canales, A. R., & Jahren, C. (2008). Successful multi-national workforce integration program to improve construction site performance. *Safety Science*, 46(4), 603–618.

Kjellén, U. (2012). Managing safety in hydropower projects in emerging markets – Experiences in developing from a reactive to a proactive approach. *Safety Science*, 50(10), 1941–1951.

Loosemore, M., & Lee, P. (2002). Communication problems with ethnic minorities in the construction industry. *International Journal of Project Management*, 20(7), 517–524.

Lu, W., Li, H., Shen, L., & Huang, T. (2009). Strengths, weaknesses, opportunities, and threats analysis of Chinese construction companies in the global market. *Journal of Management in Engineering*, 25(4), 166–176.

Mahalingam, A., & Levitt, R. (2007). Safety issues on global projects. *Journal of Construction Engineering and Management*, 133(7), 506–516.

Mearns, K., & Yule, S. (2009). The role of national culture in determining safety performance: Challenges for the global oil and gas industry. *Safety Science*, 47(6), 777–785.

Mohamed, S., Ali, T. H., & Tam, W. Y. V. (2009). National culture and safe work behaviour of construction workers in Pakistan. *Safety Science*, 47(1), 29–35.

National Bureau of Statistics. (2015). *China Statistical Yearbook 2015*. Beijing: China Statistics Press.

Reina, P., & Tulacz, G. J. (2016). Top 250 International Contractors: Seeking Stable Markets. *ENR* (*Engineering News-Record*). https://www.enr.com/articles/40110-top-250-international-contractors-seeking-stable-markets?v=preview

Schubert, U., & Dijkstra, J. J. (2009). Working safely with foreign contractors and personnel. *Safety Science*, 47(6), 786–793.

Shan, M., Chan, A. P. C., Le, Y., Hu, Y., & Xia, B. (2016). Understanding collusive practices in Chinese construction projects. *Journal of Professional Issues in Engineering Education & Practice*, 143(3), 05016012.

Spangenberg, S., Baarts, C., Dyreborg, J., Jensen, L., Kines, P., & Mikkelsen, K. L. (2003). Factors contributing to the differences in work related injury rates between Danish and Swedish construction workers. *Safety Science*, 41(6), 517–530.

Starren, A., Hornikx, J., & Luijters, K. (2013). Occupational safety in multicultural teams and organizations: A research agenda. *Safety Science*, 52, 43–49.

Suraji, A., Duff, A., & Peckitt, S. (2001). Development of causal model of construction accident causation. *Journal of Construction Engineering and Management*, 127(4), 337–344.

Swuste, P., Frijters, A., & Guldenmund, F. (2012). Is it possible to influence safety in the building sector? A literature review extending from 1980 until the present. *Safety Science*, 50(5), 1333–1343.

Tam, C. M., Zeng, S. X., & Deng, Z. M. (2004). Identifying elements of poor construction safety management in China. *Safety Science*, 42(7), 569–586.

Zhang, M., & Fang, D. (2013). A continuous behavior-based safety strategy for persistent safety improvement in construction industry. *Automation in Construction*, 34, 101–107.

Zhao, Z. Y., & Shen, L. Y. (2008). Are Chinese contractors competitive in international markets? *Construction Management and Economics*, 26(3), 225–236.

Zilke, J. P., & Taylor, J. E. (2014). Evaluating the suitability of using international market analyses to characterize the global construction industry. *Journal of Management in Engineering*, 31(5), 04014078.

15 Learning from failures (LFF)
A multi-level conceptual model for changing safety culture in the Nigerian construction industry

Clara Man Cheung, Akilu Yunusa-Kaltungo, Obuks Ejohwomu and Rita Peihua Zhang

Summary

In this theory-driven literature review, we proposed to examine how the national, industrial, and organisational level of safety culture may inform our understanding of human behaviours that affect safety outcomes. According to safety science theory, the level of safety culture significantly affects safety outcomes. However, from the emerging research on safety performance in Africa, it has not put much focus on studying the effect of safety culture. From this we constructed a conceptual model based on learning from failures (LFF) that depicted how a multi-level safety culture of the nation, industry and organisation could affect safety outcomes.

Introduction

The general concept of safety in Africa is not novel as it dates back to pre-colonial times, precisely 1789 through the implementation of the medical examination board of the Liverpool infantry that oversaw the safety and health of British slave masters in Africa (Kalejaiye, 2013). Despite this seemingly long affiliation with fundamental safety principles, most African countries still possess some of the worst safety records in recent decades (see Figure 15.1). In addition, in spite of the enactment of several country-specific labour and factories acts in the major economies of Africa such as Nigeria, Hämäläinen and colleagues (2009) and Abubakar (2015) found that fatality rates per 100,000 workers in those countries has not been significantly improved. Against this backdrop, existing studies (Nnaji et al., 2017) that focused on studying how to improve safety performance through increasing regulations in Africa seem to provide a missing picture on the issue. However, safety culture at national, industrial and organizational levels that could affect the effectiveness of implementing those regulations has been overlooked. For example, a recent 35-year systematic review (1983–2018) of construction management research revealed that only three studies to date considered safety culture and climate in the African context (Umeokafor, 2018).

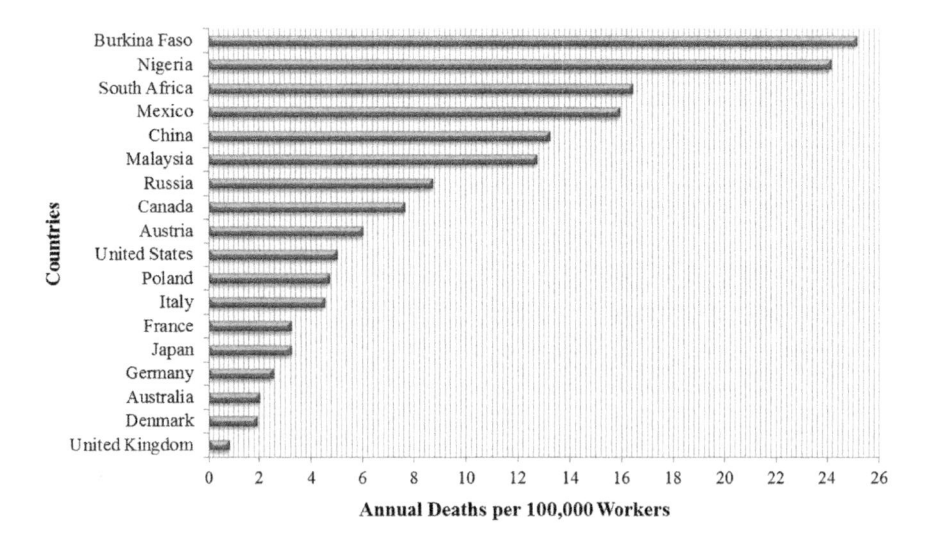

Figure 15.1 Comparison of workplace fatalities for selected countries.

(Data from Abubakar, U., 2015, An overview of the occupational safety and health systems of Nigeria, UK, USA, Australia and China: Nigeria being the reference case study. *American Journal of Educational Research*, 3(11), pp. 1350–1358.)

In addition, every society possesses its own peculiar sets of attributes, defined as national culture, which directly or indirectly influence the decision-making process of its dwellers on handling safety matters. While it is typical for most workers to depend on both formal and informal workplace instructions that are affected by the related industrial standards when performing their tasks, studies have also indicated that they tend to incorporate elements of national cultural values to interpret as well as evaluate such tasks (Power et al., 2015). Using the principles of uncertainty avoidance (UA) as a reference case, it can be inferred that most African countries function within low UA cultures, which leads to intuition-based and informal decision-making approaches (Power et al., 2015), which in turn limits their learning from failure (LFF) abilities.

While a multi-level safety culture is an important aspect to look into for the sake of improving safety performance in Africa, the majority of the safety culture studies in the construction industry have used a quantitative approach in which survey methods are widely adopted (Umeokafor, 2018). This method allows researchers to evaluate general features of a very large group or groups. However, since the basic assumptions underlying culture are so deeply rooted, culture research seeks to understand a culture in depth rather than assessing it. Guldenmund (2000) states that culture is typically investigated using qualitative methods such as in-depth interviews and observations, field research, and ethnography. This conclusion should also be applicable for studying safety culture, as it aims to reveal deep basic assumptions (often developed from past events in a country, industry and organisation) and provide rich information about how culture influences safety. From the interpretivist standpoint, this study aims to understand the cultural influence of the individual, organisation, industry and environment (national and international) on safety performance in the construction industry in Nigeria, which had the second highest annual occupational deaths per 100,000 workers in the world (Abubakar, 2015).

Literature review

Evolution of safety: technical and human factors, safety management systems and safety culture

Scholars and practitioners have been making continuous efforts in improving the safety performance of the construction industry. Approaches to reducing workplace incidents and injuries have progressed through a few discernible historical ages. Hale and Hovden (2003) identified that safety has evolved through three "ages of safety": First was the technical age, which lasted from the 19th century until after the Second World War. The technical age was concerned with developing technical measures to guard machinery, stop explosions and prevent structures from collapsing. Second was the human factors age, which was ushered in by the realisation that a technical approach to safety could not resolve all problems. Meanwhile, early research into personnel selection, training and motivation as prevention measures informed an alternative approach to safety, i.e. human-focused. From the 1960s to 1970s, the two approaches were merged in the human factors age, and it was recognised that the main source of incidents was human error arising from the interaction between human and technical factors. The third age was the management systems age, which started to develop from 1980s onward after it was realised that matching the individual to technology was not successful in resolving all safety problems either. The management systems age focused on management principles and practices applied to safety and gave considerations to organisational factors, e.g. safety-related policies and procedures, responsibilities, planning, implementation, monitoring and review.

Hudson (2007) proposed slightly different stages in the evolution of health, safety and environment as shown in Figure 15.2, which illustrates how each of the technology and management systems approaches reduces incidents and then reaches a plateau in safety improvement. Hudson (2007) argued that after the "systems stage", a cultural approach should be used to drive safety improvement. This was based on the recognition that people, rather than technology or systems, were the missing component in organisations' safety management processes (Hudson, 2007). It is expected that in a positive safety culture, individuals would be intrinsically motivated to develop a strong commitment for the implementations of formal safety policies and plans to achieve high safety standards.

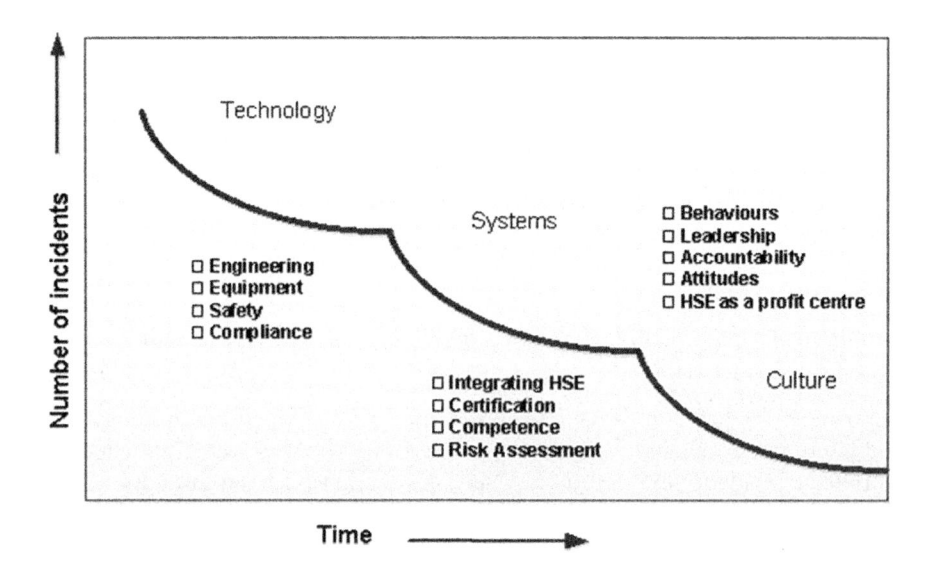

Figure 15.2 Evolving stages of health, safety and environment.

(From Hudson, P., 2007, Implementing safety culture in a major multi-national, *Safety Science*, 45(6), p. 700.)

Safety culture

The term "safety culture" has gained attention since it appeared in investigation reports of major safety catastrophes, e.g. the Chernobyl nuclear accident (IAEA, 1986), the Piper Alpha oil platform explosion in the North Sea (Hidden, 1989) and the Clapham Junction rail disaster (Cullen, 1990). Inquiries into those disasters revealed serious problems inherent in the prevailing organisational cultures, which were claimed to create the preconditions that allowed the accidents to develop and ultimately occur. An organisation's prevailing culture has fundamental influence on safety outcomes because it determines how safety is prioritized in an organisation and how organisational members enact safety systems. It permeates all parts of an organisational system and influences safety through various mechanisms. Reason describes culture as the "engine" that drives the organisational system towards the goal of sustaining the maximum resistance towards hazards (Reason, 1998).

However, the concept of safety culture has been inconsistently defined. Researchers have studied safety culture using different perspectives. For example, some definitions treat safety culture as an entity that an organisation either has or does not (Hale, 2000). It is assumed that that the existence of such an entity ensures that an organisation performs well in safety, otherwise the organisation will perform poorly. Consistent with this perspective, the UK Health and Safety Executive's (HSE) Advisory Committee on the Safety of Nuclear Installations (ACSNI, 1993) offers a widely accepted definition of a safety culture as "the product of individual and group values, attitudes, competencies, and patterns of behaviour that determine the commitment to, and the style and proficiency of, an organisation's health and safety management". However, such a perspective has been questioned by researchers, who raise concerns about the idea of separating safety culture from the broader operating culture of an organisation. For example, Guldenmund (2000) questioned whether in fact there is such an entity or object that we can call safety culture.

Alternatively, safety culture is viewed as the outcome of the broader organisational culture. This view assumes that organisational cultures have characteristics that impact how safety is prioritized and enacted. Guldenmund (2000) claims that the basic assumptions underpinning the operation of an organisation are from the underlying organisational culture and have a profound influence on the effectiveness of safety management. For example, the way that safety is understood and enacted within an organisation is shaped by basic assumptions about what causes incidents. Some basic assumptions may not be specifically related to safety but still have an impact on safety. For example, a basic assumption in an organisation that written rules and procedures are futile is not directly relate to safety but will impact how individuals respond to safety rules and procedures (Guldenmund, 2000). From this view, safety culture is defined as "those aspects of the organisational culture which will impact on attitudes and behaviours related to increasing or decreasing risk" (Guldenmund, 2000, p. 251). Similarly, Antonsen (2009) argues "there is no such thing as a 'safety culture' but rather there are different traits of larger organisational culture that can affect the organisations' safety levels" (p. 184).

Safety cultural and safety climate

Safety culture and safety climate are two related but theoretically different concepts. Safety culture is a multi-level phenomenon. Based on Schein's (2010) organisational culture model, Glendon and Stanton (2000) suggested that cultural influences on safety are manifested at three different levels: (1) at the deepest level, as mentioned earlier, the basic assumptions underlying the operation of an organisation influence safety; (2) at the intermediate level, the beliefs and espoused values regarding safety that have a safety impact; and (3) at the surface level, observable artefacts (e.g. safety policy documents, rules and procedures, supervisory actions) that are the most evident expression of cultural influences on safety. Guldenmund (2000) suggested that the outer layers (i.e. the beliefs and values, and artefacts) are rooted in and logically flow from the basic assumptions underpinning the organisational culture. The basic assumptions are so deeply rooted and the 'truth' about them is so self-evident that it is very difficult for people who hold such assumptions to recognise or express them (Guldenmund, 2000). However, the beliefs and values as well as the artefacts can be revealed by measuring workers' attitudes or perceptions. It is commonly accepted that the safety-related aspects of the two outer layers are reflected in the safety climate (Clarke, 2000; Glendon and Stanton, 2000; Guldenmund, 2000). In other words, safety climate captures the "surface features of the safety culture discerned from the workforce's attitudes and perceptions at a given point in time" (Flin et al., 2000, p. 178). It describes a "snapshot" assessment of cultural influences on safety that prevail in an organisation (Mearns and Flin, 1999).

The need to understand safety culture using a qualitative approach

Culture is developed over a long period of time and is relatively enduring, while safety climate is a temporary state and subject to change. Capturing the state of the safety climate provides important information about what is happening in an organisation at a specific point in time, but the exploration of culture can explain why safety is enacted in a specific way.

Safety climate is often assessed using quantitative approaches, e.g. measured with a questionnaire designed to capture individuals' attitudes and perceptions in relation to specific aspects of safety culture. Measuring the safety climate assumes that safety culture is expressed in different ways within an organisation. Numerous efforts have been made by researchers to identify aspects or components of safety culture for safety climate measurement in the construction industry (see for example Fang et al., 2006, and Larsson et al., 2008).

In contrast, due to that the basic assumptions underlying culture are so deeply rooted, culture research seeks to understand a culture in depth rather than assessing it. Safety culture studies aim to reveal deep basic assumptions (often developed from past events in an organisation) and provide rich information about an organisation's cultural influence on safety. Guldenmund (2000) states that culture is typically investigated using qualitative methods such as in-depth interviews and observations, field research and ethnography.

Current status of safety culture in Nigeria

National safety culture in Nigeria

Knowledge of the effects of national culture on safety performance is quite crucial to the understanding of how global safety policies might perform in multi-cultural and cross-cultural environments. While it is typical for most workers to depend on both formal and informal workplace instructions when performing their tasks, studies have also indicated that they tend to incorporate elements of national cultural values to interpret as well as evaluate such tasks (Power et al., 2015). Using the principles of uncertainty avoidance (UA) as a reference case, it can be inferred that most African countries function within low UA cultures, which leads to intuition-based and informal decision-making approaches (Power et al., 2015), which in turn limits their LFF abilities. Using the largest economy and most populated nation in Africa (i.e. Nigeria) as a case study, Abubakar (2015) found that the core national culture issues responsible for the dwindling safety performances can be summarised into the following categories:

1. National financial commitment to regulatory agencies
   Table 15.1 indicates that the Nigerian Ministry of Labour and Productivity (NMLP), which is responsible for regulating health and safety performance across the country, is grossly underfunded when compared to regulators in some of the countries presented in Figure 15.1. To further conceptualise the magnitude of the disparity, the United Kingdom's HSE, which performs

Table 15.1 Comparison of the Nigerian safety regulatory agency to other countries

Country	National regulatory agency	National workforce	National financial commitment
United Kingdom	HSE	32.2m [3]	£150m [1]
United States of America	OSHA	156m [4]	£433m [5]
Australia	SWA	12.4m [4]	£18.3m [6]
Nigeria	NMLP	55m [4]	£11m [2]

Source: 1 = The Health and Safety Executive (HSE), 2014, Annual report and accounts 2013/14, available at: www.hse.gov.uk/aboutus/reports/1314/ar1314.pdf; 2 = Federal Government of Nigeria, 2014, 2014 FGN budget proposal – Summary by MDAs, available at: www.budgetoffice.gov.ng/pdfs/2014_budget_proposals/23.%20Summary_Labour%20&%20Prod.pdf; 3 = Office for National Statistics, 2014, Statistical Bulletin: Labour market statistics, available at: www.ons.gov.uk/ons/dcp171778_363998.pdf; 4 = The World Fact Book, 2014, Country comparison: labour force, available at: www.cia.gov/library/publications/the-worldfactbook/rankorder/2095rank.html; 5 = Occupational Safety and Health Administration (OSHA), 2014, Commonly Used Statistics (FY 2014), available at: https://www.osha.gov/oshstats/commonstats.html; 6 = Safe Work Australia, 2015, Agency Resources and Planned Performance (2013/14). Available at: https://docs.employment.gov.au/system/files/doc/other/safe_work_australia_pbs_2014-15.pdf. Accessed September, 2015.

a similar function as NMLP, had 2,621 employees in 2014 and a budget of approximately £150m (The Health and Safety Executive, 2014). In the same year, NMLP only managed £11m (Federal Government of Nigeria, 2014) despite Nigeria's labour force exceeding that of the UK by approximately 42%. This invariably indicates that NMLP is significantly incapacitated and neither able to monitor safety-related activities across all worksites nor influence the abilities of workers to learn from previous failures.

2. Adequacy of technical safety and risk management expertise

Traditionally, operational management of health and safety in most countries is overseen by inspectors. Currently, the most universally accepted index for measuring the adequacy of factory inspectors in any given country as recommended by the International Labour Organisation (ILO) is the national workforce size per inspector ratio (NWSPIR). There is no definitive scale for NWSPIR that would guarantee successful safety performance, owing to the influence of other factors such as the number of workplaces to be considered by the regulators, physical sizes and spread of the workplaces, types of hazards to be managed, and the maturity level of the particular country in health and safety management. However, ILO recommends a minimum NWSPIR of 10,000/1 and 40,000/1 for industrial market–based economies and developing nations respectively. According to data from 31 March 2014, UK HSE had 3,081 employees (45% of which are inspectors) responsible for a workforce of 32,000,000 spread over 242,495 km^2. NMLP in the same period was estimated to have approximately 5,500 employees (including inspectors, administrators, apprentices, accountants, engineers, surveyors, drivers, security guards, etc.) to monitor the activities of the 55,000,000 workforce across 923,768 km^2 (The World Fact Book, 2014).

212 Cheung et al.

3. Efficiency of the legal process

    A legal framework is established to offer safety inspectors the authority they need to lawfully execute their responsibilities as well as prescribe commensurate penalties to defaulters. In Nigeria, the National Industrial Court of Nigeria (NICN) was established through the National Industrial Court Act of 2006 to serve as the sole custodians of health and safety laws. Unfortunately, the efficiency of NICN is far from impressive owing to the extremely long life cycle of dispute resolutions. Previous studies such as Amadi (2009) indicate that a typical Nigerian court takes more than five years to complete a case. A superficial comparison of the performances of NICN and the US Occupational Safety and Health Regulatory Commission (OSHRC) in Figure 15.3 indicates that the NICN is 66% less efficient. These delays are closely related to delayed or lack of accurate evidence from workplaces due to inadequate safety inspectors.

4. Accidents and incidents history

    A fundamental success factor for LFF is the availability of accurate historical data of events, as these are often used to reconstruct scenarios that led to initial occurrence, so as to generate effective mitigating actions. Unfortunately, data availability and accuracy has always posed challenges on countries especially amongst the African countries including Nigeria (Mbakwe, 2011).

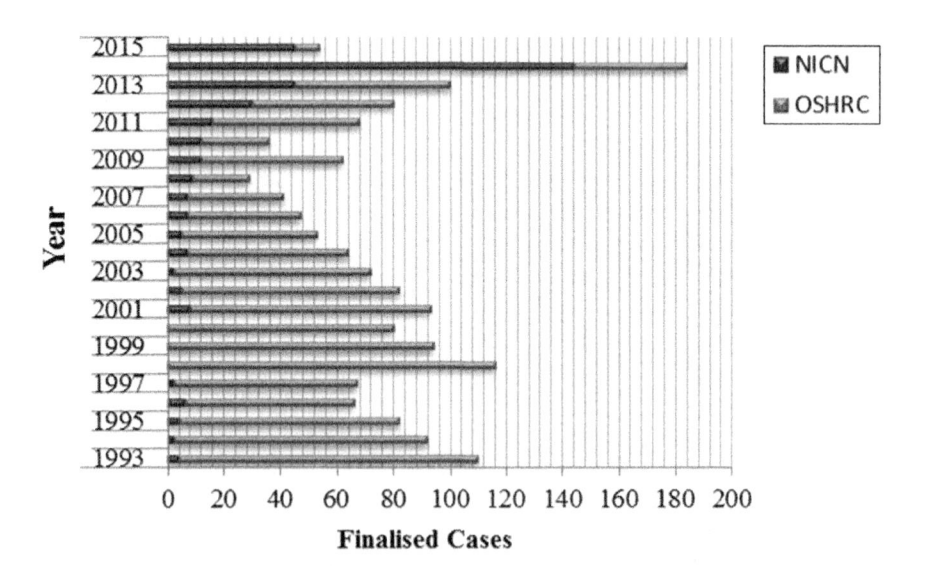

Figure 15.3 Comparing the performance of NICN (Nigeria) and OSHRC (USA) in terms of finalised cases between 1993 and 2015.

(Data Source: http://judgment.nicn.gov.ng/courtruling.php and www.oshrc.gov/decisions/).

Industrial safety culture in Nigeria

Nigeria has been described as one of the fastest urbanising nations within the sub-Saharan Africa region (Oluwakiyesi, 2011), which makes its construction industry immensely vital to economic development (Aibinu and Odeyinka, 2006). Nigeria's construction industry is highly multi-cultural and lacks a unique national identity, based on the premise that it was originally modelled after the British system but is significantly influenced by Germany, China, France and Italy (Aibinu and Odeyinka, 2006; Idoro, 2004). In fact, virtually all of the industrial safety regulations have their origins from foreign countries. For instance, the very popular 1987 Factory Act, 1988 Control of Substances Hazardous to Health Regulations, 1992 Personal Protective Equipment at Work Regulations, and 1999 Management of Health & Safety at Work Regulations were all adopted from British laws, while the 1970 Occupational Safety and Health Act was adopted from the US Department of Labour (Belel and Mahmud, 2012). Similarly, more recent laws such as the 1992 Manual Handling Operations Regulations as well as the 1994 New Design and Management Regulations all have their roots from different foreign countries (Belel and Mahmud, 2012). While this fusion of laws has brought about some degree of good practices over the decades, a combination of the very wide disparity that is inherent in the safety cultures of individual countries, especially China (see Figure 15.1 for safety performance) and the absence of a well-developed national safety framework will continue to pose great challenges to practical safety excellence in Nigeria.

Organisational safety culture in Nigeria

While the influence of different safety cultures within the Nigerian construction industry has been discussed on the basis of its negative impacts, there also seems to be a lost opportunity with regard to capitalising on some of its strengths. For instance, cement is one of the most vital construction materials in Nigeria and Africa as a whole, and some of the large cement manufacturing organisations have a well-established industrial safety framework. One of such organisations is LafargeHolcim PLC, a global leader of building materials manufacturing. In LafargeHolcim, one of the three core areas of its code of business conduct is health and safety (Schönsleben et al., 2016; Labib and Read, 2015), which is perhaps why it has very stringent and functional in-house safety standards/advisories such as work at height (WAH), personal protective equipment (PPE), accident investigation and near miss reporting (AINMR), contractor safety management (CSM), energy isolation (EI), hot work, journey management, etc.

Conceptual model

The proposed conceptual model for enhancing safety performance in Nigeria is based on the concept of LFF originally developed by Labib and Read (2015) but has been enhanced in the current study to incorporate more health and

214 Cheung et al.

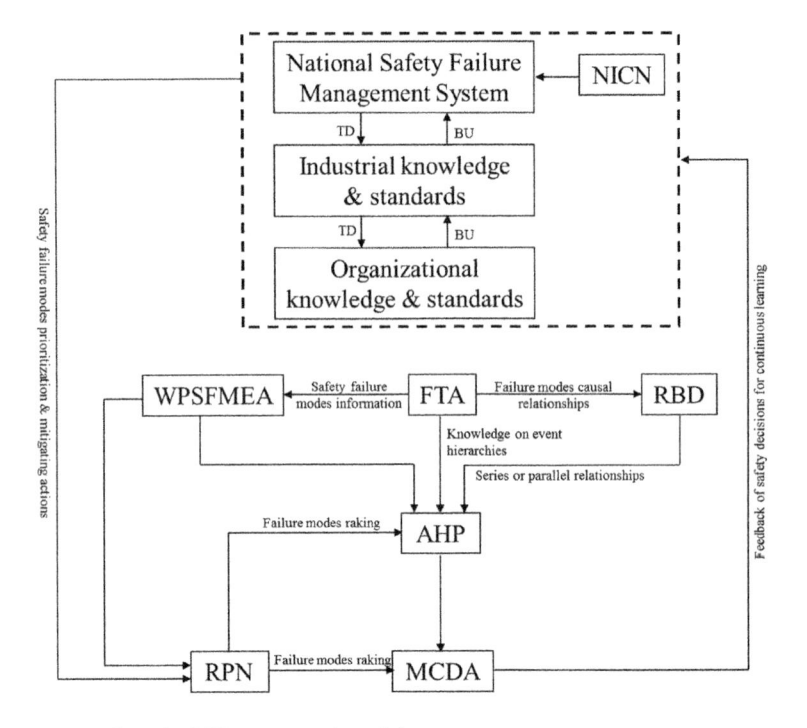

Figure 15.4 An LFF conceptual model.

safety-specific parameters (see Figure 15.4). Just as in the original, the current model is also a fusion of several top-down and bottom-up tools including failure ranking, fault tree analysis (FTA), workplace safety failure modes effects analysis (WPSFMEA), reliability block diagrams (RBDs), analytical hierarchy process (AHP) and multiple criteria decision making (MCDM). The fundamental rationale behind the fusion of several tools lies in the fact that it paves way for the integration of national regulations (top-down approach) and technical expertise from within the industries (bottom-up approach). It is also envisaged that the harmonised framework could improve robustness, whereby the limitations of certain tools might be compensated by the strengths of others and vice versa. The proposed LFF model is implemented in the following stages:

Step 1: Owing to the multi-cultural influence on safety practices within the Nigerian construction industry, the model commences with the development of a universal failure capturing system which involves an integration of well-defined national safety regulations (e.g. from NICN) and the existing best practices within multi-national industries (e.g. LafargeHolcim PLC industrial safety standards).

Step 2: The next stage of the model is to ascertain the root causes of major safety failures using FTA. With complex safety systems whereby several basic events could lead to the top event, it is often difficult to visualise the causal relationships that exist. Hence, the FTA is used to construct a resultant RBD that simplifies the relationship.

Step 3: The determined root causes (often referred to as basic events) from the FTA will then be used to furnish a WPSFMEA so that all stakeholders within the industry develop a common understanding of the general causes of failures.

Step 4: The individual failure modes from WPSFMEA are then ranked according to severities/impacts using pre-defined risk priority numbers (RPN).

Step 5: This stage combines the information obtained from WPSFMEA, RPN, FTA and RBD to design the AHP, which allows for the comparison of alternatives that would later form the basis for decision-making in the MCDM stage.

Conclusions and future research

A key objective of future research in this area would be to gain a fuller understanding of how a multi-level safety culture could inform our understanding on its effect on individual's safety behaviours and performance in Africa. More specifically, the issue will be to examine if the relationships among the individual, organisational, industrial and environmental (national and international) levels of safety culture could influence individual's safety decision-making and related outcomes in construction firms in Nigeria. It is particularly relevant if the national and industrial level of safety culture could affect the formation of safety culture at the organisational level that creates work culture where concern for employees' safety is a priority, even in the face of conflicting demands between productivity and safety. Under this multi-level safety culture influence, individuals could understand the importance of work safety and comply with it accordingly. In addition, articulating the multi-level nature of such a model (see Figure 15.4) would be very important. As Schein (2010) and Glendon and Stanton (2000) have suggested, safety culture is a multi-level phenomenon, although most of the existing studies have focused on the organisational level. Against this backdrop, the proposed LFF multi-level model could address the limitation.

Another suggestion for further research is methodological. Although survey designs and independent outcome measures have been commonly used to quantify the status of safety culture, this quantitative approach, in fact, is often criticised as it fails to depict the reasons that create a certain safety culture. Since the basic assumptions underlying culture are so deeply rooted, culture research seeks to understand a culture in depth rather than assessing it. Under the circumstances, Guldenmund (2000) proposed to study culture using qualitative methods. Along with his line of thought, this study proposed to use multiple qualitative method approaches such as individual and focus group interviews or case studies that could yield unique new information and reduce problems associated with common method bias (Podsakoff et al., 2003).

Based on this review, we believe that it would represent an innovative and potentially critical advancement in safety science by constructing a model that could depict the impact of multi-level safety culture on safety outcomes in the construction industry of Nigeria.

References

Abubakar, U., 2015. An overview of the occupational safety and health systems of Nigeria, UK, USA, Australia and China: Nigeria being the reference case study. *American Journal of Educational Research*, 3(11), pp. 1350–1358.

Advisory Committee on the Safety of Nuclear Installations, 1993. *ACSNI Study Group on Human Factors: Third report*, Her Majesty's Stationary Office, London.

Amadi, J., 2009. Enhancing access to justice in Nigeria with judicial case management: An evolving norm in common law Countries. Available at SSRN 1366943.

Antonsen, S., 2009. Safety culture and the issue of power. *Safety Science*, 47(2), pp. 183–191.

Aibinu, A. A. and Odeyinka, H. A., 2006. Construction delays and their causative factors in Nigeria. *Journal of Construction Engineering and Management*, 132(7), pp. 667–677.

Belel, Z. A. and Mahmud, H., 2012. Safety culture of Nigerian construction workers – A case study of Yola. *International Journal of Scientific & Engineering Research*, 3(9), pp. 1–5.

Clarke, S., 2000. Safety culture: Under-specified and overrated? *International Journal of Management Reviews*, 2(1), pp. 65–90.

Cullen, D., 1990. *The public inquiry into the Piper Alpha disaster*. Her Majesty's Stationary Office, London.

Fang, D., Chen, Y. and Wong, L., 2006. Safety climate in construction industry: A case study in Hong Kong. *Journal of Construction Engineering & Management*, 132(6), pp. 573–584.

Federal Government of Nigeria, 2014. 2014 FGN budget proposal – Summary by MDAs. Available at: www.budgetoffice.gov.ng/pdfs/2014_budget_proposals/23.%20 Summary_Labour%20&%20Prod.pdf. Accessed September 2015.

Flin, R., Mearns, K., O'Connor, P. and Bryden, R., 2000. Measuring safety climate: Identifying the common features. *Safety Science*, 34(1–3), pp. 177–192.

Glendon, I. and Stanton, N. A., 2000. Perspectives on safety culture. *Safety Science*, 34, pp. 193–214.

Guldenmund, F. W., 2000. The nature of safety culture: A review of theory and research. *Safety Science*, 34(1–3), pp. 215–257.

Hämäläinen, P., Saarela, K. L. and Takala, J., 2009. Global trend according to estimated number of occupational accidents and fatal work-related diseases at region and country level. *Journal of Safety Research*, 40(2), pp. 125–139.

Hale, A. R., 2000. Culture's confusions. *Safety Science*, 34 (1–3), pp. 1–14.

Hale, A. R. and Hovden, J., 2003. Management and culture: The third age of safety. A review of approaches to organizational aspects of safety, health and environment. In A. M. Feyer and A. Williamson (Eds), *Occupational Injury: Risk Prevention and Intervention*, Taylor & Francis, London.

Hidden, A., 1989. *Investigation into the Clapham Junction Railway Accident*. Her Majesty's Stationary Office, London.

Hudson, P., 2007. Implementing safety culture in a major multi-national. *Safety Science*, 45(6), pp. 697–722.

Idoro, G. I., 2004. November. The effect of globalisation on safety in the construction industry in Nigeria. In *Proceedings of International Symposium on Globalisation and Construction*, Vol. 17, No. 18, pp. 817–826.

International Atomic Energy Agency, 1986. Summary Report on the Post-Accident Review Meeting on the Chernobyl Accident. Safety Series 75-INSAG1, International Safety Advisory Group, International Atomic Energy Agency, Vienna.

Kalejaiye, P., 2013. Occupational health and safety: Issues, challenges and compensation in Nigeria. *Peak Journal of Public Health and Management*, 1(2), pp. 16–23.

Labib, A. and Read, M., 2015. A hybrid model for learning from failures: The Hurricane Katrina disaster. *Expert Systems with Applications*, 42(21), pp. 7869–7881.

Larsson, S., Pousette, A. and Törner, M., 2008. Psychological climate and safety in the construction industry-mediated influence on safety behaviour. *Safety Science*, 46(3), pp. 405–12.

Mbakwe, A. C., 2011. Modeling highway traffic safety in Nigeria using Bayesian network.Morgan State University.

Mearns, K. and Flin, R., 1999. Assessing the state of organisational safety – Culture or climate? *Current Psychology*, 18(1), pp. 5–17.

Nnaji, C., Gambatese, J., Awolusi, I. and Oyeyipo, O. 2017. Construction safety innovation awareness and adoption in Nigeria: A mixed method approach. In *Joint CIB W099 and TG59 International Construction Conference*, Cape Town, South Africa, pp. 361–372.

Office for National Statistics, 2014. Statistical Bulletin: Labour market statistics. Available at: www.ons.gov.uk/ons/dcp171778_363998.pdf. Accessed September 2015.

Oluwakiyesi, T., 2011. Construction industry report: A haven of opportunities. OSHA UD. Commonly Used Statistics, 2014. Available at: www.osha.gov/oshstats/commonstats.html. Accessed September 2015.

Podsakoff, P.M., MacKenzie, S.B., Lee, J.Y. and Podsakoff, N.P. 2003. Common method biases in behavioral research: A critical review of the literature and recommended remedies. *Journal of applied psychology*, 88(5), p. 879..

Power, D., Klassen, R., Kull, T. J. and Simpson, D., 2015. Competitive goals and plant investment in environment and safety practices: Moderating effect of national culture. *Decision Sciences*, 46(1), pp. 63–100.

Reason, J., 1998. Achieving a safe culture: Theory and practice. *Work & Stress*, 12(3), pp. 293–306.

Schein, E. H., 2010. *Organisational Culture and Leadership* (4th ed.), Jossey-Bass, San Francisco, CA.

Schönsleben, P., Friemann, F. and Rippel, M., 2016. Managing the socially sustainable global manufacturing network. In *IFIP International Conference on Advances in Production Management Systems* (pp. 884–891), Springer, Cham.

The Health and Safety Executive (HSE), 2014. Annual report and accounts 2013/14. Available at: www.hse.gov.uk/aboutus/reports/1314/ar1314.pdf. Accessed September 2015.

The World Fact Book, 2014. Country comparison: labour force. Available at: www.cia.gov/library/publications/the-world-factbook/geos/ni.html. Accessed September 2015.

Umeokafor, N. I., 2018. Construction health and safety research in Nigeria: Towards a sustainable future. In *Joint CIB W099 and TG59 International Safety, Health, and People in Construction Conference*, Salvador, Brazil, pp. 213–221.

16 Overview of safety behaviour and safety culture in the Malaysian construction industry

Mazlina Zaira Mohammad and Bonaventura H. W. Hadikusumo

Summary

Malaysia has seen rapid growth in construction in the 2000s. The Malaysian construction industry is heavily dependent on foreign labour. Indonesians, Filipinos, Nepalese and Bangladesh labourers are commonly hired in Malaysia. Undeniably, Malaysian construction safety performance has not improved due to unsafe behaviour and a weak safety culture. Awareness of safe behaviour among labourers varied and the social impacts were related to labourers' health and safety, productivity and social well-being. There is a lack of systematic reviews for understanding the concepts of safety behaviour and safety culture as applied in the Malaysian construction industry. Hence, concepts related to specific safety behaviour and safety culture issues in the Malaysian construction industry have been explained to promote and enhance good practices of occupational safety and health (OSH) at the workplace. The main source of information discussed in this chapter is a case study. The results from site observations found that the safety practices conducted with empowerment and two-way communication with labour led to zero accidents and allowed labourers to work safely. The case study's observational findings can help safety personnel establish a safe working environment and increase productivity at construction projects.

Introduction

Since the 1980s, the growth of the Malaysian construction industry has contributed great economic value in terms of Malaysia's employment and developmental plans. The construction company always aims for projects to be completed on time, leading to hectic work schedules, which make daily construction site operations more dangerous. The construction industry has dynamic and complex working environments. The complexity of work activities and procedures remain unchanged and are critical in order to maintain the safety and health of labourers. On top of that, the construction building design itself involves architectural, structural, electrical and mechanical work exposed to hazards in the workplace, leading to a high likelihood of accidents and incidents. In addition to labourers in an unsafe working environment, unsafe actions and behaviours are some of the main causes of accidents (Zhang & Fang, 2013).

The Department of Occupational Safety and Health (DOSH) is the authority in charge of humanitarian and environmental protection across all industrial sectors in Malaysia. Based on a statistics report by DOSH Malaysia, the construction industry has the highest number of fatalities in the workplace in the country. Figure 16.1 shows the occupational accidents statistics by sector for the year 2017. Although the manufacturing sector contributes the highest total number of the occupational accidents including non-permanent disability (NPD), permanent disability (PD) and death (D), with a total of 2178 cases, only 68 death cases were recorded as compared to the construction sector with a death toll of 111 cases. Table 16.1 shows the number of NPD, PD and D cases per sector for 2017.

An investigation must be conducted after an accident occurs to identify the cause. The employer is always asked common questions by DOSH, such as "Has the employer checked the workplace recently?" "How does the employer make sure the safety of their workplace?" and "How often does the employer perform an inspection to secure their workplace from any unwanted incidents or accidents?" Accidents do not just happen. Accidents are caused by those who are not doing a job properly or who tend to take shortcuts somewhere along the process to accelerate the work. Occasionally, workers do things that they are not supposed to do and in a field in which they are not well trained. It is a well-established hypothesis that safety behaviour and safety culture are the key factors to improve safety at the workplace, as stated in previous research by Ismail et al. (2012) and Misnan et al. (2008). This chapter reviews this aspect in the context

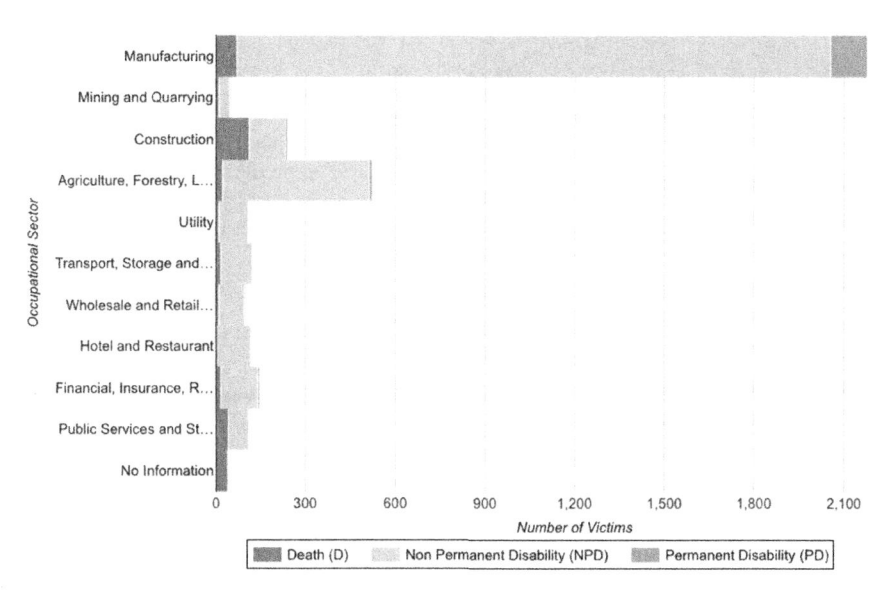

Figure 16.1 Total occupational accidents for all sectors in Malaysia for 2017. (From DOSH, 2018.)

Table 16.1 NPD, PD and D for occupational accidents in Malaysia for 2017

Sector	NPD	PD	Death	Total
Manufacturing	1985	125	68	2178
Mining and quarrying	37	1	8	46
Construction	123	6	111	240
Agriculture, forestry and fishery	488	11	23	522
Utilities (electricity, gas, water and sanitary service)	90	4	10	104
Transport, storage and communication	105	1	16	122
Wholesale and retail trades	86	1	10	97
Hotels and restaurants	110	1	3	114
Finance, insurance, real estate and business services	124	6	16	146
Public services and statutory authorities	64	0	2	66
No information	0	0	0	0
Total	3212	156	267	3635

Note: NPD, non-permanent disability; PD, permanent disability.

of Malaysian construction industry conditions. The knowledge gap is to compare practices on site with previous research findings. Hence, the objective of this chapter is to define and describe worker safety behaviour and safety culture in the Malaysian construction industry in an overview based on a selected case study.

Perspectives and concepts

In the early 1930s, Heinrich found that 88 percent of accidents are caused by unsafe acts of labour, followed by 10 percent unsafe conditions. Two percent of accidents are unavoidably known by 'act of God' (Seo, 2005). Forty years later in the early 1970s, Bird's loss control theory identified basic causes of accidents at workplaces (Abdelhamid & Everett, 2000). The theory posits that the lack of management control and poor management decisions are main causes of accidents. Figure 16.2, which draws on the accident causation study by Abdelhamid and Everett (2000), shows basic causes of accidents (e.g. lack of management control). Unsafe acts and unsafe conditions can also be triggered by the basic causes, which then lead to the occurrence of an accident. This shows the relevance of safety behaviour as one of the focuses to overcome accidents.

In essence, unsafe acts of a person or unsafe workplace conditions are considered as the primary causes of accidents. Wrong use of equipment, negligence during work, failure to warn others of danger and working without authority are described as unsafe acts. Hamid, Majid and Singh (2008) stated that using defective tools, poor ventilation and lighting in the workplace, and inadequate or missing machine guards are examples of unsafe conditions in the workplace.

```
┌─────────────────────────────────┐
│         Basic causes            │
├─────────────────────────────────┤
│ • Lack of management control    │
│ • Personal or job factor        │
│ • Environment factor            │
└─────────────────────────────────┘
        │              │
        ▼              ▼
 ┌────────────┐  ┌──────────────────┐
 │ Unsafe act │  │ Unsafe condition │
 └────────────┘  └──────────────────┘
        │              │
        ▼              ▼
┌─────────────────────────────────┐
│            Accident             │
├─────────────────────────────────┤
│ • Personal injury               │
│ • Property damage               │
│ • Damage to environment         │
└─────────────────────────────────┘
```

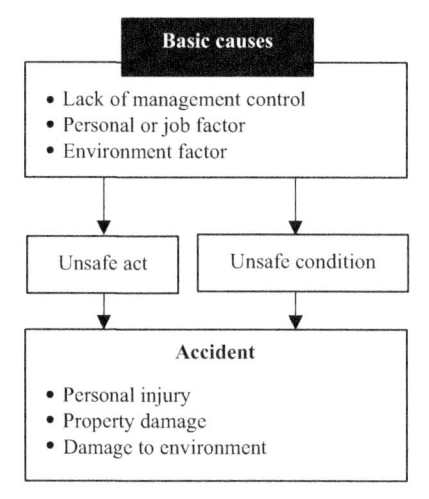

Figure 16.2 A depiction of accident causation.

In addition, safety behaviour is interrelated with the development of safety culture. Based on Geller's safety triad theory, total safety culture involves three domains: behaviour factors, person factors and environment factors (Geller, 2001). These domains are dynamically interactive, as changes within one domain's factor ultimately affects the others. An example for person factor is whether the labourer attended safety training on housekeeping and practicing the concept on site shows they have safety knowledge. This is indirectly associated with the behaviour factor, such as influencing the labourer's attitude. For instance, actively caring for and always cleaning a site for everyone's safety may result in an environmental factor such as a safe layout.

In the Malaysian construction industry, there have been several research works associated with safety behaviour and safety culture (Abdul Rahim, Zaimi & Singh, 2008; Aziz, & Osman, 2019; Ibrahim et al., 2018; Ishak & Azizan, 2018; Majid, 2010; Ismail, Doostdar & Harun, 2012; Nawi et al., 2017). The Malaysian construction industry's labourers are typically not local. A big challenge is to develop the safety culture at the workplace with various nationalities (Mohammad & Hadikusumo, 2015).

This chapter presents a comprehensive review of the literature and case study of three construction companies on safety behaviour and safety culture in the Malaysian construction industry. The main focus of this chapter is to understand the perspectives of safety behaviour and safety culture in the Malaysian construction company, and to provide a better insight on recent construction safety practices that encourage companies to move forward. The results of this literature synthesis and observation case study will help form ideas for applying the safety concept on what is needed in order to improve safety performance in the construction industry.

Methodology

This chapter has adopted a narrative review approach and is based on case study observations. Figure 16.3 shows a summary of the three sources from which information was gathered to give an overview on safety behaviour and safety culture in the Malaysian construction industry. The first source is for statistical data information. The recent statistical data is required to review the current construction industry situation. Some of the relevant secondary data was obtained from Malaysian government websites such as data.gov.my, and from safety-related regulatory bodies websites such as the official DOSH site.

The second source is the full text of other studies. Research journals were selected and collected through a database search of Google Scholar and ScienceDirect. The key words *safety behavior, safety culture, construction industry* and *Malaysia* were used when searching the journals. Journal selection was filtered by recent years, from 2000 to 2018. Geographically, the research locations are all in Malaysia. A number of research papers related to safety behaviour and safety culture were also reviewed.

The Occupational Safety and Health Act (OSHA) 1994 is the main Malaysian practice safety act regulation. The construction industry should comply with OSHA 1994 regulations. Based on OSHA 1994, safety and health officers (SHOs) are required in the construction industry for a project value of more

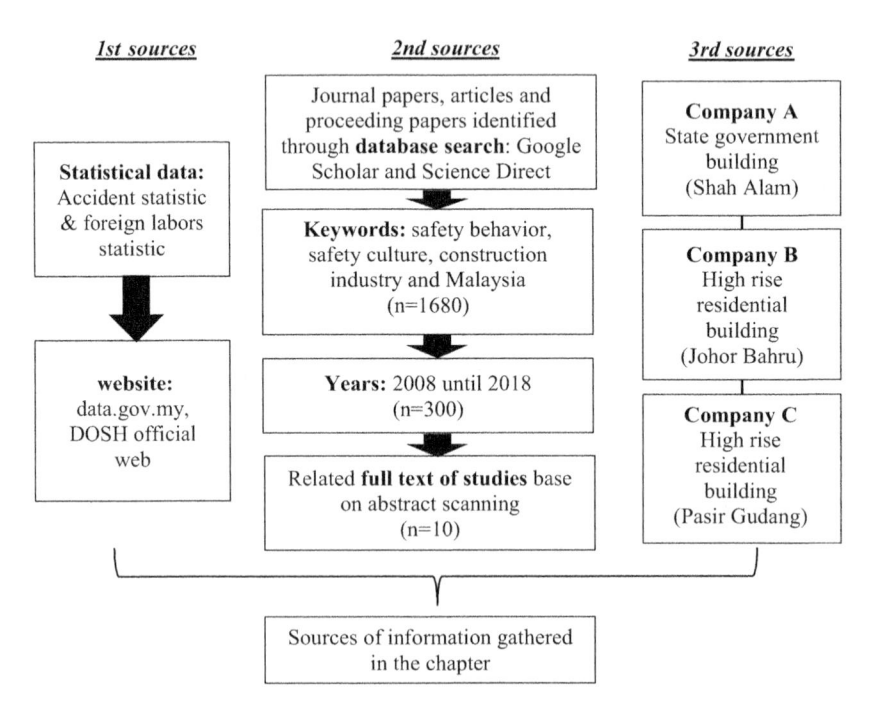

Figure 16.3 Summary of information gathered.

than 20 million Malaysian ringgit, which is classified as a Grade 7 (G7). Hence, a company under G7 has been selected for the case study. The third source includes three construction companies selected for the case study, labelled as Company A, Company B and Company C for private and confidential purposes. These construction projects were selected at different locations including one in Selangor state (central part of Peninsular Malaysia) and two in Johor state (southern part of Peninsular Malaysia).

The cases were evaluated based on two groups of safety behaviour: top management and labourers. There are three selected criteria to be observed for top management, which are leadership, management commitment and budget for safety management. There are six criteria under the labourers group to be observed, which are wearing personal protective equipment (PPE), reward or penalty system, conducting toolbox meeting, safety signage, housekeeping practice and safety training. The criteria have been chosen based on the previous research findings and the common safety effort practiced in the Malaysian construction industry.

The author (Mohammad) visited each construction site project for one whole day. Some of the basic information such as total labour on the site, basic safety practices provided and other necessary information were asked from the safety officer and site safety supervisor (SSS). These two safety personnel were competent persons registered under DOSH.

Results and discussion

This section describes the summary from comprehensive literature, results of onsite case study observations, and discussion on the findings from the observations. Safety behaviour and safety culture are discussed. Definitions and descriptions will follow.

Comprehensive overview of previous research

A few studies have been associated with safety behaviour and safety culture in the Malaysian construction industry. Hui-Nee (2014) indicated the high number of accidents is due to lack of safety culture and non-compliance of the requirements of OSHA 1994. The objective of Ismail and colleagues's (2012) study was to identify the behavioural factors of safety culture among Malaysian construction companies. Their study described OSHA 1994 as a self-regulation enforcement reflective to promote safety culture. The goal of safety performance is to achieve zero hazards in the construction project (Ishak & Azizan, 2018). This goal is in parallel with the authority bodies' (National Institute Occupational Safety and Health [NIOSH] and DOSH) targets by strengthening the safety culture.

According to research studies by (Ibrahim et al., 2018; Majid, 2010), similar findings have shown that management commitment contributes to the implementation of safety practices. Without management support, safety practices cannot be implemented because they are required for the company to provide a safe working environment, as in OSHA 1994 under Section 14 makes safety the responsibility of employers towards employees. Basic safety practices such as providing PPE, conducting toolbox meetings, updating safety boards and installing safety signboards are compulsory practices to comply with Building Operations and Works of Engineering Construction (BOWEC) under the Factories and Machinery Act (FMA) 1967.

Hamid, Majid and Singh (2008) performed research related with the Malaysian construction sites. They classified causes of accidents into six factors: unsafe equipment, job site conditions, unique nature of the industry, unsafe method, human element and management. The findings exhibited that unsafe methods (incorrect procedure, knowledge level and fail to obey work procedure) and human elements (negligence, body effort, experience, PPE, self-esteem/motivation, attitude) were the most common. These factors are related to the negative attitudes of labour as mentioned by Chong and Low (2014).

Most of the safety professionals notice that a good safety program is an attempt to change behaviour and to encourage safe behaviour. The Malaysian-based research study by Ismail et al. (2012) identified the components to evaluate safety culture, which include leadership, management commitment, safety training, and reward or penalty system. Li and colleagues's (2015) study found that complimenting and encouraging labourers' safety behaviour is preferable to pushing labourers' unsafe behaviour through proactive behaviour-based safety (PBBS) approach.

Case study analysis

The authors conducted certain observations for better overview of safety behaviour and safety culture at three different construction sites. Table 16.2 summarizes the details of the observations for each construction company's site. The companies have been named Company A, Company B and Company C. These three companies were observed based on safety behaviours among both labourers and top management. Safety culture is developed from the overall safety behaviour elements.

The case study was conducted in two states. Company A's site is a state government building project located in Selangor state. Company B and Company C are high-rise residential building projects located in Johor state. The highest number of labourers at a site project was at Company B with 590 labourers, followed by Company C with 410 labourers and Company A with 300 labourers. Accident records are confidential for these companies; however, they provided information for fatality and serious injury accidents but not in detail. Company A and Company C reported no fatal accidents. Unfortunately, Company B had one fatality and also serious injuries.

Detailed descriptions of observations are listed in Table 16.2. According to Fung and colleagues (2005), safety culture demonstrates management and worker values. Therefore, the case study analysis–related safety culture involves safety behaviour of the labour and safety behaviour among top management. Safety behaviour involves

Table 16.2 Case study observation information at three construction sites

Company	A	B	C
Type of project Location | Period | # of labourers	State government building Shah Alam, Selangor | 3 years | 300	High rise residential building Johor Bahru, Johor | 3 years | 590	Pasir Gudang, Johor | 4 years | 410
Labour nationalities	Indonesia 60%, Bangladesh 32%, Pakistan 6%, local 2%	Indonesia 55%, Bangladesh 35%, Nepal 5%, the Philippines 3%, local 2%	Indonesia 45%, Bangladesh 45%, Nepal 7%, local 3%
Accident	No fatal but one serious injury	One fatality and serious injury	No fatal, no serious injury
Safety culture			
Safety behaviour (Labourers)			
Wearing PPE	80% wore PPE	60% wore PPE	90% wore PPE
Reward/Penalty system	Reward	Penalty	Reward & Penalty
Conducting Toolbox meeting	Talk given by the SSS	Talk given by the safety officer	Talk and safety sharing session by the labour's leader
Safety signage	• Basic safety signage visible at the entrance • Some at the site walkway	• Attractive and basic safety signage at the security post	• Basic safety signage visible at the entrance • Some at the site walkway
Housekeeping	• Green housekeeping, 3R concept, 5S concept • The labourers were committed	• Improper material arrangement and storage	• 5S concept • The labourers were committed • Proper layout work site
Safety training	Safety induction once in a year and compulsory for all new labourers to attend the safety training		
Safety culture			
Safety behaviour (Labourers)			
Wearing PPE	80% wore PPE	60% wore PPE	90% wore PPE
Reward/Penalty system	Reward	Penalty	Reward & Penalty
Conducting Toolbox meeting	Talk given by the SSS	Talk given by the safety officer	Talk and safety sharing session by the labour's leader
Safety signage	• Basic safety signage visible at the entrance • Some at the site walkway	• Attractive and basic safety signage at the security post	• Basic safety signage visible at the entrance • Some at the site walkway

(Continued)

Table 16.2 Continued

Company	A	B	C
Housekeeping	• Green housekeeping, 3R concept, 5S concept • The labourers were committed	• Improper material arrangement and storage	• 5S concept • The labourers were committed • Proper layout work site
Safety training	Safety induction once in a year and compulsory for all new labourers to attend the safety training		
Safety behaviour (top management) Leadership	SHO and SSS are • Concerned with the labour • Visible at work site • Committed	• SHO strict with the labour • SHO & SSS are committed	SHO & SSS are • Friendly • Role models • Supportive • Committed • Visible at work site
Management commitment	Safety and health committee established (organization with more than 40 employees)	Private and confidential	
Budget for safety management	1–2% from the total project cost has been allocated		3–4% from the total project cost has been allocated

safety participation and safety compliance (Hedlund et al., 2016). Hence, analysis is based on how well these three companies participate and comply. References from previous researchers (Iyer et al., 2004; Oyewole et al., 2010; Nielsen, 2014; Teo & Ling, 2006; Zaira & Hadikusumo, 2017) have listed various safety aspects related to the safety behaviour of labourers. In this case study, the safety behaviour of labourers is a key aspect and includes the wearing of PPE, reward or penalty systems, toolbox meetings, safety signage, housekeeping and safety training. These selected aspects are based on the references and practices implemented on site in the Malaysian construction industry, which is relevant for the improvement of foreign labourers' safety behaviour. From the observation of wearing PPE, Company C shows the highest percentage, followed by Company A and then Company B. Company C allows empowerment of their labour leaders by letting them conduct a toolbox meeting, but not the other two companies.

A construction site project is known as a 3D sector, meaning dangerous, dirty and difficult. Hence, housekeeping is important in part to observe whether site safety is implemented, because housekeeping will indirectly reduce site hazards and make the site less dangerous, less dirty and less difficult. Housekeeping is also implemented in a different way for each company. Company B does not practice housekeeping at its site. Company A practices the Sort, Set in order, Shine, Standardize and Sustain (5S), and Reduce, Reuse and Recycle (3R) concepts of housekeeping. Site observation findings at this company were neat, proper layout and well-organized materials. This is similar to Company C; however, it does not practice the 3R concept.

Key safety aspects regarding the safety behaviour of top management include the leadership elements of safety personnel and inquiring about the management commitment and safety budget. These aspects are based on references (Nielsen, 2014; Teo & Ling, 2006) and the current practices implemented in the selected three companies. Luria and Morag (2012) also found that interaction about safety between managers and employees is increased and improved with leadership intervention of safety management by walking around. The relationship between safety personnel and labourers for Company C is good because the SHO and SSS are friendly, supportive, committed, visible at the construction site and role models. Company B has different elements in terms of leadership, which is strict with labour but committed. Company A's leadership style shows concern for its labourers, including taking care of their duties and applying the buddy system. The relationship between safety personnel and labourers is also important because communication in delivering safety message is crucial among multinational labourers.

As stated in OSHA 1994 Section 30 and Safety and Health Committee Regulation 1996, it is compulsory for an employer to establish a safety committee if there are more than 40 employees and if DOSH advises them to establish one. Therefore, the company must be responsible for and show commitment to safety management. This legislation indirectly encourages all construction companies to make positive efforts towards improving safety behaviour and strengthening safety culture.

In terms of the information on safety budget, the authors inquired directly to the SHO. This is important to gather the information to evaluate top management's commitment to developing the safety culture. Company C had the highest safety budget with three to four percent of the total project cost allocated. This was followed by Company A with one to two percent; Company B did not reveal its information. Keng and Razak (2014) highlighted that a construction project must be allocated a sufficient safety budget to assist safety committees and safety officers in improving the safety culture at the workplace.

Comparing these three companies, Company C has a better safety behaviour based on the overall elements. The authors believed that Company C has the strongest safety culture, followed by Company A and Company B. Following the aforementioned observational information, Company C had zero serious accidents, with safety commitments which are greater than the other two companies. For instance, in terms of the safety behaviour of labourers, those working for Company C were 90% compliant in wearing PPE, which is the highest. The company also implemented both rewards and penalties for the improvement of labourers' safety behaviour. Empowerment was given to labour leaders during the toolbox meeting, which contributes to trust among the labourers. In terms of safety behaviour among top management, Company C shows that the leadership is closer with the labourers as compared with the other two companies. The safety budget of Company C is also higher than those of Companies A and B.

Per Geller's triad safety theory, communicating is a basic factor under the behaviour domain. According to Mohammad and Hadikusumo (2017), in order to implement safety practices, effective communication among foreign labourers is required. Hence, construction companies should implement safety practices by empowerment and two-way communication with labourers, especially when they are foreigners. The ultimate vision of a safety improvement mission is a total safety culture. A total safety culture is achieved when everyone acts safely and feels responsible for safety practices on a daily basis. Employees should be willing to identify at-risk behaviours and environmental hazards and intervene to correct problems.

Defining and describing safety behaviour and safety culture

Accidents are defined as unplanned or unwanted events that cause injuries, death, property damage or loss of production. It is extremely hard to prevent an accident if there is no understanding of hazards and causes of accidents. Everyone always wants to feel safe and healthy in every workplace. Accident rates could be reduced if there is strong safety behaviour and eventually a safety culture at the workplace.

The authors' point of view on safety behaviour is that individual attitudes and concerns on safety, willingness to do any action safely and a well-established safety mindset allows workers to be always careful. In a construction organization, this cannot be quickly developed in each individual. It is a long-term process which requires an effort from all management levels. Safety behaviour involves only one level, which is directly that of the individual, while safety culture involves all levels of an organization. Labour might cooperate if management is committed to safety implementation. Safety behaviour is considered a subset of safety culture. Hence, safety culture plays a big role in any organization which cares about safety performance.

The meaning of safety culture includes mutual values and beliefs in safety within an organization and is a control system to produce a distinctive behavioural standard. Therefore, everyone must acknowledge the importance of safety and health in their workplace and always remind each other. The perspective of the authors on safety culture is a collective behaviour and shared similar value on safety in any size group of people, especially in an organization.

Poor system and operation, defective or poorly designed equipment, and poor workplace conditions could encourage unsafe behaviour; however, these behaviours are not inevitable. It is difficult to change the attitude and belief of labourers through direct strife, yet by showing an example of acting safely from the management, labourers could begin to always think about safety. This practice has led to the development of a safety behaviour approach. A poor culture of safety and health may lead to weaknesses because problems for people who work the interface may be due to poor communication or training. A poor culture encourages situations in which non-compliance with safe work practices is acceptable, and does not help the organization to take effective action to address health and safety issues.

Remember that safety culture is a long-term process which develops slowly for the fundamental change and requires time. Initiative and leading from the top are required for a safety culture to develop and succeed. Decision makers on safety must be embraced within an organization. Labour should be empowered to challenge unsafe behaviour of others and have a positive attitude towards safety if leaders are able to nurture an environment by developing a culture of belief, and supporting open and informed conversation about safety.

Implications for practice and research

The Occupational Safety and Health Master Plan for Malaysia 2015 (OSH-MP15) had the target that towards the end of 2015 labour in Malaysia should be ready to be at the level of preventive culture. OSH-MP15 implementation is the second phase of the 15-year national strategy and plan for continuous OSH improvement in Malaysian industry. OSH-MP15 was focused on the self-regulation concept, which was executed from 2011 until 2015. The recent master plan for 2016 until 2020 is called OSH-MP20 and is focused on preventive culture. To attain this goal requires cultivating and enhancing a safe and healthy work culture in the Malaysian construction industry. According to DOSH, the cooperation by industry on this OSH-MP20 until 2018 has been positive, with specific safety programs conducted on site. Therefore, in order to achieve this, workplace safety and health culture should be nurtured and considered important to all employers.

NIOSH Malaysia promoted the good practices of occupational safety and health at the workplace by hosting the 18th Conference of Occupational Safety and Health (COSH). "Fostering an OSH Culture at the Workplace" was the theme for the 18th COSH. This shows the efforts of the Malaysian government, employers and employees together emphasizing greater cooperation to reduce workplace accidents and work towards the goal of zero accidents at the workplace.

The working culture and habits of construction labourers are the most typical elements when visiting a construction site. A normal practice when entering a construction site is compulsory PPE. This practice is enforced by DOSH and the Construction Industry Development Board (CIDB) due to legal requirements. Nonetheless, in daily working routines, labourers would forget or ignore basic PPE which they must comply with.

Accidents happen either due to unsafe acts or exposure to hazards. A hazard could be known as an unsafe condition. The definition of hazard is a source or situation with a potential harm in terms of human injury or ill health, damage to property, damage to the environment or a combination of these. Therefore, safety behaviour among labourers is crucial for them to understand hazards, hence labourers are able to identify and avoid hazards. Why do we need to prevent accidents? To prevent employees and the public from being injured. Moreover, accident prevention in the workplace allows the avoidance of legal implications, complies with human rights and is in the interest of business sustainability.

This overview could raise safety awareness and provide useful references for further effective safety precautions and management plans (understanding of safety behaviour and safety culture). These case study observation findings can help safety personnel to improve the working environment and productivity in the Malaysian construction industry. Developing safe behaviour among labourers requires creativity from safety personnel and top management of the construction company to conduct attractive and effective safety programs. The author (Mohammad) has experience at construction sites for the data collection on safety research, and she found that construction companies have different safety practice implementations and safety culture styles on site.

Safety culture is not for the sake of policies or safety management requirements. It must include the willingness for everyone in the organization to have a safety concern. Safety practices such as toolbox meetings by allowing the labourers to do the sharing information session and housekeeping with various concepts and managing a positive relationship with labourers could be components for creating a safety culture. Two general factors also could be considered for the best fit: management's commitment to safety, and labourers' involvement in safety. The elements that create and nurture a safety culture are as follows:

1. All levels are committed to embracing safety.
2. Acknowledge safety as an investment and not a cost.
3. Apply safety as a continuous improvement process.
4. Deliver training and information for all related to safety.
5. Implementing a system to prevent and control hazards.

Future studies may include research on these specific parts, components and elements associated with safety behaviour and safety culture using a quantitative method. The results may further explain the issues in an accurate and precise manner along with statistical data.

Conclusions

The contents of this chapter were compiled using three sources. The first is statistical data from official websites of authorities. The second included journal papers from search engines. The third included research findings described in this section as derived from observations at three different construction sites. The overview discussed and supported three categories of interest: defining and describing safety behaviour and safety culture, comprehensive overview on previous research and a case study.

Safety culture does not involve only one person. It requires effort from an entire organization, including all levels of the department. In order to maintain a safe workplace, all leaders are responsible for developing and shaping a safety culture in their organization, which is a critical role for them. This will be beneficial to the employer, here construction companies, as an established safety culture will improve safety performance and increase the productivity of the project. Moreover, this will also be beneficial to society, as the safety behaviour will encourage the labourers to work safely, which could reduce the number of accidents and reduce ill health and injuries.

Acknowledgement

The authors would like to thank the editors for the opportunity to contribute a chapter to this book.

References

Abdelhamid, T. S. & Everett, J. G. (2000) Identifying root causes of construction accidents. *Journal of Construction Engineering and Management*, 126(1), pp. 52–60.

Aziz, S. F. A. & Osman, F. (2019) Does compulsory training improve occupational safety and health implementation? The case of Malaysian. *Safety Science*, 111, pp. 205–212.

Chong, H. Y. & Low, T. S. (2014) Accidents in Malaysian construction industry: Statistical data and court cases. *International Journal of Occupational Safety and Ergonomics*, 20(3), pp. 503–513.

DOSH (2018) Occupational Accidents Statistics by Sector 2017. DOSH. Available at: http://www.dosh.gov.my/index.php/en/archive-statistics/2017/2003-occupational-accidents-statistics-by-sector-2017

Fung, I. W., Tam, C. M., Tung, K. C. & Man, A. S. (2005) Safety cultural divergences among management, supervisory and worker groups in Hong Kong construction industry. *International Journal of Project Management*, 23(7), pp. 504–512.

Geller, E. S. (2001) *Working safe: How to help people actively care for health and safety*, 2nd edn. Boca Racon, Florida: CRC Press.

Hamid, A. R. A., Majid, M. Z. A. & Singh, B. (2008) Causes of accidents at construction sites. *Malaysian Journal of Civil Engineering*, 20(2), pp. 242–259.

Hedlund, A., Gummesson, K., Rydell, A. & Andersson, M. (2016) Safety motivation at work: Evaluation of changes from six interventions. *Safety Science*, 82, pp. 155–163.

Hui-Nee, A. (2014) Safety culture in Malaysian workplace: An analysis of occupational accidents. *Health and the Environment Journal*, 5(3), pp. 32–43.

Ibrahim, I. I., Noor, S. M., Nasirun, N. & Ahmad, Z. (2018) Favorable working environment in promoting safety at workplace. *Journal of ASIAN Behavioral Studies*, 3(8), pp. 71–78.

Ishak, N. & Azizan, M. A. (2018) A review on the benchmarking concept in Malaysian construction safety performance. *AIP Publishing*, 1930, pp. 20–24.

Ismail, F., Ahmad, N., Janipha, N. A. I. & Ismail, R. (2012) Assessing the behavioral factors' of safety culture for the Malaysian construction companies. *Procedia – Social and Behavioral Sciences*, 36, pp. 573–582.

Ismail, Z., Doostdar, S. & Harun, Z. (2012) Factors influencing the implementation of a safety management system for construction sites. *Safety Science*, 50(3), pp. 418–423.

Iyer, P. S., Haight, J. M., Del Castillo, E., Tink, B. W. & Hawkins, P. W. (2004) Intervention effectiveness research: Understanding and optimizing industrial safety programs using leading indicators. *Chemical Health and Safety*, 11(2), pp. 9–19.

Keng, T. C. & Razak, N. A. (2014) Case studies on the safety management at construction site. *Journal of Sustainability Science and Management*, 9(2), pp. 90–108.

Li, H., Lu, M., Hsu, S. C., Gray, M. & Huang, T. (2015) Proactive behavior-based safety management for construction safety improvement. *Safety Science*, 75, pp. 107–117.

Luria, G. & Morag, I. (2012) Safety management by walking around (SMBWA): A safety intervention program based on both peer and manager participation. *Accident Analysis & Prevention*, 45, pp. 248–257.

Majid, A. (2010) A framework of safety culture for the Malaysian construction companies: A methodological development. *Pertanika Journal of Social Sciences & Humanities*, 18, pp. 45–54.

Misnan, M.S. and Mohammed, A.H. (2007) Development of safety culture in the construction industry: a conceptual framework. Boyd, D. (ed.) *Proceedings of 23rd Annual ARCOM Conference*, 3–5 September 2007, Belfast, UK. Association of Researchers in Construction Management.

Mohammad, M. Z. & Hadikusumo, B. H. (2015) A model of integrated multi-level safety intervention practices in construction industry. *Proceedings CIB W099*, Belfast, 10–11 September 2015, pp. 49–61.

Mohammad, M. Z. & Hadikusumo, B. H. (2017) A model of integrated multilevel safety intervention practices in Malaysian construction industry. *Procedia Engineering*, 171, pp. 396–404.

Nawi, M. N. M., Ibrahim, S. H., Affandi, R., Rosli, N. A. & Basri, F. M. (2017) Factor affecting safety performance construction industry. *International Review of Management and Marketing*, 6(8S), pp. 280–285.

Nielsen, K. J. (2014) Improving safety culture through the health and safety organization: A case study. *Journal of Safety Research*, 48, pp. 7–17.

Occupational Safety and Health Act (OSHA) 1994 is an Act available at http://www.dosh.gov.my/index.php/en/legislation/acts. The web link can be inserted in the sentence, "The Occupational Safety and Health Act (OSHA) 1994 (see http://www.dosh.gov.my/index.php/en/legislation/acts) is the main Malaysian practice safety act"

Oyewole, S. A., Haight, J. M., Freivalds, A., Cannon, D. J. & Rothrock, L. (2010) Statistical evaluation and analysis of safety intervention in the determination of an effective resource allocation strategy. *Journal of Loss Prevention in the Process Industries*, 23(5), pp. 585–593.

Seo, D. C. (2005) An explicative model of unsafe work behavior. *Safety Science*, 43(3), pp. 187–211.

Teo, E. A. L. & Ling, F. Y. Y. (2006) Developing a model to measure the effectiveness of safety management systems of construction sites. *Building and Environment*, 41(11), pp. 1584–1592.

Zaira, M. M. & Hadikusumo, B. H. (2017) Structural equation model of integrated safety intervention practices affecting the safety behavior of workers in the construction industry. *Safety Science*, 98, pp. 124–135.

Zhang, M. & Fang, D. (2013) A continuous Behavior-Based Safety strategy for persistent safety improvement in construction industry. *Automation in Construction*, 34, pp. 101–107.

Part IV
Construction workers' well-being

17 Determinants of risky sexual behaviour by South African construction workers

Paul A. Bowen and Peter J. Edwards

Summary

Risky sexual behaviour is a significant contributor to HIV infection and re-infection. In South Africa, HIV prevalence is among the highest in the world. The construction industry is particularly affected, but little is known about risky sexual behaviour among construction workers and the factors that influence it. To explore this, data were gathered from 512 site-based workers in the Western Cape. A theoretical model postulating the determinants of risky sexual behaviour was proposed and tested using regression analysis and structural equation modelling. The findings indicated that age, gender, acquaintance with an HIV+ person, drug usage, and level of AIDS-related knowledge were direct determinants of risky sexual behaviour, with higher levels of education, ethnicity, and acquaintance with an HIV+ person each predicting better AIDS-related knowledge. Alcohol consumption and drug usage were positively associated. Recommendations for more targeted workplace interventions are proposed.

The prevalence of HIV/AIDS in sub-Saharan Africa

Eastern and Southern Africa have only 6.2% of the world's population but are home to half of the world's people living with HIV. This sub-Saharan region continues to be the hardest hit by the HIV epidemic, with 46% of the world's new HIV infections in 2015 and nearly 40% of those being in South Africa (UNAIDS, 2016). The latest national prevalence, incidence, and behaviour survey indicated that the national HIV prevalence among South Africans in 2012 was 12.2% (6.4 million persons), up from the 2008 national estimate (10.6%, or 5.2 million) (Shisana et al., 2014). The HIV/AIDS pandemic is considered one of the main health challenges facing South Africa (Mayosi & Benetar, 2014). Evidence indicates that the burden of HIV in South Africa is predominantly driven by heterosexual transmission (Fraser-Hurt et al., 2011). The rate of new HIV infections is related to biological vulnerabilities such as low rates of medical circumcision and sexually transmitted diseases (STDs); socio-behavioural factors including multiple and concurrent sexual partners, transactional sex, unprotected sexual intercourse, and alcohol use before sex; and structural factors such as poverty, wealth disparities, and migration (Zuma et al., 2016).

Risky sexual behaviour has been defined by the U.S. Centers for Disease Control and Prevention (CDC, 2010) as behaviour that heightens the risk of contracting STDs (including HIV) and unintended pregnancies. In sub-Saharan Africa, poor and inconsistent use of male condoms has been identified as a key driver of HIV infection (Hargreaves et al., 2007), and also that transactional sex is associated with HIV among women (Wamoyi et al., 2016). Adolescents who consume alcohol are more likely to engage in sex than are their counterparts, to have experienced sexual debut at a younger age, and to engage in sex with multiple partners (Morojele et al., 2013).

A cornerstone in the response to the South African pandemic is the National Strategic Plan (NSP) for HIV, TB and STIs (SANAC, 2017), currently in its fourth iteration. Goal 5 of NSP 2017–2022 proposes 'deeper involvement of the private sector and capacitation of civil society sectors and community networks' (SANAC, 2017: 5). Greater involvement of the private sector is necessitated by the inability of the public health system to adequately provide the requisite care. Gilbert (2006) and Wouters and colleagues (2009) emphasise overburdened healthcare staff and an over-stretched public health system.

Compared with other economic sectors and industries, the construction industry is disproportionately adversely affected by the HIV/AIDS pandemic (Bureau for Economic Research [BER]; South African Business Coalition on HIV/AIDS [SABCOHA], 2004). In their analysis of data collected from 10,243 construction workers between 2002 and 2005, Bowen and colleagues (2008) estimated the industry infection level to be 14%; a level above national prevalence. Despite this increased susceptibility, the sector has been one of the slowest to respond to the HIV pandemic (Meintjes et al., 2007). Notwithstanding considerable research in South Africa regarding risky sexual behaviour, and its determinants, research relating directly to on-site construction workers is sparse. This study, located within the broader public health discourse, addresses that shortcoming. The aim is to gain a better understanding of how demographic characteristics, alcohol consumption, drug usage, and AIDS-related knowledge are associated with the risky sexual behaviours of construction workers.

Risky sexual behaviour and associated factors

Risky sexual behaviour – condom use, promiscuity, and transactional sex

UNAIDS (2013) reports that the correct and consistent use of condoms reduces the risk of sexually transmitted infection (STI)/HIV transmission by over 90%. Shisana and colleagues (2014) noted that, overall, use of condoms at last sexual intercourse occasion increased significantly between 2002 and 2008, but then significantly decreased in 2012 across all age groups and for both genders except among females aged 50 years and older. They found that approximately 1 quarter of all sexually active respondents of 15 years and older had consistently used a condom in the preceding 12 months with their most recent sexual partner, but that just over half of the survey respondents had not used a condom over the same period.

McPhail and Campbell (2001) identified six factors that adversely influence condom use: low perceptions of risk; peer group norms and expectations; relative lack of condom availability (or failure to ensure appropriate availability); adult attitudes and preferences about condoms and sex; male-skewed gender power relations; and affordability of condoms for adolescents. Sexual promiscuity has been found to be a contributing factor to the spread of HIV/AIDS (Smallwood & Venter, 2001), as has sexual concurrency (more than one sexual partner overlapping in time) (Morris & Kretzschmar, 1997; Eaton et al., 2011; Mah & Shelton, 2011; Fox, 2014).

Transactional sex (intercourse with a non-primary partner in exchange for money, housing, or material goods), often driven by a survival imperative, places women at increased risk of HIV infection (Dunkle et al., 2004). The term 'transactional sex' differentiates 'sex work' from the exchange practices embedded in many relationships in contexts outside of the West (Wamoyi et al., 2016). Transactional sex relationships are non-commercial: participants have an intimate relationship, the exchange embedded in these relationships is implicit, and it is not formally negotiated (Stoebenau et al., 2016). Transactional sex is associated with sexual coercion and HIV risk behaviours such as multiple concurrent sexual partnerships (Choudry et al., 2015). Rates of HIV among young South African women in the age group of 15–24 years are disproportionately high, approximately four times that of young men and, in 2007, accounted for 90% of new infections in that age group (Rehle et al., 2007).

Alcohol consumption and drug usage

Drug use and heavy use of alcohol before sexual intercourse are each associated with risky sexual behaviours and with lack of condom use (Eich-Hochli et al., 1998; Shisana et al., 2004; Cook & Clarke, 2005; Parry et al., 2005; Kalichman et al., 2007; Peltzer et al., 2011; Seth et al., 2011).

Morojele *et al.* (2006) and Kalichman *et al.* (2008) highlighted the association between alcohol consumption, sexual promiscuity and lack of condom use at informal drinking establishments (known as 'shebeens') in townships. Shisana and colleagues (2004) and Seth and colleagues (2011) showed a strong link between sexual risk-taking behaviour and alcohol use. Shisana and colleagues (2004) also established a positive link between alcohol consumption and having multiple sex partners, and that condom use was negatively associated with frequency of alcohol use. Kalichman and colleagues (2006) identified alcohol as a major factor in risky sexual behaviour because intoxication often leads to casual sex and inconsistent condom use. They also found that men were more likely than women to have used drugs and to have had multiple sex partners. Kirby and Barry (2012) claim that alcohol represents the 'gateway' drug, leading to the use of tobacco, 'dagga' (marijuana), and other illicit substances. Moreover, alcohol users exhibited a greater likelihood of using both licit and illicit drugs. Van Heerden et al. (2009) note significant associations between male gender and alcohol, tobacco, dagga and other drug use, and that 'coloureds' and 'whites' are more likely than 'blacks' to have used alcohol, tobacco, and other drugs. (It is important to understand that these ethnic categorisations, prescribed under the former apartheid regime in South Africa, are now used in a non-pejorative way to address particular concerns pro-actively and affirmatively among previously disadvantaged population groups.)

AIDS-related knowledge

Knowledge about HIV transmission is a pre-requisite to practising safer behaviours to prevent HIV infections and re-infections. Scott-Sheldon and colleagues (2013), in a study of 'Black' African men drawn from four townships outside Cape Town, found that, for men who had tested HIV-negative, knowledge of HIV transmission was inversely related to sexual risk behaviours.

In a study of adolescents in Botswana, Letamo (2018) found that, despite major interventions aimed at behavioural change, myths and misconceptions about HIV transmission remained. Such misconceptions were associated with having primary or lower level education and with being male. Hong et al. (2012), in a study in rural Kenya, reported low HIV knowledge to be associated with lower education, lower household finances, sex without a condom, and being HIV+. Faimau and colleagues (2016) established that, although more than 90% of college students in Botswana identified routes of HIV transmission correctly, misconceptions regarding HIV/AIDS still existed. These included the belief that people can be infected with HIV because of witchcraft and that only people who have sex with gay or homosexual partners can be infected with HIV. Apart from their independent effects, alcohol consumption, drug usage, and HIV transmission knowledge have a *combined* effect on risky sexual behaviour.

Given the associations between alcohol and drug use, AIDS-related knowledge, and risky sexual behaviour, further investigation of these multi-variate relationships is important not only in terms of understanding the motivation for such behaviour, but also for informing the management interventions provided by employer organisations as their contribution to the public health response to the HIV/AIDS pandemic. These multi-variate relationships have been investigated in this study in the context of workers in the construction industry in South Africa.

A conceptual model of risky sexual behaviour

Following a review of the relevant literature, a *conceptual model* of factors predicting risky sexual behaviour was postulated (see Figure 17.1).

In the conceptual model, it was proposed that demographic, behavioural, and cognitive factors predict risky sexual behaviour. Specifically, it was

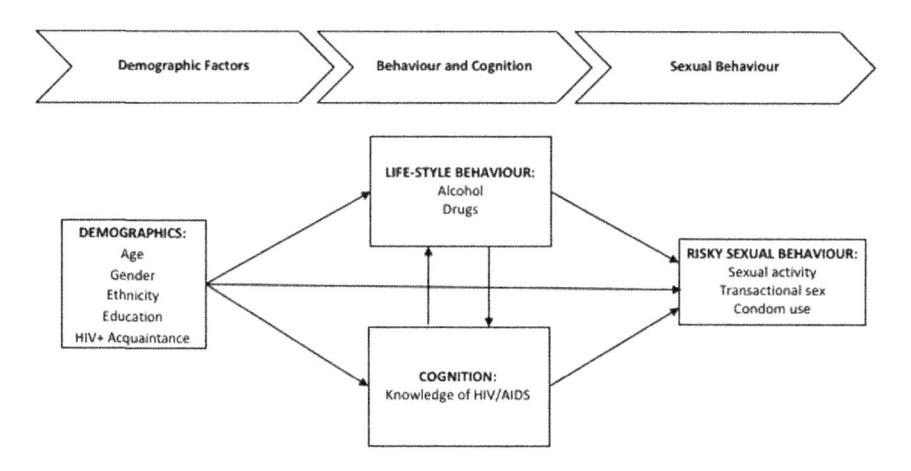

Figure 17.1 Graphical overview of the conceptual model.

proposed that age, gender, ethnicity, level of education, and acquaintance with an HIV+ person could be regarded as exogenous variables and hence do not require explanation. These exogenous variables were hypothesised to explain alcohol and drug use (behavioural factors) and AIDS-related knowledge (cognition). Alcohol and drug usage and AIDS-related knowledge were hypothesised to interact, and each to explain risky sexual behaviour and, together with the exogenous variables, to explain risky sexual behaviour by construction workers. Multiple sex partners, transactional sex, and condom use were used to represent risky sexual behaviour.

Research method

Participants and setting

Ethical clearance was obtained from the University of Cape Town. The survey was undertaken on 18 construction sites, using a self-administered but supervised questionnaire. Convenience sampling was used for the selection of construction firms and sites, as well as the workers interviewed. The six companies had previously participated in an investigation into HIV/AIDS policies and treatment programmes implemented by Western Cape construction firms (Bowen et al., 2010, 2014). The sample frame consisted of all employees present when researchers visited the sites by prior arrangement. For logistical reasons, the geographical scope of the study was restricted to the Western Cape region of South Africa.

Survey participants ($n = 512$) were site-based workers in a range of occupations. The questionnaires were available in English, Afrikaans and isiXhosa (an indigenous African language), as these are the most commonly spoken languages in the province. Workers were briefed about the nature of the study. They were not paid or rewarded and were assured that their participation was entirely voluntary and anonymous. Participants who provided informed consent then proceeded to complete the questionnaires. This took place in large converted shipping-container offices equipped with tables and chairs. Proficiency in all three languages was available through the attending field researchers, whose assistance was limited to clarifying the meaning of particular questions. One field researcher was female, and the other two were male. The time taken to complete the questionnaires ranged from 30 minutes to 1 hour, depending on the participant's literacy level.

Measures

The questionnaire was based on items previously employed to survey the general population in South Africa (Kalichman & Simbayi, 2003, 2004). These comprise-validated instruments were developed especially for application in South Africa. The questions and measures are depicted in Table 17.1.

Demographic characteristics: Participants provided personal information including age, gender, ethnicity, level of education, and whether they were acquainted with an HIV+ person. Ethnicity was captured with the following classifications: 'Black' African, 'Coloured' (mixed race), 'Indian', and 'White', with the latter three classes being combined as 'Others' in the statistical analysis. Participants were asked also whether they had been tested for HIV and about their sero-status.

Alcohol consumption: Consumption in the preceding three months was ascertained. Alcohol was not distinguished by type or amount. Higher frequency scores indicated higher levels of alcohol consumption.

Drug use: Four items, drawn from Kalichman and Simbayi (2003, 2004), were used to create a scalar measure for frequency (but not amount) and type of illegal drug use in the preceding three months.

AIDS-related knowledge: Carey and Schroder (2002) and Kalichman and Simbayi (2003, 2004) provided the basis for the seven items used to create a scalar measure of AIDS-related knowledge. The scale was scored for the number of correct responses, with higher scores indicating higher levels of AIDS-related knowledge. 'Do not know' responses were treated as incorrect.

Risky sexual behaviour: Four items, based on Kalichman and Simbayi (2003, 2004), formed the basis for a scalar measure for risky sexual behaviour. Respondents were asked about their sexual activity in the previous three months; whether or not they had ever exchanged sex for money, housing, gifts, or food ('transactional sex'); and if a condom was used at last coital activity. The scale was scored so that higher scores indicated higher levels of risky sexual behaviour.

Table 17.1 Demographic and scale items (*n* = 512)

Items	Response options
1. Demographic variables	
Age	Age in years
Gender	Male = 1; Female = 2
Ethnicity	'Others' ('Coloured', Indian, 'White') = 1; 'Black' African = 2
Education	Primary or less = 1; Secondary = 2; Tertiary or higher = 3
Knowing at least one HIV+ person	None = 0; At least 1 = 1
HIV testing	Not tested = 0; Tested = 1
HIV status	Negative = 0; Positive = 1
2. AIDS-related knowledge (AK): (*'Agree'*; *'Disagree'*; or *'Do not know'*)	Correct response = 1; Incorrect response = 0; Do not know = 0
AK1. Can men give AIDS to women? (*Agree*)	
AK2. Can women give AIDS to men? (*Agree*)	
AK3. Must a person have many different sex partners to get AIDS? (*Disagree*)	
AK4. Does washing after sex help protect someone from getting AIDS? (*Disagree*)	
AK5. Can a pregnant woman give AIDS to her baby? (*Agree*)	
AK6. Can the use of vitamins and healthy foods cure AIDS? (*Disagree*)	
AK7. Can traditional African medicines cure AIDS? (*Disagree*)	
3. Alcohol (AU1.) – in the past 3 months, how often have you used alcohol?	Never = 0; Once only = 1; More than once = 2
4. Drugs (DU) – in the past 3 months, how often have you used:	Never = 0; Once only = 1; More than once = 2
DU1. 'Dagga' (cannabis)	
DU2. 'Tik' (crystal methadone)	
DU3. Cocaine	
DU4. Mandrax (Methaqualone)	
5. Risky sexual behaviour (RS)	No = 0; Yes = 1
RS1. Have you had two or more sex partners in the last 3 months?	
RS2. Have you ever received money, housing, gifts or food for sex?	
RS3. Have you ever given money, housing, gifts or food for sex?	
RS4. Did you use a condom the last time you had sex?	

Note: For AIDS-related knowledge, correct responses are indicated in parentheses.

Statistical analysis

Using IBM SPSS version 25 for Macintosh, a variety of descriptive and bi-variate statistical analyses was performed. To verify the factorial structure of all composite scales, confirmatory factor analysis (CFA) using structural equation modeling (SEM) was conducted on the items measuring drug usage, AIDS-related knowledge, and risky sexual behaviour. The CFA, conducted using IBM AMOS version 24.0 for Windows, employed maximum likelihood estimation to evaluate model fit. Four critical fit indices were applied to determine the degree of fit of the structural equation models as follows (with index values reflecting good model fit indicated in parenthesis): χ^2/df ratio (less than 4); Comparative Fit Index (CFI of 0.90 and greater); Root Mean Square Error of Approximation (RMSEA 0.06 and less); and Hoelter critical N (CN index 200 and greater). Sample size was deemed sufficient for CFA, and modification indices, available within AMOS, were used to guide the model revision process.

Given the categorical nature of the data, Bayesian estimation (the methodological approach available within AMOS for analysing categorical data using the Markov Chain Monte Carlo [MCMC] algorithm) was used to compare the parameter estimates derived from both the ML and Bayesian approaches. Specifically, in Bayesian estimation, the mean of the posterior distribution can be reported as the parameter estimate (regression weight), and the standard deviation of the posterior distribution serves as an analogue to the standard error in ML estimation (Byrne, 2010). In this instance, Bayesian estimation was used to verify the parameter estimates derived from the ML estimation approach. Sampling convergence was deemed to have occurred when the convergence statistic was less than 1.002.

Once the factorial structure had been validated, unweighted scale scores were created by summating the scores of their respective constituent items with reverse scoring of individual items where appropriate. To facilitate the specification of a structural model, multiple linear regression analysis was used to identify significant predictors of alcohol consumption, drug use, AIDS-related knowledge, and risky sexual behaviour. Variables entered into the regression models were selected on the basis of evidence from the extant literature. Following the regression analyses, an integrated theoretical model, to examine the direct and indirect determinants of risky sexual behaviour, was then specified and tested using structural equation modelling.

Results

Missing-value analysis

The proportion of the sample with missing values on all of the items of interest was less than 3%, with most items having less than 2% missing values. Listwise deletion was therefore appropriate (Graham, 2012).

Participant characteristics

Survey participants were predominantly male (91%; n = 461). Ages ranged between 18 and 69 years old (mean = 36, SD = 10.86), with most respondents in the age group of 21–30 years (34%; n = 168). Almost two-thirds (62%; n = 313) were 'Black' African (as distinct from the other ethnic groups). Over a quarter (29%; n = 144) had primary-level education at most, whilst 52% (n = 260) had secondary-level education. Forty-nine per cent (n = 234) reported not knowing any HIV+ persons. With regard to sero-status, a quarter (26%; n = 131) claimed not to have been tested for HIV, and 10% (n = 34) of those who had tested reported that they were HIV+.

Confirmatory factor analysis

The initial factorial model was specified and tested using CFA. Correlation was indicated between the error terms of the men-to-women and women-to-men transmission knowledge items (r = 0.76, p < 0.01). With this path specified, model fit proved to be excellent ($\chi2/df$ ratio = 1.430, CFI = 0.977, RMSEA = 0.029, and Hoelter [95%] = 452).

The convergence statistic cut-off point was 1.0009. The parameter estimates were very close to each other. Inspection of the Bayesian SEM diagnostic first and last combined polygon plot for each item indicated that AMOS successfully identified salient features of the posterior distribution for each item. The Bayesian SEM diagnostic trace plots indicated convergence in distribution occurred rapidly, a clear indicator that the SEM model was specified correctly (Byrne, 2010). The psychometric validity of the individual scalar instruments was thus confirmed.

Developing the theoretical model

Multiple linear regression analysis was used, first, to explore demographic characteristics as predictors of each of alcohol consumption, drug usage, AIDS-related knowledge, and risky sexual behaviour; and, second, demographic characteristics (in various combinations), alcohol consumption, drug use, and AIDS-related knowledge as predictors of risky sexual behaviour. The full set of models is shown in Table 17.2.

Predictors of alcohol consumption

Model 1 examined the relationship between the five demographic factors and alcohol consumption. This model was significant: F (5, 443) = 14.18, p < 0.001, R^2 = 0.14. Age (β = –0.19, p < 0.001), gender (β = –0.17, p < 0.001), ethnicity (β = –0.22, p < 0.001), and being acquainted with an HIV+ person (β = 0.13, p < 0.01) each proved to be significant predictors of alcohol consumption (see Table 17.2). Younger workers, males, workers other than 'Black' African, and

Table 17.2 Regression models of demographic characteristics, behavioural and cognitive factors, and risky sexual behaviour

Model	Dependent variable	Independent variable	B	S.E.	β	t	p-value	R	R^2	F
Model 1: Demographic predictors of alcohol consumption	Alcohol use	(Constant)	2.714	.317		8.567	.000***	.371	.138	14.176***
		Age	-.017	.004	-.194	-4.221	.000***			
		Gender	-.554	.151	-.169	-3.670	.000***			
		Ethnicity	-.429	.092	-.222	-4.671	.000***			
		Education	.057	.069	.042	.831	.407			
		HIV+ acquaintance	.252	.084	.133	2.984	.003**			
Model 2: Demographic and alcohol consumption predictors of drug use	Drug use	(Constant)	1.799	.377		4.775	.000***	.286	.082	6.462***
		Age	-.015	.005	-.165	-3.386	.001**			
		Gender	-.174	.169	-.050	-1.035	.301			
		Ethnicity	-.328	.104	-.160	-3.152	.002**			
		Education	-.175	.076	-.121	-2.317	.021*			
		HIV+ acquaintance	-.050	.094	-.025	-.535	.593			
		Alcohol	.126	.052	.119	2.403	.017*			
Model 3: Demographic, alcohol consumption, and drug use predictors of AIDS-related knowledge	AIDS-related knowledge	(Constant)	4.309	.704		6.120	.000***	.506	.256	20.814***
		Age	-.002	.008	-.009	-.205	.838			
		Gender	.260	.306	.038	.849	.396			
		Ethnicity	-1.043	.193	-.254	-5.414	.000***			

		Education	.765	.140	.261	5.450	.000***	
		HIV+ acquaintance	.855	.172	.213	4.958	.000***	
		Alcohol	.039	.097	.018	.403	.687	
		Drugs	.110	.087	.056	1.270	.205	
		(Constant)	1.762	.303		5.822	.000***	
Model 4 Conceptual Model: Demographic, alcohol consumption, drug use, and AIDS-related knowledge predictors of risky sexual behaviour	Risky sexual behaviour						.268	.072	4.002***
		Age	-.006	.003	-.086	-1.690	.092	
		Gender	-.320	.128	-.126	-2.504	.013*	
		Ethnicity	-.213	.082	-.142	-2.601	.010*	
		Education	-.002	.060	-.002	-.037	.970	
		HIV+ acquaintance	.069	.073	.047	.947	.344	
		Alcohol	.044	.040	.057	1.112	.267	
		Drugs	.078	.036	.108	2.184	.029*	
		AIDS knowledge	-.028	.020	-.077	-1.392	.165	

Note: *$p < 0.05$; **$p < 0.01$; ***$p < 0.001$

workers being acquainted with an HIV+ person exhibited significantly higher levels of alcohol consumption than did their fellow workers.

Predictors of drug use

Model 2 examined the relationship between the five demographic factors, alcohol consumption, and drug use. This model was significant: $F(6, 436) = 6.46, p < 0.001, R^2 = 0.08$. Age ($\beta = -0.17, p < 0.01$), ethnicity ($\beta = -0.16, p < 0.01$), level of education ($\beta = -0.12, p < 0.05$), and alcohol consumption ($\beta = 0.12, p < 0.05$) each proved to be significant predictors of drug use (see Table 17.2). Younger workers, workers other than 'Black' Africans, workers with higher levels of education, and workers consuming more alcohol reported significantly higher levels of drug use than did their counterparts.

Predictors of AIDS-related knowledge

Model 3 explored the five demographic factors, alcohol consumption, and drug use as determinants of AIDS-related knowledge. This model was significant: $F(7, 423) = 20.81, p < 0.001, R^2 = 0.26$; with ethnicity ($\beta = -0.25, p < 0.001$), education ($\beta = 0.26, p < 0.001$), and acquaintance with an HIV+ person ($\beta = 0.21, p < 0.001$) as significant determinants of AIDS-related knowledge (see Table 17.2). Higher levels of AIDS-related knowledge were positively associated with ethnicity being other than 'Black' African, possessing higher levels of education, and knowing an HIV+ person.

Predictors of risky sexual behaviour

Model 4 (the conceptual model) explored the demographic variables, alcohol consumption, drug use, and AIDS-related knowledge as predictors of risky sexual behaviour. This model was significant: $F(8, 414) = 4.00, p < 0.001, R^2 = 0.07$. Gender ($\beta = -0.13, p < 0.05$), ethnicity ($\beta = -0.14, p < 0.05$), and drug use ($\beta = 0.11, p < 0.05$) each proved significant predictors of risky sexual behaviour (see Table 17.2). Male workers, workers other than 'Black' African, and workers reporting higher levels of drug use, reported significantly higher levels of risky sexual behaviour than did workers in other categories.

An initial structural model specifying the antecedents of risky sexual behaviour was then postulated.

Testing the structural model

An initial structural model (with correlated error terms for the men-to-women and women-to-men knowledge items as indicated by the CFA) was specified and tested. A number of paths were not significant and changes to the model were indicated. The paths between age and drug use, between ethnicity and risky sexual behaviour, and between acquaintance with an HIV+ person and alcohol were omitted. Paths between knowing an HIV+ person and risky sexual behaviour, and between age and risky sexual behaviour were added.

With these modifications, the model was a good fit to the data ($\chi 2/df$ ratio = 2.073, p = 0.000, CFI = 0.914, RMSEA = 0.046, and Hoelter [95%] = 293). All paths were then significant, most at $p < 0.01$ or $p < 0.001$. The final structural model, regression weights, and associated levels of significance are shown in Figure 17.2.

A number of significant *direct* pathways were identified in the SEM. Gender was significant in predicting levels of alcohol consumption (β = –0.19, $p < 0.001$) and risky sexual behaviour (β = –0.12, $p < 0.05$). Female workers were less likely than males to report higher levels of alcohol consumption or to engage in risky sexual behaviour. Level of education had a direct role in predicting AIDS-related knowledge (β = 0.32, $p < 0.001$) and drug usage (β = –0.18, $p < 0.001$), with better-educated workers being more likely to possess better AIDS-related knowledge and less likely to use illegal drugs. Acquaintance with an HIV+ person was a direct predictor of AIDS-related knowledge (β = 0.23, $p < 0.001$) and risky sexual behaviour (β = 0.19, $p < 0.001$). Specifically, workers knowing an HIV+ person were more likely to possess higher levels of AIDS-related knowledge than were their counterparts and more likely to engage in risky sexual behaviour. Age was found to be significant in predicting alcohol consumption (β = –0.19, $p < 0.001$) and risky sexual behaviour (β = –0.19, $p < 0.01$). Older workers were less likely

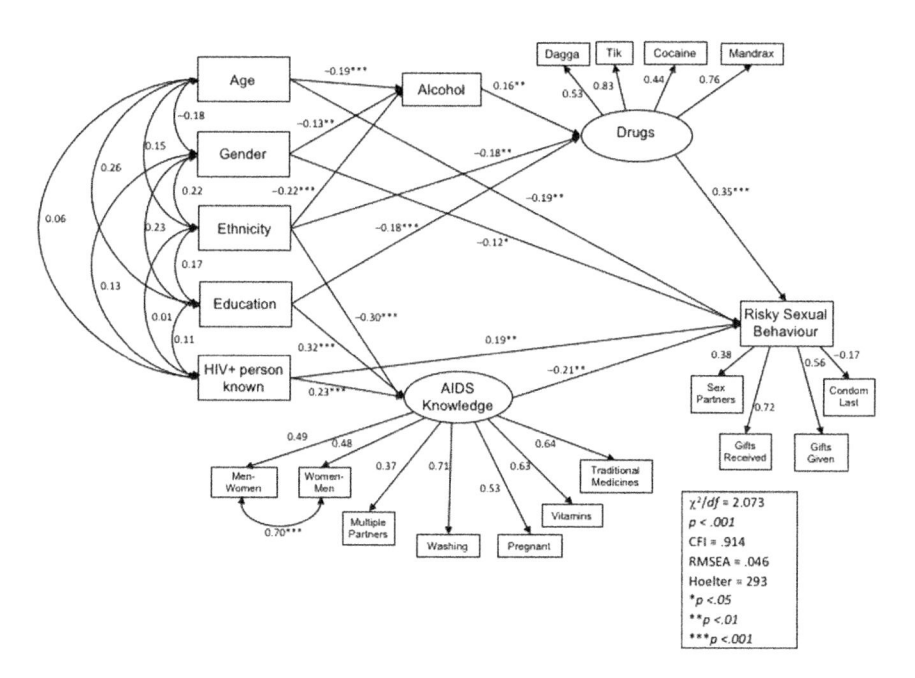

Figure 17.2 The structural model.

than younger workers to engage in either alcohol consumption or risky sexual behaviour.

Risky sexual behaviour was also predicted by drug use ($\beta = 0.35$, $p < 0.001$) and level of AIDS-related knowledge ($\beta = -0.21$, $p < 0.01$). Workers with higher levels of drug use and lower levels of AIDS-related knowledge were more likely to indulge in risky sexual behaviour.

The regression analyses had determined that AIDS-related knowledge was not a direct predictor of either alcohol consumption or drug use (or vice versa), and that education was not a direct predictor of risky sexual behaviour. Notably, in the SEM model, education predicted risky sexual behaviour *indirectly* via its direct relationship with AIDS-related knowledge (itself a direct predictor of risky sexual behaviour). The SEM found no relationship (direct or indirect) between AIDS-related knowledge and either alcohol consumption or drug usage. In the SEM model, no direct relationship was found between ethnicity and risky sexual behaviour. Rather, ethnicity was found to have an indirect effect on risky sexual behaviour via its direct effect on AIDS-related knowledge; itself the direct predictor of risky sexual behaviour.

The final structural model validated the inclusion of the path between AIDS-related knowledge and risky sexual behaviour. This path was not identified in the regression analysis.

Discussion

Risky sexual behaviour by construction workers was explored as a function of demographic factors, alcohol consumption, drug usage, and AIDS-related knowledge. The importance of focusing on the construction industry was underscored by the disproportionately higher impact of the pandemic on this sector, the industry-specific characteristics contributing to this estimated higher rate, and the importance of accurate and comprehensive AIDS-related knowledge as an effective barrier to the spread of infection.

The results showed that risky sexual behaviour of construction workers is directly predicted by age, gender, acquaintance with an HIV+ person, drug usage, and level of AIDS-related knowledge. Alcohol consumption and level of education are indirect predictors of risky sexual behaviour through drug use and AIDS-related knowledge, respectively. These findings agreed with the findings in the literature (Stein & Nyamathi, 2000; Kalichman et al., 2006; Shisana et al., 2014). The findings were also in accord with intuition, as a less-educated person is less likely to properly comprehend communication messages. However, a less obvious link is related to the correspondence between lower educational attainment and likely adherence to traditional beliefs about the cause of HIV/AIDS. In this regard, the observed relationship between lower education and poor HIV knowledge indicates another considerable challenge for public health efforts to improve levels of knowledge transmission, particularly in societies and communities where non-scientific, traditional beliefs about HIV and AIDS are common (Yamba, 1997; Kalichman & Simbayi, 2004, Govender et al., 2016).

Alcohol consumption was predicted by age, gender, and ethnicity, whilst drug usage was determined by alcohol consumption, ethnicity, and education level. Older, female, and workers other than 'Black' African reported significantly lower levels of alcohol consumption than did younger, male, and 'Black' African workers. Workers other than 'Black' African and workers with lower levels of education reported significantly higher levels of drug usage than did their colleagues. This finding partly supports Kader and colleagues (2014), who found that males were significantly more likely than females to engage in hazardous and harmful use of alcohol and drugs. Kader and colleagues (2014) also reported that alcohol use was predicted by age, as did the study by Peltzer and colleagues (2010), where alcohol and drug use were found to be more likely amongst predominantly younger persons. Drug use in the survey for the present study was not predicted directly by either age or gender, but indirectly via their direct associations with alcohol consumption. In this respect, alcohol can be considered to be a gateway drug to illegal drugs such as dagga, tik, mandrax, and cocaine (Kirby & Barry, 2012).

Acquaintance with an HIV+ person was positively associated with both AIDS-related knowledge and risky sexual behaviour. The first relationship is unsurprising. One would expect that greater exposure to, and interaction with, infected persons would be associated with greater understanding of the disease and its spread.

The positive association between acquaintance with an HIV+ person and greater risky sexual behaviour is less clear and is not found in the literature. One would expect that such proximity to the illness and its effects would actually encourage positive behaviour change. The survey data were not sufficiently incisive to explain the apparent anomaly, but a possible reason might lie in the disparity in the levels of AIDS-related knowledge held by construction workers acquainted with HIV+ persons and those who are not. Workers knowing infected persons demonstrate significantly higher levels of AIDS-related knowledge than do those not so acquainted. Being more knowledgeable may lure the former into a false confidence regarding the risk of infection and lead them towards more risky behaviours. This corresponds with the concept of 'lower perceptions of risk' noted by McPhail and Campbell (2001) and Kabwama and Berg-Beckoff (2015), whereby increased knowledge results in greater perceptions of risk up to a point but, thereafter, further knowledge instills a greater (but false) sense of confidence and actually decreases the risk perception. Other explanations could include adherence to traditional beliefs, feelings of personal invincibility, addictive behaviour towards sexual intercourse, or perceptions that the HIV+ acquaintance is in recovery. However, the psycho-social implications of all of these require future research.

Finally, the associations between alcohol consumption, drug usage, and risky sexual behaviour are supported by previous research (Scott-Sheldon et al., 2009; Shisana et al., 2014). The findings of the present study reinforce the argument that alcohol and drug use have an adverse effect on risky sexual behaviour in general and condom use in particular.

Limitations

The survey findings were limited by the cross-sectional nature of the survey, the regional bias of the sample (Western Cape), and the potential under-reporting of risky behaviours. Care must be taken in attempting to generalise the results to other regions, other demographic groups, or other industries.

Conclusions

The purpose of this study was to explore factors associated with South African construction workers' engagement in risky sexual behaviour in the context of the HIV/AIDS pandemic. A field survey was conducted among 512 workers in the Western Province. Regression analysis and structural equation modelling were used to analyse the data.

The contribution of the research findings expands current knowledge about construction workers' knowledge and behaviours relating to the risk of contracting HIV/AIDS. Direct predictors of workers' knowledge were confirmed as gender, acquaintance with an HIV+ person, and drug usage behaviours. Ethnicity and level of education were found to be indirect determinants of risky sexual behaviour, with higher levels of education predicting better AIDS-related knowledge. Greater knowledge was inversely associated with risky behaviour; with more knowledge leading to less propensity to engage in risky sex. Construction workers reporting higher use of alcohol and drugs exhibited higher levels of risky sexual behaviour. The positive association between alcohol consumption and drug use was re-affirmed. Surprisingly, however, acquaintance with an HIV+ person was also found to be associated with greater risky sexual behaviour. This needs further investigation.

The findings should help to inform proactive management of HIV/AIDS intervention by construction firms. First, AIDS-related knowledge clearly has a strong influence on risky sexual behaviour, but knowledge acquisition is influenced by literacy and education, and the ways in which knowledge messages are communicated. This presents a challenge for the construction industry where the majority of the workforce is still poorly educated or has low levels of literacy, and this is likely to be the case in all developing countries. The South African construction industry needs to consider the efficacy of media and communication messages on a continuous basis, with less reliance on written text and more emphasis on visual images and live performance. Language issues must be addressed, in a participatory process that should truly engage with the culture of the target audience. Traditional beliefs, myths, and customs should be addressed in non-confrontational ways, ideally with peer-group help. In this study, the significant, positive causal relationship between alcohol and drug use and risky sexual behaviour has been shown. This presents additional challenges for workplace programmes seeking to promote more frequent and consistent condom use. Passive initiatives such as the provision of free condoms on site should be reinforced by more direct and active participatory activities with workers.

Intervention programmes cannot ignore or discount worker habits with regard to alcohol and drug use. The interventions do not need to be framed more aggressively, given that 'lectures' about personal behaviour are notoriously ineffective. More sophisticated and more worker-focused means must be sought. The solution probably lies in 'testimony' approaches, whereby the personal change stories of individual workers are presented (with suitable coaching) at workplace toolbox meetings. This would directly engage positive messages *from* workers and direct them *to* workers. Beyond the workplace environment, the content and targeting of messages aimed at behavioural change must be carefully designed and promoted, especially for interventions intended for venues where alcohol is consumed such as taverns and beer halls. This means going beyond the construction industry environment itself. More prudent alcohol and drug use can lead to safer sexual behaviours and lower risk of HIV infection. Integrating alcohol and drug use issues into worker education also has wider positive impacts such as reduced worker absenteeism, enhanced job performance, and improved safety records.

Enabling the construction industry to play its part in dealing with the HIV/AIDS pandemic not only in South Africa but also in other developing countries does not involve offering prescriptive advice to workers, but addressing issues effectively that directly and indirectly influence the behaviours that put them at risk.

Funding

This study was based on research supported by the National Research Foundation (NRF) of South Africa (grant-specific, unique reference number [UID] 85376). The grantholder acknowledges that opinions, findings, and conclusions or recommendations expressed in any publication generated by research supported the NRF are those of the authors, and that the NRF accepts no liability whatsoever in this regard.

References

Bowen, P. A., Allen, Y., Edwards, P. J., Cattell, K. S. & Simbayi, L. C. (2014), Guidelines for effective workplace HIV/AIDS intervention management by construction firms, *Construction Management and Economics*, Vol. 32(4), pp. 62–81.

Bowen, P. A., Cattell, K. S., Edwards, P. J. & Marks, J. (2010), Perceptions of HIV/AIDS policies and treatment programmes by Western Cape construction firms, *Construction Management and Economics*, Vol. 28(9), pp. 997–1006.

Bowen, P. A., Dorrington, R., Distiller, G., Lake, H. & Besesar, S. (2008), HIV/AIDS in the South African construction industry: An empirical study, *Construction Management and Economics*, Vol. 26(8), pp. 827–839.

Bureau for Economic Research/South African Business Coalition on HIV/AIDS. (2004), *The Economic Impact of HIV/AIDS on Business in South Africa, 2003*. Stellenbosch: Bureau for Economic Research.

Byrne, B. (2010), *Structural Equation Modeling with AMOS: Basic Concepts, Applications, and Programming*, 2nd ed. New York: Taylor & Francis.

Carey, M. P. & Schroder, K. E. (2002), Development and psychometric evaluation of the brief HIV knowledge questionnaire (NIV-KQ-18), *AIDS Education and Prevention*, Vol. 14(2), pp. 174–184.

Centers for Disease Control and Prevention (2010). Youth risk behavior surveillance - United States, 2009. *Morbidity & Mortality Weekly Report*, Vol. 59, (No. SS-5), pp. 1–143.

Choudhry, V., Ambresin, A-E., Nyakato, V. N. & Agardh, A. (2015), Transactional sex and HIV risks – Evidence from a cross-sectional national survey among young people in Uganda, *Global Health Action*, Vol. 8, pp. 1–11.

Cook, R. L. & Clark, D. B. (2005), Is there an association between alcohol consumption and sexually transmitted diseases? A systematic review, *Sexually Transmitted Diseases*, Vol. 32(3), pp. 156–164.

Dunkle, K. L., Jewkes, R. K., Brown, H. C., Gray, G. E., McIntyre, J. A. & Harlow, S. D. (2004), Transactional sex among women in Soweto, South Africa: Prevalence, risk factors and association with HIV infection, *Social Science & Medicine*, Vol. 59(8), pp. 1581–1592.

Eaton, J. W., Hallett, T. B. & Garnett, G. P. (2011), Concurrent sexual partnerships and primary HIV infection: A critical interaction, *AIDS and Behavior*, Vol. 15(4), pp. 687–692.

Eich-Hochli, D., Niklowitz, M., Clement, U., Luthy, R. & Opravil, M. (1998), Predictors of unprotected sexual contacts in HIV infected persons in Switzerland, *Archives of Sexual Behavior*, Vol. 27(1), pp. 77–90.

Faimau, G., Maunganidze, L., Tapera, R., Mosomane, L. C. K. & Apau, S. (2016), Knowledge of HIV/AIDS, attitudes towards sexual risk behaviour and perceived behavioural control among college students in Botswana, *Cogent Social Sciences*, Vol. 2(1), pp. 1–16.

Fox, A. M. (2014), Marital concurrency and HIV risk in 16 African countries, *AIDS and Behaviour*, Vol. 18(4), pp. 791–800.

Fraser-Hurt, N., Zuma, K., Njuho, P., Chikwava, F., Slaymaker, E., Hosegood, V. & Gorgens, M. (2011), *The HIV Epidemic in South Africa: What Do We Know and How Has It Changed?* Pretoria: SANAC.

Gilbert, L. (2006), Delivery of health care in a time of AIDS: The impact of HIV/AIDS on the nature and practice of health professionals in South Africa. Presented at XVIth Congress of the International Sociological Association, The Quality of Social Existence in a Globalising World, Session RC15_03, 23–27 July, Durban, University of KwaZulu-Natal.

Graham, J. (2012), *Missing Data: Analysis and Design*. New York: Springer.

Govender, R., Bowen, P., Edwards, P. & Cattell, K. (2016), AIDS-related knowledge, stigma and customary beliefs of South African construction workers, *AIDS Care*, Vol. 29(6), pp. 711–717.

Hargreaves, J. R., Bonell, C. P., Morison, L. A., Kim, J. C., Phetla, G., Porter, J. D., Watts, C. & Pronyk, P. M. (2007), Explaining continued high HIV prevalence in South Africa: Socioeconomic factors, HIV incidence and sexual behaviour change among a rural cohort, 2001–2004, *AIDS*, Vol. 21(Supplement 7), pp. S39–S48.

Hong, S. Y., Thompson, D., Wanke, C., Omosa, G., Jordan, M. R., Tang, A. M., Patta, S., Mwero, B., Mjomba, I. & Mwamburi, M. (2012), Knowledge of HIV transmission and associated factors among HIV-positive and HIV-negative patients in rural Kenya, *Journal of AIDS Clinical Research*, Vol. 3(7). doi:10.4172/2155-6113.1000170.

Kabwama, S. N. & Berg-Beckhoff, G. (2015), The association between HIV/AIDS-related knowledge and perception of risk for infection: A systematic review, *Perspectives in Public Health*, Vol. 135(6), pp. 299–308.

Kader, R., Seedat, S., Govender, R., Koch, J. R. & Parry, C. D. H. (2014), Hazardous and harmful use of alcohol and/or other drugs and health status among South African patients attending HIV clinics, *AIDS and Behaviour*, Vol. 18(3), pp. 526–534.

Kalichman, S. C. & Simbayi, L. C. (2003), HIV testing attitudes, AIDS stigma, and voluntary HIV counseling and testing in a black township in Cape Town, South Africa, *Sexually Transmitted Infections*, Vol. 79(6), pp. 442–447.

Kalichman, S. C. & Simbayi, L. C. (2004), Traditional beliefs about the cause of AIDS and AIDS-related stigma in South Africa, *AIDS Care*, Vol. 16(5), pp. 572–580.

Kalichman, S. C., Simbayi, L. C., Kagee, A., Toefy, Y., Cain, D. & Cherry, C. (2006), Association of poverty, substance use, and HIV transmission risk behaviours in three South African communities, *Social Science & Medicine*, Vol. 62(7), pp. 1641–1649.

Kalichman, S. C., Simbayi, L. C., Vermaak, R., Cain, D., Jooste, S. & Peltzer, K. (2007), HIV/AIDS risk reduction counseling for alcohol using sexually transmitted infections clinic patients in Cape Town, South Africa, *Journal of Acquired Immune Deficiency Syndrome*, Vol. 44(5), pp. 594–600.

Kalichman, S. C., Simbayi, L. C., Vermaak, R., Jooste, S. & Cain, D. (2008), HIV/AIDS risks among men and women who drink at informal alcohol serving establishments (shebeens) in Cape Town, South Africa. *Prevention Science*, Vol. 9(1), pp. 55–62.

Kirby, T. & Barry, A. E. (2012), Alcohol as a gateway drug: A study of US 12th graders, *Journal of School Health*, Vol. 82(8), pp. 371–379.

Letamo, G. (2018), Misconceptions about HIV transmission among adolescents: Levels, trends and correlates from the Botswana AIDS impact surveys, 2001–2013: A short report, *AIDS Care*, Vol. 31(1), pp. 48–52.

MacPhail, C. & Campbell, C. (2001), 'I think condoms are good but, aai, I hate those things': Condom use among adolescents and young people in a Southern African township, *Social Science & Medicine*, Vol. 52(11), pp. 1613–1627.

Mah, T. L. & Shelton, J. D. (2011), Concurrency revisited: Increasing and compelling epidemiological evidence, *Journal of the International AIDS Society*, Vol. 14, pp. 33–41.

Mayosi, B. M. & Benatar, S. R. (2014), Health and health care in South Africa: 20 years after Mandela, *New England Journal of Medicine*, Vol. 371(14), pp. 1344–1353.

Meintjes, I., Bowen, P. A. & Root, D. (2007), HIV/AIDS in the South African construction industry: Understanding the HIV/AIDS discourse for a sector specific response, *Construction Management and Economics*, Vol. 25(3), pp. 255–266.

Morojele, N. K., Kachieng'a, M. A., Mokoko, E., Nkoko, M. A., Parry, C. D. H., Nkowane, A. M., Moshia, K. M. & Saxena, S. (2006), Alcohol use and sexual behaviour among risky drinkers and bar and shebeen patrons in Gauteng province, South Africa, *Social Science & Medicine*, Vol. 62(1), pp. 217–227.

Morojele, N. K., Nkosi, S., Kekwaletswe, C. T., Saban, A. & Parry, C. D. H. (2013), Review of research on alcohol and HIV in Sub-Saharan Africa. Policy brief, Alcohol and Drug Abuse Research Unit. February. Pretoria: South African Medical Research Council.

Morris, M. & Kretzschmar, M. (1997), Concurrent partnerships and the spread of HIV, *AIDS*, Vol. 11(5), pp. 641–648.

Parry, C. D. H., Plüddemann, A., Steyn, K., Bradshaw, D. Norman, R. & Laubscher, R. (2005), Alcohol use in South Africa: Findings from the first demographic and health survey (1998), *Journal of Studies on Alcohol*, Vol. 66(1), pp. 91–97.

Peltzer, K., Davids, A. & Njuho, P. (2011), Alcohol use and problem drinking in South Africa: Findings from a national population-based survey, *African Journal of Psychiatry*, Vol. 14(1), pp. 30–37.

Peltzer, K., Ramlagan, S., Johnson, B. D. & Phaswana-Mafuya, N. (2010), Illicit drug use and treatment in South Africa: A review. *Substance Use and Misuse*, Vol. 45(13), pp. 2221–2243.

Rehle, T., Shisana, O., Pillay, V., Zuma, K., Puren, A. & Parker, W. (2007), National HIV incidence measures: New insights into the South African epidemic. *South African Medical Journal*, Vol. 97(3), pp. 194–199.

Scott-Sheldon, L. A. J., Carey, M. P., Vanable, P. A., Senn, T. E., Coury-Doniger, P. & Urban, M. A. (2009), Alcohol consumption, drug use, and condom use among STD clinic patients, *Journal of Studies on Alcohol and Drugs*, Vol. 70(5), pp. 762–770.

Scott-Sheldon, L. A. J., Carey, M. P., Carey, K. B., Cain, D., Simbayi, L. C., Mehlomakhulu, V. & Kalichman, S. C. (2013), HIV testing is associated with increased knowledge and reductions in sexual risk behaviours among men in Cape Town, South Africa, *African Journal of AIDS Research*, Vol. 12(4), pp. 195–201.

Seth, P., Wingood, G. M., DiClemente, R. J. & Robinson, L. S. (2011), Alcohol use as a marker for risky sexual practices and biologically-confirmed sexually transmitted infections among young adult African-American women, *Women's Health Issues*, Vol. 21(2), pp. 130–135.

Shisana, O., Simbayi, L. C., Rehle, T., Onoya, D., Jooste, S., Zungu, N., Labadarios, D. & Zuma, K. (2014), *South African National HIV Prevalence, Incidence and Behaviour Survey, 2012*. Pretoria: HSRC Press.

Shisana, O., Stoker, D., Simbayi, L. C., Orkin, O., Bezuidenhout, F., Jooste, S .E., Colvin, M. & van Zyl, J. (2004), South African National Household Survey of HIV/AIDS prevalence, behavioural risks and mass media impact: Detailed methodology and response rate results, *South African Medical Journal*, Vol. 94(4), pp. 283–288.

Smallwood, J. & Venter, D. (2001), Feedback report on a HIV/AIDS, STDs, and TB Study conducted among general contractors. Report, Department of Construction Management, University of Port Elizabeth, 8 August, p. 9.

South African National AIDS Council (SANAC). (2017), South African National Strategic Plan on HIV, TB and STIs 2017–2022. Draft 1.0, NSP Steering Committee Review, 30 January 2017. Pretoria: SANAC.

Stein, J. A. & Nyamathi, A. (2000), Gender differences in behavioural and psychosocial predictors of HIV testing and return for test results in a high-risk population, *AIDS Care*, Vol. 12(3), pp. 343–356.

Stoebenau, K., Heise, L., Wamoyi, J. & Bobrova, N. (2016), Revisiting the understanding of "transactional sex" in sub-Saharan Africa: A review and synthesis of the literature, *Social Science & Medicine*, Vol. 168, pp. 186–197.

UNAIDS. (2013), *UNAIDS Report on the Global AIDS Epidemic 2013*. Geneva: UNAIDS.

UNAIDS. (2016), *Prevention Gap Report*. Geneva: UNAIDS.

Van Heerden, M. S., Grimsrud, A. T., Seedat, S., Meyer, L., Williams, D. R. & Stein, D. J. (2009), Patterns of substance use in South Africa: Results from the South African Stress and Health study, *South African Medical Journal*, Vol. 99(5 Part 2), pp. 358–366.

Wamoyi, J., Stobeanau, K., Bobrova, N., Abramsky, T. & Watts, C. (2016), Transactional sex and risk for HIV infection in sub-Saharan Africa: A systematic review and meta-analysis, *Journal of the International AIDS Society*, Vol. 19, pp. 1–16.

Wouters, E., Van Damme, W., Van Loon, F., van Rensburg, D. & Meulemans, H. (2009), Public-sector ART in the Free State Province, South Africa: Community support as an important determinant of outcome, *Social Science & Medicine*, Vol. 69(8), pp. 1177–1185.

Yamba, C. B. (1997), Cosmologies in turmoil: Witch-finding and AIDS in Chiawa, Zambia, *Africa: Journal of the International African Institute*, Vol. 67(2), pp. 200–223.

Zuma, K., Shisana, O., Rehle, T. M., Simbayi, L. C., Jooste, S., Zungu, N., Labadarios, D., Onoya, D., Evans, M., Moyo, S. & Abdullah, F. (2016), New insights into HIV epidemic in South Africa: Key findings from the National HIV Prevalence, Incidence and Behaviour Survey, 2012, *African Journal of AIDS Research*, Vol. 15(1), pp. 67–75.

18 Construction migrants from 'developing' countries in 'developed' hosts

David Oswald

Summary

The aim of this study was to explore the experiences of migrant workers from developing countries when working in a more developed host country. Construction workers from developing countries that migrate to developed countries should benefit from remittances, savings, new skills and international contacts. In terms of their health and safety (H&S), they should also benefit from more advanced H&S regulations and ways of working. Arguably, such migrant workers (from developing countries in developed countries) should receive even more health and safety support than local workers, as they have to quickly adapt to new ways of working, systems, rules, regulations and policies. This chapter draws upon ethnographic research undertaken within the United Kingdom (UK) with migrant workers from lower-income countries in Eastern Europe that included Romania, Bulgaria and Croatia. The findings suggest that while migrant workers from lower-income countries received certain H&S benefits, they were also at risk of exploitation risks in terms of being paid less than the local workers; accidents due to arriving without adequate training or competency checks; precarious employment arrangements; extra demands and restrictions on travel home if they were bilingual; and gaining acceptance within the workforce due to inadequate communication strategies. Most construction migrant worker H&S research has focused on the physical accident risks, but this chapter also draws attention to the mental health risks that can be associated with these exploitations.

Introduction

The construction sector is one that is very sensitive to even the smallest changes in the global economy. It is a sector that thrives during acute economic peaks and suffers during economic downturn. In countries that are experiencing economic growth, migrant construction workers are often crucial, by filling job vacancies and providing the necessary skills. Hence, migrant workers are an important resource for the construction industry, which is dynamic, transient, multi-tiered and very susceptible to the economic climates of countries. Migrant workers from developing countries may have the opportunity to temporarily work in a more developed host country. Much of the previous research in the health and safety space of migrant workers has focused on the accident rates of migrants in comparison to local workers. However, within the construction industry there has been less focus on the migrant worker experience and the risks posed to their mental health, as well as physical health and safety. Migrant workers from lower-income countries can take great risks leaving their home country. The purpose of this study was to provide insights on potential mental, as well as physical, H&S risks to migrant workers from developing countries when working in a more developed host country.

Globalisation has supported movement of migrant workers across the globe, including the movement of workers from so-called developing countries to developed countries. The use of foreign labour on construction project is a strategy that can be explained by factors such as immobility of the construction site, the labour-intensive work and the fragmented nature of the production process (Bobek et al., 2018). This is an attractive strategy as it provides such flexibility while also lowering costs (Fellini et al., 2007; Meardi et al., 2012; Oswald et al., 2018). For the migrant employees from countries with poorer wage structures, they often aim to maximise their income within a short time frame. This income is typically sent home to family, as their centre of interest remains their country of origin, and there could be no further work for them once the construction project is completed (Bobek et al., 2018; Caro et al., 2015; Datta et al., 2007; Trevena, 2013). The economic differences between the countries central to this study (Bulgaria, Romania, Croatia and the UK) can be interpreted through national minimum wages. At the time the data was collected, the national average monthly wage in Bulgaria was €184; Romania, €217; Croatia, €395; and the UK, €1378 (Eurostat, 2015). Hence, the free movement opportunities within the European Union can allow workers from Eastern European countries to earn several times their monthly wage.

The regulation of labour immigration is an issue that is one of the most controversial and important public policies in high-income countries (Ruhs & Anderson, 2010). The debate in the UK has been obvious for over the last decade. The same government in 2007 which stated that expanded labour immigration would bring 'obvious enormous economic benefits' (see Byrne, 2007), would later state in 2008 that 'it's been too easy to get into this country in the past and it's going to get harder' (see Woolas, 2008). Between 1993 and 2015 the number of foreign-born immigrants in the UK more than doubled from 3.7 million to over 8.7 million (Vargas-Silva & Rienzo, 2017). As numbers have risen, so has the hostility towards migrants (Marino et al., 2017), and the Brexit vote of 2016 has brought great uncertainty for current and prospective UK immigrants.

Since 1993, the greatest rise of foreign-born immigrants into the UK was between 2005 and 2008, which coincided with the expansion of the European Union in 2004 (Vargas-Silva & Rienzo, 2017). Poland was one of the countries that joined in 2004 and is the most common country that immigrants to the UK are from (Vargas-Silva & Rienzo, 2017). Cyprus, Czech Republic, Estonia, Hungary, Latvia, Lithuania, Malta, Slovakia and Slovenia also joined the European Union in 2004. According to the World Bank list of economies, all of these countries are categorised as 'high-income' countries. Later in 2007, Bulgaria and Romania joined the European Union, and they were followed by Croatia in 2013, with the World Bank list of economies categorising these countries as 'upper middle'. Hence, while arguably a blunt indicator, Bulgaria, Romania and Croatia fall into the group of middle-income economies, and therefore are 'developing' countries, as per the definition within the theme of this book. Their status as 'upper middle' and given that they were accepted within the European Union does suggest that while they are perhaps less advanced than some other countries within the European Union, they are more advanced than many other countries in the world. Nevertheless, workers migrating from less-advanced countries within the European Union should benefit from the structures in place in more advanced countries European countries. It is through this lens that this chapter unpacks the challenges that migrant workers from Croatia, Bulgaria and Romania face in the UK.

Construction migrant worker H&S

It is widely regarded in research evidence around the world that migrant workers are more likely to have workplace accidents than local workers. For example, Meardi and colleagues (2012:14) found research participants 'overwhelming[ly] confirm' that construction migrants were at higher H&S risk. Hassan and Houdmont (2014) reported that 40% of their 651 surveyed migrant workers reported having had experienced a workplace accident; and in the UK, migrant workers make up only 8% of the UK construction labour population, but they account for 17% of total fatalities (CCA, 2009). These findings should raise questions about the management of the health and safety of migrant workers globally.

As well as potentially facing increased risk of injury, migrant workers are subject to other risks. Those that move to developing countries can risk poor, unsanitary living conditions, being 'trapped' with passports withheld and having wages withdrawn, as is reportedly the case in Qatar (see Millward, 2016). In more developed countries, risks of not having such basic human rights or violations are reduced; but nevertheless there are risks that migrant workers will not reach the same levels of rights and opportunities as local workers. Murie (2007) pointed out that workers should have an expectation of 'decent' work, which applies to all workers, including those on daily wages and in very temporary, informal employment. 'Decent' work is undertaken in a safe physical environment under conditions that respect the rights of workers as defined in national law and international conventions. However, there are risks of indecency within the construction industry, especially amongst migrant workers, as work can be low-paid, short-term, unregistered, informal and hazardous (Murie, 2007). Construction projects are also typically characterised by high demand and low job control, which are known to contribute to mental health problems. Migrant workers are typically casually employed and without access to paid sick leave, both of which pose risks to mental health. The mental health problem in the industry can be interpreted through suicide rates, where construction workers in the UK have a high rate in comparison to other occupations (Meltzer et al., 2008).

The Health and Safety Executive (HSE, n.d.) highlighted four areas to ensure migrant worker health and safety in the UK:

- *Training*, as workers may be 'completely unfamiliar with workplace risks', and may 'have never done the sort of work' required;
- *Communication*, as there may have problems communicating in English;
- *Competence*, which may be unclear, so before work commences assessments should be made; and
- *Attitude to health and safety*, as there may be different expectations about health and safety responsibilities.

In terms of communication, there has been a wide range of approaches implemented and/or recommended such as encouraging workers to learn the local language (Chan et al., 2016); having a bi-lingual speaking workgroup leader to act as a translator (Oswald et al., 2019); a 'buddy' system where a foreign worker is paired with a colleague of the same nationality who can speak English (Bust et al., 2008); pictorial aids or visuals (Hare et al., 2013); and even the creation of a workgroup's 'own language' (see Tutt et al., 2013). Yet, a clear assessment on the most effective strategies is still somewhat unclear, and further research in this area is required. The HSE acknowledges that workers may require training, as migrant workers could be unfamiliar with risks and not undertaken similar work before; and also encourages competency checks prior to the workers starting. McKay and colleagues (2006) reported that two-thirds of migrant workers received no health and safety training, and the other third typically received a short site induction, which was not communicated effectively. The different expectations and ways of working, have led to recommendations of fully flexible approaches that fit the different cultural characteristics of specific project teams, while harmonizing with the host-country legislative requirements (Oswald et al., 2018). These aspects have been explored as areas that could affect physical injury to migrant worker, but also could affect their mental health, if the workers are untrained, unable to effectively communicate, feel incompetent and have different conflicting views on the way health and safety should be performed.

Ethnography

The complexities surrounding multi-national workforces have been unpacked using ethnographic approaches (see, for example, Peltonen, 2007, and Sharpe, 2006). Ethnography is 'the art and science of describing a group or culture' (Fetterman, 1998:1), and holds considerable promise for addressing practical, problem-based research concerned with construction projects (Pink et al., 2010). During a three-year period, the researcher spent approximately two days per week (8 am to 5 pm) on a large multinational construction project in the UK. There were many different nationalities present, most of which were from countries that are considered 'developed', such as Germany and the United States of America; but others, within the definition of this book, can be considered 'developing', namely Bulgaria, Croatia and Romania. The Romanian workers typically shared workgroups with the Spanish and Portuguese nationals, whereas the Croatian workers were a homogeneous national workgroup and there were only a few Bulgarian nationals present within larger Polish workgroups. The workgroups were typically on the project for a short period, which is not uncommon within the construction industry; and there was a high turnover of foreign workers. The group in focus is those workers considered from 'developing' countries: Bulgaria, Croatia and Romania. Hence, while a range of themes developed within the wider study, the findings relevant to describing the migrant worker experience from these countries are those that are communicated within this chapter.

The researcher was a member of the H&S department on the project, adopting the role as a student. An ethnographic researcher's main tool is often the use of participant observation. Participant observation 'is a unique method for investigating human existence whereby the researcher more or less actively participates with people in commonplace situations and everyday life settings while observing and otherwise collecting information' (Jorgensen, 1989:1). This role included site walks, ad hoc discussions with employees, and being present at H&S meetings and at accident and incident responses. Note-taking of observations and informal discussion were captured where possible in the field, and then more detailed notes were constructed as soon as possible after leaving the action (Pole & Morrison, 2003). This collected data was analysed in a concurrent process (Silverman, 2013) until a point of saturation was reached (Kumar, 2005).

Risks for migrants from lower-income countries

The following results and discussion section unpacks the key areas of risks where migrant workers can be exploited in developed countries.

Financial disparity between migrant and local workforces

Organisations in developed host countries can reduce their direct wage costs by employing migrant labour. Hence, while migrant workers may earn much more than they would in their home countries, they may risk exploitation through not earning as much as the local workers. For example, referring to a workgroup of Romanian and Portuguese construction workers, a UK safety representative and operator on the project explained: 'They get paid through an agency, which must allow some loopholes, as they are on less than the [UK] minimum wage'.

Considering Romania's monthly minimum wage (€217) was over five times less than the UKs and Portugal's monthly salary was €589, less than half of the UKs (Eurostat, 2015), it is conceivable that even being paid less than the minimum wage would still have been attractive to workers, as they could have still been earning significantly more than they would in their home countries. Considering Portuguese workers migrated, it cannot be assumed that the complexity of the migrant worker phenomenon can be simplified into two categories of developed and undeveloped countries, as countries such as Portugal, which is considered developed, still had economic differences that can encourage migration. The movement of these workers resonates with Dainty et al.'s (2007) report that financial incentive was the single most common reason for migrant construction workers wishing to work in the UK, with some workers stating they could earn six times what they could in their own countries. However, Bobak and colleagues (2018) pointed out that financial motivations are not the only factor for such migrants in the European Union, but also other factors such as better career prospects, improving English language competence and new experiences. Massey (1990) also explained that the situation in their home country, in terms of risk, can also contribute to the decision.

Fellini et al. (2007) previously reported that construction companies recruit foreign workers to reduce labour costs, and can do so through use of informal and sometimes illegal strategies. In some places around the world these illegal strategies can be very lucrative for recruiters, who can charge over 6,000% more than the legally permitted (Buckley, 2012). The mental well-being of migrant workers, who have to come to the terms they are being paid less than local workers, and potentially have costs associated with paying recruiters and relocation, is a risk for migrants from lower-income countries. Fellini and colleagues (2007) explained that one reason for foreign immigrants being paid less than local workers is that they have less bargaining power. In this case, they were probably more likely to accept lesser terms as well due to the economic differences across the countries. Migrant workers from developing countries should carefully consider the potential financial risks involved in relocation to what is usually temporary and precarious employment.

In terms of physical health and safety, it has been argued that organisations should consider alternative strategies of keeping migrant workers healthy and safe (Bust et al., 2008). Multinational projects should understand that this may entail initial extra costs through the use of appropriate management approaches, which may incorporate further training or translators. This shouldn't be perceived as going above and beyond basic health and safety requirements, as the appropriate risks with migrant workers, such as communication barriers, should be managed by the employers. In the case of migrant workers from developing countries moving to developed countries, it is possible that the organisation will be making a direct labour-saving cost, and therefore there should be no excuse for reinvesting this into the health and safety management of the migrant workers.

Inappropriate competency assessments leaving migrants at risk

The informal nature of recruitment of workers from less developed countries raises questions regarding competency and training levels – two areas the HSE (n.d) draws attention to. On this large project, there was confusion over the accepted levels of competence, as different nations have different levels of qualification. Many of these qualifications were not known in the UK or could be understood as reciprocal to the UK standards. Evidence of competency came in a variety of forms, some of which were more robust than others, as one H&S manager explained after the arrival of a workgroup of Croatian workers: 'We have had managers signing letters to say the [migrant] workers are competent, and they have never even met them'.

In another study, Davis and Gibb (2009) found that most construction migrants to the UK had very limited construction experience and understanding of the UK health and safety system. Hence, workers may arrive with more significant H&S risks, as they are unaware of their surroundings and have not undergone appropriate training to gather the required competencies. This is especially relevant to migrant workers that have decided on a career change to work in the construction industry. There should be raised awareness of this dangerous current practice, where migrant workers are operating in a high-risk environment without having adequate competency checks or training provided.

The levels of competency, expectations and standard in H&S vary across countries, and different ways of working were evident from on-site observations. For example, during the summer, a Bulgarian worker arrived at a site induction with no steel-toe cap boots and in shorts. Both of these actions broke the construction site rules, yet this was how work was undertaken in his home country. Another Bulgarian worker refused to wear safety gloves (another mandatory site rule) for months, as he 'did not need them'. The health and safety advisor in the area eventually explained he had found a pair the worker liked, and the worker thereafter used them to comply with the basic H&S standards expected on the site. The foreign migrant workers mostly appreciated the increased attention on H&S. For example, in regard to the H&S system in place, a UK site operative stated: 'The [Romanian and Portuguese] workers think it's great that there is much more focus on safety, it's their management that are more concerned about profits'.

The differences in ways of working have been highlighted as a challenge in previous studies, with Bust et al. (2008) calling for a new approach to H&S management of migrant workers. This arguably begins before migrant workers set foot on site, considering Dainty and colleagues (2007) identified there were ineffective control measures in place to assess the competence of migrant construction workers. The on-site observations in this study confirmed that due to a lack of competency many workers were deemed unsuitable for the job, and were removed from site. This also suggests there are still inadequate control measures in place for deeming competency prior to arrival. The method of assessing competency really began when the workers arrived on-site, with workers having three months to demonstrate their skills were up to standard. This was similar to findings of Davis and Gibb (2009), who reported that employers undertook appraisals of the migrants' own practical work before offering them a job. Those that didn't pass this probation period were removed from site, which did create some security-related issues, as the security manager explained:

> Their passes have been revoked but the dismissed Romanians have been seen loitering around the work site. Apparently they have got into the site

Health and safety of migrant workers 265

once before. I'm not sure if they are after their job back, or are going to try and steal belongings or equipment, but it has become a security concern.

The turnover was not only a problem for the organisation, but also an example of potential outcomes for migrant workers, who could lose their jobs very quickly and have to pay for an expensive flight to relocate home sooner than expected. This risk could have significant implications on migrant workers, in terms of their mental health, as they may not have the finances to return home or adequately support their families from losing their job swiftly after starting. In an H&S meeting it was highlighted that in a workgroup of ten Romanian workers, eight were sent home: 'There are only two of the Romanians left; one of them had an accident on his shoulder but couldn't communicate it, so kept working'.

Workers involved in accidents were either too afraid to raise it or did not have the language skills to do so, meaning they could continue to work while injured. Hence, communication was an issue for direct physical health and safety accidents, and also had implications on migrant workers' mental health, as discussed in more detail next.

Risks of increased demands and isolation from inadequate communication strategies

Communication has been highlighted in many previous studies as important for reducing accidents, and without a robust communication strategy, migrant workers are at an additional risk of accidents. This work also found that they are also at risk of increased demands if they are bilingual and/or divided from others in the organising. Both of these scenarios could have implications for migrant workers' mental health.

One particular group that was reliable, flexible and low-cost was bilingual supervisors, whom were very important for communication of operations. They would get favouritism during the disciplinary process, as the project did not want to lose these workers, and were difficult to cover, which meant many stayed long hours and over holiday periods away from their families. This affected the mental health of some of the bilingual construction workers, as one H&S advisor explained:

> I asked him when his last [alcoholic] drink was; and he responded 'this morning'. He would have been way over the limit. He was depressed, it's Christmas time; he sees everyone around him going back to their families, and he kept mentioning 'Everyone goes home except me'. Technically he should be red carded (fired) but as a bilingual supervisor, we will see what happens. I wouldn't be surprised if it was rescinded to a yellow (warning).

Bilingual workers were essential for the communication abilities. Indeed, some local workers, recognised non-English speaking workers as being an additional H&S risk. For example, the researcher observed a local UK scaffolder state in a site induction: 'Foreign workers who cannot speak fluent English should not be allowed to work on site in the UK as it is a safety hazard'.

The view that non-English speaking migrant workers 'should not be allowed on site' is a very blunt and controversial point that could be interpreted as resentment. Dainty and colleagues (2007) found that there was a degree of resentment from some UK workers towards the influx of migrant workers, due to cultural differences and under-cutting of work tenders. The argument from the scaffolder was that, communication, or lack of due to language barriers, can raise safety issues. The risk of non-acceptance within the organisation is another mental health risk that migrant workers potentially face. According to Loosemore et al. (2010), segregation can be caused by language and cultural barriers, and on this project there were suggestions of division and discrimination. The researcher observed the following conversation between the local works manager and a Croatian foreman:

WORKS MANAGER: 'Some guys [migrant workers] want to move from one area of the site to another. They say they are feeling intimidated'.
FOREMAN (MIGRANT WORKER): 'There has been some discrimination. But it has been coming from all sides. I think it's a problem, some guys say it is a joke, but it is not always seen that way'.

The Eastern European workers complained they sometimes referred to as 'Poles', as Polish people are one of the most common immigrant countries in the UK. Yet the Eastern European workers, from various countries, were very proud of their different nationalities, and therefore it is very understandable why they felt these comments were unacceptable. On the other side, local workers complained that: 'The foreign workers have been taking any equipment they require to use, despite it belonging to other subcontractors. Equipment is going missing from all over the site and this is just the start'.

As well as saving costs, the sharing of equipment and machinery amongst subcontractors can be regarded as a good safety practice, as there is less equipment that needs to be checked, and less chance of equipment obstructing works in tight areas. The lack of effective communication was arguably creating a barrier to shared understanding, instead creating division and perceptions of misconduct, again raising mental health risks due to a lack of acceptance on the project.

Conclusions

Workers from less developed countries that migrate to developed countries should benefit from the H&S systems in place in their host country. The migrant workers in this study were impressed with the H&S focus on the project, but they were also subject to exploitation and additional risks through their migration. In terms of both their physical and mental H&S, the migrant workers risked:

- being paid less than the local workers, which could affect their mental health;
- accidents, due to arriving without adequate training or competency checks;

- losing their job soon after starting, due to precarious employment arrangements;
- extra demands and restrictions on travel home if they were bilingual; and
- gaining acceptance within the workforce, partly due to inadequate communication strategies.

Migrant workers from low-income countries can take risks by leaving their home to move to a more developed host country. However, unfortunately there are risks financially. As after a potentially costly relocation, they could lose their job in the host country very quickly if deemed incompetent. Hence, there is a need to develop a more sophisticated approach to adequately assess competency prior to arrival. During their work period, the workers' mental health risks can be increased by extra demands placed upon them and restrictions on holidays, especially if they are bilingual.

While, the risks of corruption and modern-day slavery to the migrant worker are less likely to occur in developed countries such as the UK, the research suggests that workers are still not treated on par to local workers. Host countries could be doing more to ensure the H&S of migrant workers is not at greater risk by investing in more efficient systems for additional challenges, such as H&S communication. Established migration channels between developing and developed countries should have more robust systems in place for assessing competence, and further research should explore the most appropriate method to avoid situations of precarious employment, high turnover, and incompetent workers being in a high-risk environment.

References

Bobek, A., Wickham, J., Moriarty, E. & Salamon, S. (2018) Is money always the most important thing? Polish construction workers in Ireland, *Irish Journal of Sociology*, 26(2), 162–182.

Buckley, M. (2012) From Kerala to Dubai and back again: Construction migrant workers and the global economic crisis, *Geoforum*, 43, 250–259.

Bust, P. D., Gibb, A. G. & Pink, S. (2008) Managing construction health and safety: Migrant workers and communicating safety messages, *Safety Science*, 46(4), 585–602.

Byrne, L. (2007) Home affairs – Minutes of evidence, 27 November 2007. Available at: https://publications.parliament.uk/pa/ld200708/ldselect/ldeconaf/82/8206.htm.

Caro, E., Bernsten, L., Lillie, N. & Wagner, I. (2015) Posted migration and segregation in the European construction sector, *Journal of Ethnic and Migration Studies*, 41(10), 1600–1620.

CCA, Centre for Corporate Accountability. (2009) Migrants' workplace deaths in Britain, commissioned and jointly published by Irwin Mitchell and the Centre for Corporate Accountability, London.

Chan, A., Javed, A., Lyu, S. & Hon, C. (2016) Strategies for improving safety and health of ethnic minority construction workers, *Journal of Construction Engineering & Management*, 142(9), 05016007.

Dainty, A. Gibb, A., Bust, P. & Goodier, C. (2007) Health, safety and welfare of migrant construction workers in the South East of England. Report for the Institution of Civil Engineers.

Datta, A., McIlwaine, C., Wills, J., Evans, Y., Herbert, J. & May, J. (2007) The new development finance or exploiting migrant labour? Remittance sending among low-paid migrant workers in London, *International Development Planning Review*, 29(1), 43–67.

Davis, N. & Gibb, A. (2009) Factors that increase health and safety risks for migrant construction workers. In *CIB W099 International Conference*, Melbourne, Australia, pp. 41–49.

Eurostat. (2015) Minimum wage in Europe, European Commission. Available at: https://ec.europa.eu/eurostat/en/web/labour-market/statistics-illustrated.

Fellini, I., Ferro, A. & Fullin, G., (2007) Recruitment processes and labour mobility: The construction industry in Europe, *Work, Employment and Society*, 21(2), 277–298.

Fetterman, D. (1998) *Ethnography: Step-by-Step*, 2nd edn. Thousand Oaks, CA: Sage Publications.

Hare, B., Cameron, I., Real, K. J. & Maloney, W. F. (2013) Exploratory case study of pictorial aids for communicating health and safety for migrant construction workers, *Journal of Construction Engineering and Management*, 139(July), 818–825.

Hassan, H. & Houdmont, J. (2014) Health and safety implications of recruitment payments in migrant construction workers, *Occupational Medicine*, 64(5), 331–336.

HSE. (n.d.) Migrant workers. Available at: http://www.hse.gov.uk/toolbox/workers/migrant.htm.

Jorgensen, D. L. (1989) *Participant Observation: A Methodology for Human Studies*. Newbury Park, CA: Sage Publications.

Kumar, R. (2005) *Research Methodology*, 2nd edn. London: Sage Publications Limited.

Loosemore, M., Phua, F., Dunn, K. & Ozguc, U. (2010) Managing cultural diversity in Australia construction sites, *Construction Management and Economics*, 28(2), 177–188.

Marino, S., Roosblad, J. & Penninx, R. (2017) *Trade Unions and Migrant Workers: New Contexts and Challenges in Europe*. ILERA Publication series. Cheltenham, UK: Edward Elgar Publishing.

Massey, D. (1990) The social and economic origins of immigration, *The Annals of the American Academy of Political and Social Science*, 510 (World Population: Approaching the Year 2000 (Jul., 1990), 60–72.

McKay, S., Craw, M., & Chopra, D. (2006) *Migrant Workers in England and Wales, An Assessment of Migrant Worker Health and Safety Risks*. HSE Publication.

Meardi, G., Martin, A. & Lozano Riera, M., (2012) Constructing uncertainty: Unions and migrant labour in construction in Spain and the UK, *Journal of Industrial Relations*, 54(1), 5–21.

Meltzer, H., Griffiths, C., Brock, A., Rooney, C. & Jenkins, R. (2008) Patterns of suicide by occupation in England and Wales: 2001–2005, *British Journal of Psychiatry*, 193(1), 73–76.

Murie, F. (2007) Building safety – An international perspective, *International Journal of Occupational and Environmental Health*, 13(1), 5–11.

Millward, P. (2016) World Cup 2022 and Qatar's construction projects: Relational power in networks and relational responsibilities to migrant workers, *Current Sociology*, 65(5), 756–776 (2017).

Oswald, D., Sherratt, F., Smith, S. & Hallowell, M. (2018) Exploring safety management challenges for multi-national construction workforces: A UK case study, *Construction Management & Economics*, 36(5), 291–301.

Oswald, D., Wade, F., Sherratt, F. & Smith, S. (2019) Communicating health and safety on a multinational construction project: challenges and strategies, *Journal of Construction Engineering & Management*, 145(4).

Ruhs, M. & Anderson, B. (2010) *Who Needs Migrant Workers?: Labour Shortages, Immigration, and Public Policy*. Oxford: Oxford University Press.

Sharpe, D. (2006) Shop floor practices under changing forms of managerial control: A comparative ethnographic study of micro-politics, control and resistance within a Japanese multinational, *Journal of International Management*, 12(3), 318–339.

Silverman, D. (2013) *Doing Qualitative Research: A Practical Handbook*. London: SAGE Publications Limited.

Trevena, P. (2013) Why do highly educated migrants go for low-skilled jobs? A case study of Polish graduates working in London. In: B. Glorius, I. Grabowska-Lusinska & A. Rindoks (Eds), *Mobility in Transition: Migration Patterns after EU Enlargement*. Amsterdam: Amsterdam University Press.

Tutt, D., Pink, S., Dainty, A. R. J. & Gibb, A. (2013) "In the air" and below the horizon: Migrant workers in UK construction and the practice-based nature of learning and communicating OHS, *Construction Management and Economics*, 31(6), 515–527.

Peltonen, T. (2007) In the middle of managers occupational communities, global ethnography and the multinationals, *Ethnography*, 8(3), 347–360.

Pink, S., Tutt, D., Dainty, A. & Gibb, A. (2010) Ethnographic methodologies for construction research: knowing, practice and interventions, *Building Research & Information*, 38(6), 647–659.

Pole, C. & Morrison, M. (2003) *Ethnography for Education*. Maidenhead, UK: McGraw-Hill Education.

Woolas, P. (2008) Migrant numbers 'must be reduced', BBC News. Available at: http://news.bbc.co.uk/2/hi/uk_news/politics/7677419.stm.

Vargas-Silva, C. & Rienzo, C. (2017) Migrants in the UK: An overview. The Migration Observatory, at the University of Oxford COMPAS (Centre on Migration, Policy and Society). Available at: http://migrationobservatory.ox.ac.uk/resources/briefings/migrants-in-the-uk-an-overview/.

19 Occupational accidents predicting early retirement of construction workers in South Africa

Justus N. Agumba and Innocent Musonda

Summary

In 2018, the construction industry in South Africa experienced 5441 accidents, of which 49 were fatal accidents and 3 workers were forced into early retirement. The construction industry continues to lose workers not only to fatalities but also to early retirement, and this poses a problem to an industry faced with a shortage of skilled people. In this chapter, the types of accidents that result in permanent disability of construction workers leading to early retirement are discussed. Data gathered over ten years (2007–2016) were obtained from the Federated Employers Mutual Assurance (FEM) database. The data were analysed using an Excel statistical package. The analysis established that there were 91,197 accidents in the construction industry between 2007 and 2016. Further, the data revealed that there was a 15% chance of a construction worker being permanently disabled from a construction-related accident while at work. The major cause of injuries was being struck by an object. The most fatal injuries were caused by motor vehicles. Furthermore, the data showed almost a 2% chance of injuries leading to early retirement of construction workers. Despite the low probability of accidents leading to workers' early retirement, it is imperative for construction workers to receive training in best practices for health and safety when they work on any construction project. This will reduce permanent disability further that leads to early retirement among construction workers in South Africa.

The human and social costs of the construction industry

The construction industry is hazardous and dangerous. The hazardous and dangerous conditions of the industry are experienced when workers manually handle heavy materials and are exposed to malfunctioning and vibrating equipment. Furthermore, the noise and dust pollution of the work environment and chemicals exacerbate the dangerous and hazardous conditions on construction sites, which might lead to accidents and ill health of the workers. Robroek and colleagues (2013) argued that the work environment is important, as physical and psycho-social demands in certain industries can make it hard for workers with chronic diseases and ill-health to work. In support of this observation, Brenner and Ahern (2000) found that over 677,000 working days were lost owing to workers being absent from work as a result of work-related injuries and sickness.

Furthermore, they established that over 24,000 potential years of quality working lives were lost owing to early retirement on health grounds. These statistics on lost man-hours are supported by the current Federated Employers Mutual Assurance (FEM, 2018) statistics in South Africa. In September 2018, a record 16,921 working days were lost in the construction sector. Undoubtedly, such a loss has a negative effect on the country's economy.

Apart from the direct, negative effect on the economy, accidents resulting in fatal and non-fatal injuries rob the construction industry of the much-needed, skilled workforce. In South Africa at least 1.5 to 2.5 fatalities are recorded per week in the construction industry (Department of Labour, 2017). This record is comparable with other parts of the world as well. In Britain, 43 fatal injuries were recorded in 2015 among construction workers, equating to a rate of 1.94 per 100,000 workers (HSE, 2016). In Hong Kong, the construction industry recorded 3723 industrial accidents in 2015 and had the highest accident rate and number of fatalities among all industry sectors in Hong Kong (Labour Department, 2016). The picture in the United States of America is no different. The Bureau of Labor Statistics (BLS, 2016) recorded 904 incidents of workers being exposed to harmful substances or environments, being in contact with objects and equipment.

Despite the high economic cost of fatal and non-fatal workplace injuries and the impact on the construction labour force, there is a dearth of research in the type of accidents that cause permanent disability of construction workers, leading to early retirement. The impact on the industry and economy is huge considering that Waehrer and colleagues (2007) argue that the total cost of fatal and non-fatal injuries in the construction industry in the US was estimated at $11.5 billion, disproportionately higher than costs for all industries by 15%.

It is imperative to prevent work disability leading to retirement because of the human and societal costs. Prevention of early retirement as a result of disability is important because the number of construction workers at risk of early retirement on health grounds is likely to rise in the near future (de Zwart, Frings-Dresen & van Duivenbooden, 1999). A study conducted by Brenner and Ahern (2000) established that the mean annual rate of early retirement on health grounds of construction workers in Ireland was 5.3/1000 workers.

Disability retirement is a significant factor affecting working life. In a European study of 4923 older people, physical and psycho-social demands led to an increased exit from paid employment owing to work disability (Robroek et al., 2013). Brenner and Ahern (2000) found that in the Irish construction industry, cardiovascular disease and musculoskeletal disorders each accounted for nearly one third of the conditions leading to permanent disability on the grounds of which early retirement was granted. Boschman and colleagues (2012) indicated that a high prevalence of musculoskeletal disorders (MSDs) among construction workers in two different construction-related occupations, bricklayers and supervisors, was evident in the Netherlands.

Chau and colleagues. (2004) also established that the causes of occupational injuries depended on the type of work, and that preventive measures against injuries from moving objects must be taken, more especially for masons, plumbers and electricians. Preventive measures against injuries from construction machinery and devices should be taken for carpenters, roofers and civil-engineering workers. In Poland, Szubert and Sobala (2005) suggested that the risk factors for early retirement were the conditions of work, i.e. the piecework system, heavy lifting at work, shortage of leisure time, disability, increased rate of absence because of sickness and alcohol abuse. A study in Sweden by Falkstedt et al. (2014) found that women had a higher rate of disability retirement than men, regardless of diagnosis, whereas men had a steeper increase in disability retirement with decrease in educational level. Claessen and colleagues (2009) established that being moderately overweight was not associated with occupational disability among construction workers, but obesity increased the risk of work disability resulting from osteoarthritis and cardiovascular disease.

Järvholm and colleagues (2014) indicated that the work environment was a predictor of disability retirement among construction workers, with those in physically heavy jobs having the highest likelihood of disability. They found that risk varied considerably among blue-collar workers. For example, rock workers had double the risk of disability retirement compared with electricians. However, Arndt et al., (2005) found that the general workforce appeared to experience a higher risk of disability than blue-collar workers in general. Furthermore, most working years lost because of disability retirements were found among men over 50 years, mainly as a result of musculoskeletal and cardiovascular diseases. Arndt et al., (2005) affirmed the sentiment of Järvholm and colleagues (2014). Furthermore, it is important to note that Arndt et al., (2005) called for further efforts to sustain the health of construction workers in their working lives. FEM (2018) statistics in South Africa indicated that by September 2018, 5,441 accidents and 49 fatal accidents had occurred, which led to 3 construction workers being forced into early retirement. However, despite this statistical evidence, there is a dearth of research in South Africa on the type of accident(s) that lead to early retirement as a result of permanent disability of construction workers in the construction industry. The purpose of this chapter is to fill this knowledge gap.

Therefore, a number of specific research questions have been posed in this chapter as follows:

- What types of accidents cause permanent disability of construction workers in South Africa?
- What types of accidents result in fatalities of construction workers in South Africa?
- What are the types of accidents that incur the highest average medical cost?
- What are the lost days in relation to the accidents incurred?
- What is the type of accident that leads to the early retirement of a construction worker?

The specific research objectives were to

- determine the type of accidents that cause permanent disability of construction workers in South Africa;

- establish the type of accidents that cause fatalities among construction workers in South Africa;
- establish the type of accident that has the highest average medical cost;
- determine the number of lost working days as a result of the accidents; and
- determine the type of accidents that lead to early retirement of construction workers.

In order to address the specific research questions and objectives, data were obtained from the Federated Employers Mutual Assurance in South Africa, which enabled empirical analysis to be undertaken. An overview of the types of accidents, permanent disability and cost of accidents in the construction industry is given in the next section.

Literature

Types of accidents

According to Agumba and Haupt (2014), frequently experienced accidents and injuries among small- and medium-sized business construction workers were workers being cut while working, struck by falling objects and workers falling from height. The injuries experienced most frequently were wounds. Hinze and McGlothlin (2002) reported that slips, trips and falls accounted for 15% to 20% of all workers' compensation cases, with older workers having a higher percentage of falls compared with younger workers.

The collapse of tower cranes was another source of danger on construction sites. When there were failures of any part of the crane or the load carry systems, they cause serious damage and injuries to crane operators, site personnel and the general public (Skinner et al., 2006). The US Occupational Safety and Health Administration (2005) posited that significant and serious injuries or fatalities might occur if cranes were not inspected before use and if they were not used properly. Often, these injuries occurred when a worker was struck by an overhead load or caught within the crane's swing radius. Accidents involving cranes emanate from erection or assembling, usage, dismantling, and supervision or inspection, and are a major threat to life of workers on any building site.

Other causes of accidents on construction sites included falls from ladders. Mitra, Cameron and Gabble (2002) reported that accidents resulting from ladder falls were a problem, as the records showed a significant increase between 2001 and 2005 in Australia. These accidents resulted in a significant rise in serious injury from ladder falls. This was evident in their investigation of 4,553 site workers with injuries from ladder falls presented to Victorian Hospital. Of these, 160 patients were classified as major trauma cases. A fall from height, of more than one metre, was the most common incidence of injury accounting for 59% of the total. It was also established that about 20% of ladder-related falls greater than one metre and major trauma cases occurred while people were working on site. It can be stated further that, despite the knowledge of the dangers of falls from ladders, there has been a significant increase in the number of casualties from ladder falls on building sites, which resulted in broken limbs, fractures and bruises.

In order to overcome some of the causes of injuries and illness among construction workers, Chau and colleagues (2004) suggest that training young workers when using hand tools is necessary. Further, the occupational physician could encourage overweight workers to reduce their weight, workers with hearing disorders or sleep disorders to consult a specialist to solve their problems and all workers to practise sporting activities regularly, as this could prevent injuries when handling and carrying objects (Chau et al., 2004).

Disability and associated costs

A disability, whether temporary or permanent, results in socio-economic loss that an individual and society sustains because of an injury, illness or condition. Temporary disability is when a worker is unable to work or cannot do all the work because of an injury or disease, but it is expected that the person will get better. In this case, workers can claim for compensation. However, the worker should be put off duty by a medical practitioner for more than three days. The worker is compensated for the duration of time that he or she was unable to work including the first three days (Western Cape Government, 2014; Compensation for Occupational Injuries and Diseases Act, [COIDA], 1997). Barth (2003/2004) indicated that in the US, states pay permanent partial disability benefits to workers because they suffer an impairment, a disability or some combination of the two. Each state's approach to compensating permanent partial disabilities differs, but for convenience the states can use an approach based on impairment, loss of earning capacity, loss of wages or one that combines features of the other approaches.

In South Africa if a worker is not able to work because of temporary disability, the worker is paid 75% (three-quarters) of the normal monthly or weekly wage. If the worker can only do some of the work, the worker will still get paid by the employer. The compensation fund will pay the worker 75% (three-quarters) of the difference between what the worker is paid and what the worker would have been paid before the injury. All medical expenses are also paid if the medical accounts are submitted to the commissioner (Western Cape Government, 2014; COIDA, 1997). A worker can claim compensation for temporary disability for one year. This can be extended to two years, after which, the commissioner may decide that the condition is permanent and grant compensation on the basis of permanent disability (Western Cape Government, 2014).

In a review of temporary and partial disability benefit programs in Australia, Germany, Great Britain, Japan, the Netherlands, Norway, South Africa, Sweden and the US by Mitra (2009), it was posited that in several countries a focus on time-limited benefits and the design of specific programmes for young adults are important recent developments. Time-limited benefits appear to offer some potential in terms of improved return to work and reduced programme costs.

In South Africa, according to the Amended Compensation for Occupational Injuries and Diseases Act (COIDA, 1997), permanent disability is an injury or illness from which the worker will never recover, for example, loss of hands, an eye, etc. The seriousness of the disability determines whether the worker will ever be able to work again or whether the work will be more difficult to undertake (Western Cape Government, 2014). Disabilities are rated from 100% to 1% depending on the seriousness. For example, 100% would be the loss of both hands or the loss of sight. The loss of one small toe is a 1% disability as indicated in COIDA (1997). If the injury leads to more than 30% disability, the worker is paid a monthly retirement. The amount of retirement depends on what the worker's salary was and on the seriousness of the disability. If a worker suffers 100% disability, the worker would be paid 75% (three-quarters) of the wages. If the disability is less serious, the commissioner would calculate the monthly payment. If the disability was less than 30%, the worker would be paid a lump sum. The lump sum payment is a once-off payment. However, the monthly payment would be paid for the rest of the worker's life (Western Cape Government, 2014; COIDA, 1997).

According to the International Labour Organization (ILO), 337 million accidents occur at work annually, while close to 2 million people suffer from work-related diseases. The ILO suggests that approximately $1.25 trillion is lost annually through costs such as lost working time, increased workers' compensation, interruption of production and medical expenses. Although work should not be a dangerous undertaking, in reality it has resulted in the deaths of more people than past wars (Al-Tuwaijri, 2008).

Construction accidents cost construction organisations, the sector and national economy a great deal annually (Griffith & Howarth, 2000). According to Haupt and Pillay (2016), the costs of construction accidents were estimated at R32,981,200.00. Of this total, R10,087,350.00 has been attributed to direct costs and R22,893,850.00 has been attributed to indirect costs. The costs of construction accidents are based on four cost components: sick pay, administrative costs, recruitment costs, and compensation and insurance costs. This reveals that construction accidents present a substantial cost to employers and to society at large. Furthermore, Haupt and Pillay (2016) established the average costs per accident type, It is evident that accidents involving falling objects (R24,900) and electrical installations (R13,428) had the highest average, while exertion/ergonomic accidents (R1,633) had the lowest average medical costs per accident. Electrical accidents (R199,306) and those involving falls (R125,257) had the highest costs per accident, while exertion/ergonomics accidents (R9, 400) had the lowest costs per accident. Cut/caught-between accidents had the lowest total average cost (R9,612).

Research methodology

To achieve the objectives of the study, a review of literature was undertaken, and an analysis was conducted of secondary data on injuries, early retirements and pension pay-outs resulting from accidents on construction projects, sourced from records published by FEM in South Africa. FEM is the only private workman's compensation insurer registered by the Department of Labour to deal with accidents involving construction workers. The data presented were for the types of accidents recorded during the period 2007 to 2016. The data were comprehensive and sufficient for analysis. It can be indicated that this information is a record of the actual data recorded by FEM. Hence, the information is reliable and indicative of a high degree of ecological validity (Gill & Johnson, 2002). Due to the nature of the study, ethical approval for the study was not necessary given that no particular party would be harmed by the disclosure of the findings of the study. No data would be traceable to a particular construction enterprise or its personnel. The data were represented using descriptive and inferential statistics, that is, percentages and disjoint probability respectively. These statistical parameters were achieved using an Excel statistical package. Disjoint probability was used to determine if an accident led to a construction worker being permanently disabled and early retirement.

Results and discussion

The data in Table 19.1 shows that the construction industry is dangerous and hazardous based on the 91,197 accidents that were experienced from 2007 to 2016. This finding is supported by the data obtained from other reputable sources such as the Bureau of Labour Statistics (BLS) in the US in 2017 and from the Labour Department (2016) in Hong Kong. The major type of accident that was experienced during that decade in the South African construction industry was workers being struck by an object. In this period, 36,396 such cases were reported out of a total of 91,197 accidents. On average, this result translates into approximately 9119.7 cases per year from 2007 to 2016. In the US, the BLS (2016) reported that 159 construction workers were in contact with equipment and objects in 2015. It can be suggested that more construction workers were injured per year in South Africa compared with the US. In view of the statistics, it might be suggested that the standard of health and safety implemented in the US is much higher than that in South Africa.

The results in Table 19.1 indicate further that the major type of accident that renders construction workers permanently disabled was workers being struck by an object while at work. Of the 5951 cases of permanent disability of construction workers reported in this period, being struck by an object contributed 1978 cases, of which 90 workers were forced to take early retirement because of their permanent disability, whereas 1888 workers did not take early retirement or receive their retirement. The probability of permanent disability occurring among construction workers in South Africa was determined to be 15% with a 6.1% chance of the construction worker being permanently disabled without leading to retirement compared with only a 0.43% chance of a worker taking early retirement. Notwithstanding, permanent disability without retirement also reduces an individual's productivity level. The 0.43% chance is also a source of concern because the industry and the country cannot afford to lose skilled individuals to early retirement.

The results in Table 19.2 indicate that the type of accident that caused high fatality among construction workers in South Africa was motor vehicle accidents. Of the 722 fatal accidents experienced in this period nearly half

Table 19.1 Pension status of occupation disabilities in construction

Cause	Number of accidents	Accidents percentage	Permanent disabilities not resulting in pension	Permanent disabilities resulting in pension	Total permanent disabilities	Ranking using number of accidents
Struck by	36396	39.91%	1888	90	1978	1
Fall onto different levels	11452	12.56%	910	104	1014	2
Striking against	10290	11.28%	507	6	513	3
Motor vehicle accident	9341	10.24%	552	111	663	4
Slip or overexertion	8020	8.79%	201	2	203	5
Caught in, on, between	6650	7.29%	777	38	815	6
Fall on to same level	4188	4.59%	203	3	206	7
Accident type (not elsewhere classified [NEC])	2038	2.24%	277	13	290	8
Contact with extreme temperature	1198	1.31%	160	8	168	9
Inhalation absorption, ingestion	1173	1.29%	16	8	24	10
Contact with electric current	451	0.50%	68	9	77	11
Total	91197	100	5559	392	5951	

Source: Federated Employers Mutual Assurance, (2018), FEM's accident stats as of September 2018, [online], available at: www.fem.co.za/Layer_SL/FEM_Home/FEM_Accident_Stats/FEM_Accident_Stats.htm.

Table 19.2 Cost of occupational accidents in construction

Cause	Number of accidents	Accidents percentage	Fatal accidents	Lost days	Cost of accident (Rand ZAR)	Ranking using number of accidents
Struck by	36396	39.91%	134	253108	136,675.00	1
Fall onto different levels	11452	12.56%	109	217670	364,652.00	2
Striking against	10290	11.28%	5	47025	72,370.00	3
Motor vehicle accident	9341	10.24%	354	151485	245,054.00	4
Slip or overexertion	8020	8.79%	3	53912	109,738.00	5
Caught in, on, between	6650	7.29%	31	83455	209,322.00	6
Fall on to same level	4188	4.59%	5	38665	151,430.00	7
Accident type (not elsewhere classified [NEC])	2038	2.24%	36	6554	302,871.00	8
Contact with extreme temperature	1198	1.31%	16	13374	444,635.00	9
Inhalation absorption, ingestion	1173	1.29%	15	1632	164,078.00	10
Contact with electric current	451	0.50%	14	9873	984,550.00	11
Total	91197	100	722	876573	3,185,375.00	

Source: Federated Employers Mutual Assurance, (2018), FEM's accident stats as of September 2018. [online], available at: www.fem.co.za/Layer_SL/FEM_Home/FEM_Accident_Stats/FEM_Accident_Stats.htm.

(354) of the fatalities was caused by motor vehicle accidents. The Bureau of Labor Statistics (2016) in the US recorded 226 incidents based on construction transportation. The HSE (2016) in Britain reported that the industry experienced 43 fatal incidents. However, both reports did not separate the type of accident that caused the fatality. The number of fatal accidents in South Africa was slightly more than three times higher than that in the US and almost 17 times higher than that of Britain. Although an occupational fatality should not be tolerated, the comparison shows that South Africa has a lot more to do to reduce the fatality rate.

The number of days that could have contributed towards economic gains in the construction industry in South Africa that was lost was 876,753 days. The study by Brenner and Ahern (2000) found that 677,000 working days are lost because of absence related to workers' occupational injuries and illness. Furthermore, Brenner and Ahern (2000) found that over 24,000 years of potential working lives were lost because of early retirement on health grounds. In view of these statistics, it is imperative that the construction industry stakeholders in South Africa emphasise the importance of construction health and safety in their projects.

It is important to note that the number of accidents associated with electrocution was 451 compared with 36,396 accidents associated with being struck by an object etc. However, the cost of accident associated with electrocution was the highest at R984,550.00. This finding corroborates the findings of Haupt and Pillay (2016).

It can be inferred that on average, the medical cost of accidental electrocution of an employee was 7.20 times higher than the medical cost of being struck by an object, even though being struck by an object occurred approximately 80.70 times more than a worker being electrocuted in the period from 2007 to 2016. This suggests the enormity of the medical cost and severity of the injury incurred by a worker when electrocuted.

The results in Table 19.3 indicate permanent disabilities that result in early retirement, i.e. an employee being relieved from work and paid his/her retirement package. It is important to note that, in order for a construction worker to be accorded early retirement in the South African context, they would have lost total eyesight, been totally paralysed or suffered other incapacitating conditions. Overall, 392 permanent disabilities resulting in retirement were reported. The type of accident with the highest frequency that led to construction workers taking early retirement was motor vehicle accidents, followed by the frequency of workers falling from different levels and then the frequency of being struck by an object. However, the lowest predictor, based on frequency, was slipping or overexertion by construction workers. The results in Table 19.3 show further that the ratio of construction workers taking early retirement overall as a result of any type of accident was 1 in 233 accidents occurring in the construction industry. However, in relation to the types of accidents, being struck by an object was ranked highest according to the number of accidents. The probability of being struck by an object resulting in workers' permanent disability

Table 19.3 Early retirement from permanent disability

Cause	Number of accidents	Accidents percentage	Ranking using number of accidents	Permanent disabilities resulting in pension	Probability % of early retirement	Ranking using probability % of early retirement
Struck by	36396	39.91%	1	90	0.25 (1:404)	8
Fall onto different levels	11452	12.56%	2	104	0.91 (1:110)	3
Striking against	10290	11.28%	3	6	0.06 (1:1715)	10
Motor vehicle accident	9341	10.24%	4	111	1.19 (1:84)	2
Slip or overexertion	8020	8.79%	5	2	0.03 (1:4010)	11
Caught in, on, between	6650	7.29%	6	38	0.57 (1:175)	7
Fall on to same level	4188	4.59%	7	3	0.07 (1:1396)	9
Accident type (not elsewhere classified [NEC])	2038	2.24%	8	13	0.64 (1:157)	6
Contact with extreme temperature	1198	1.31%	9	8	0.67 (1:150)	5
Inhalation absorption, ingestion	1173	1.29%	10	8	0.68 (1:147)	4
Contact with electric current	451	0.50%	11	9	2.00 (1:50)	1
Total	91197	100		392	0.43 (1:233)	

Source: Federated Employers Mutual Assurance, (2018), FEM's accident stats as of September 2018, [online], available at: www.fem.co.za/Layer_SL/FEM_Home/FEM_Accident_Stats/FEM_Accident_Stats.htm.

and retirement was determined to be 0.25% at a ratio of 1 in 404 accidents. In contrast to being struck by an object, electrocution accidents ranked as being the least likely according to the number of accidents. However, electrocution was ranked higher with regard to resulting in workers taking early retirement, with 2% chance at a ratio of 1 in 50 accidents. Overall, the results suggest that any type of accident occurring in the construction industry had up to 2% chance of resulting in a worker taking early retirement from permanent disability. It could be argued that this probability is low. However, the impact at the individual, company and national level is huge and workers should be informed of the hazardous nature of the construction industry.

Conclusions

In conclusion, the major type of accident in the construction industry in South Africa in the last decade (2007–2016) was workers being struck by an object while working. However, being struck by an object was not a major type of accident that resulted in construction workers being permanently disabled and receiving early retirement payout. In fact, it was evident that electrocution was more likely to result in permanent disability and early retirement. Electrocution had a 2% chance of a worker being permanently disabled and paid a pension. Furthermore, when workers were accidentally electrocuted, their medical expenses were much higher on average in relation to the other types of injuries. This suggests that particular attention should be paid to workers who work with electricity to ensure adherence to the health and safety requirements while working. It is recommended, therefore, that construction workers receive training on health and safety when they work on any project. In addition, adequate risk assessment regarding electrical work should be undertaken on every project.

With regard to fatal injuries of workers, motor vehicle accidents were the major cause. Construction employees who were involved in motor vehicle accidents and suffered injuries were also likely to take early retirement as a result of the injuries. Therefore, it can be stated that when construction workers are using any work-related vehicle, either being transported to site or using a vehicle on site, caution should be taken to prevent accidents. The government of South Africa should ensure that strict measures are taken against drivers who do not observe the traffic rules on site and on public roads. The vehicle should be roadworthy and be driven by a competent driver. This will ensure the prevention of permanent disability and premature retirement among construction workers in South Africa as a result of motor vehicle accidents.

In South Africa, occupational injuries, especially from electrocution, have resulted in workers being permanently disabled, retiring prematurely and costing the nation a great deal of money owing to poor health and safety performance. Furthermore, the industry incurs wasted days owing to the injuries sustained by workers. This is a major concern that should be addressed continuously by the stakeholders.

Limitation of the study

The FEM data lacks demographic statistics. Therefore, it is difficult to know which trades or professions are most affected.

Recommendations for further study

A comparative study on the type of occupational injuries and diseases among construction workers that result in permanent disability but not early retirement, and the type of permanent disability that results in construction employees taking early retirement should be undertaken. This recommendation would shed light on the South African public and the private sectors involved in risky trades.

Acknowledgement

The authors acknowledge the Federated Employers Mutual Assurance in South Africa for making the data available for this study.

References

Agumba, J. N. & Haupt, T. (2014), The types of accidents and injuries encountered by construction SMEs in South Africa, in *Proceedings of the Special Sessions on Sustainable Design and Construction and Resilience Engineering Application on Disaster Mitigation*, in the 5th International Conference on Sustainable Built Environment. Kandy, Sri Lanka, 12–15 December 2014, pp. 69–76.

Al-Tuwaijri, S. (2008), Promoting safe and healthy jobs: ILO Global programme on safety, health and the environment, *World of Work*, No. 63, pp. 4–10.

Arndt, V., Rothenbacher, D., Daniel, U., Zschenderlein, B., Schuberth, S. & Brenner, H. (2005), Construction work and risk of occupational disability: A ten year follow up of 14 474 male workers, *Occupational and Environmental Medicine*, Vol. 62(8), pp. 559–566.

Barth, S. P. (2003/2004), Compensating workers for permanent partial disabilities, *Social Security Bulletin*, Vol. 65(4), pp. 16–23.

Boschman, S. J., van der Molen, F. K., Sluiter, K. J. & Frings-Dresen, H. W. M. (2012), Musculoskeletal disorders among construction workers: A one-year follow-up study, *BMC Musculoskeletal Disorders*, Vol. 13, pp. 1–9.

Brenner, H. & Ahern, W. (2000), Sickness absence and early retirement on health grounds in the construction industry in Ireland, *Occupational Environmental Medicine*, Vol. 57, pp. 615–620.

Bureau of Labor Statistics. (2016), Injuries, illnesses, and fatalities. *Census of Fatal Occupational Injuries (CFOI) – Current and Revised Data.* [Online]. Available at: https://stats.bls.gov/iif/oshcfoi1.htm#2015 [accessed 15 June 2017].

Claessen, H., Arndt, V., Drath, C., and Brenner, H., (2009), Overweight obesity and risk of work disability: a cohort study of construction workers in Germany, *Occupational Environmental Medicine*, Vol. 66(6), pp. 402–409.

Chau, N., Gauchard, G.C., Siegfried, C., Benamghar, L., Dangelzer, J-L, Francais, M., Jacquin, R, Sourdot, A., Perrin, P.P and Mur, J-M., (2004) Relationships of job, age, and life conditions with the causes and severity of occupational injuries in construction workers. *International Arch Occupational Environmental Health*, Vol. 77, pp. 60–66.

Compensation for Occupational Injuries and Diseases Act and Amendments. (1997), [Online]. Available at: www.labour.gov.za/DOL/legislation/acts/compensation-for-occupational-injuries-and-diseases/compensation-for-occupational-injuries-and-diseases-act-and-amendments [accessed 3 June 2017].

Department of Labour. (2017), Injuries and fatalities in SA construction sector, Lack of management and poor workmanship leads to injuries and fatalities in SA construction sector. [Online]. Available at: www.gov.za/speeches/sa-construction-sector-9-mar-2017-0000 [accessed 21 November 2018].

de Zwart, B. C. H, Frings-Dresen, M. H. W. & van Duivenbooden, J. C. (1999), Senior workers in the Dutch construction industry: A search for age-related work and health issues, *Experimental Aging Research*, Vol. 25, pp. 385–391.

Falkstedt, D., Backhans, M., Lundin, A., Allebeck, P. & Hemmingsson, T. (2014), Do working conditions explain the increased risks of disability pension among men and women with low education, A follow-up of Swedish cohorts, *Scandinavian Journal of Work, Environment & Health*, Vol. 50(5), pp. 483–492.

Federated Employers Mutual Assurance. (2018), FEM's accident stats as of September 2018. [Online]. Available at: www.fem.co.za/Layer_SL/FEM_Home/FEM_Accident_Stats/FEM_Accident_Stats.htm [accessed 21 November 2018].

Gill, J. & Johnson, P. (2002), *Research Methods for Managers*, 3rd edn. London: Sage.

Griffith, A. & Howarth, T. (2000), *Construction Health and Safety Management*. London: Pearson Education.

Haupt, T. C. & Pillay, K. (2016), Investigating the true costs of construction accidents, *Journal of Engineering, Design and Technology*, Vol. 14(2), pp. 373–419.

Health Safety Executive (2016), Statistics on fatal injuries in the workplace in Great Britain 2016. [Online]. Available at: www.hse.gov.uk/statistics/pdf/fatalinjuries.pdf [accessed 5 May 2017].

Hinze, J. & McGlothin, J. D. (2002), Prevention of fall from elevations in the construction industry. Poster presented at America Industrial Hygiene Conference, San Diego, CA.

Järvholm, B., Stattin, M., Robroek, J. W. S., Janlert, U., Karlsson, B. & Burdorf, A. (2014), Heavy work and disability pension: A long term follow-up of Swedish construction workers, *Scandinavian Journal of Work, Environment & Health*, Vol. 40(4), pp. 335–342.

Labour Department, Occupational Safety and Health Branch (Hong Kong). (2016), *Occupational Safety and Health Statistics Bulletin*, No. 16 (August), pp. 1–8.

Mitra, B., Cameron, P. A. & Gabble, B. J. (2002), Ladder revisited, *The Medical Journal of Australia*, Vol. 186(1), pp. 31–34.

Mitra, S., (2009), Temporary and partial disability programs in nine countries. What can the United States learn from other countries? *Journal of Disability Policy Studies*, Vol. 20(1), pp. 14–27.

Occupational Safety and Health Administration. (2005), Worker Safety Series: Construction. Washington, DC: U.S. Department of Labor.

Robroek, S. J., Schuring, M., Croezen, S., Stattin, M. & Burdorf, A. (2013), Poor health, unhealthy behaviours, and unfavourable work characteristics influence pathways of exit from paid employment among older workers in Europe: A four year follow-up study, *Scandinavian Journal of Work, Environment & Health*, Vol. 39(2), pp. 125–133.

Skinner, H., Watson, T., Dunklry, B. & Blackmore, P. (2006), *Tower Crane Stability*. CIRIA C654, London: CIRIA.

Szubert, Z. & Sobala, W. (2005), Current determinants of early retirement among blue collar workers in Poland, *International Journal of Occupational Medicine and Environmental Health*, Vol. 18(2), pp. 177–184.

Waehrer, M. G., Dong, S. X., Miller, T., Haile, E. & Men, Y. (2007), Costs of occupational injuries in construction in the United States, *Accident, Analysis & Prevention*, Vol. 39(6), pp. 1258–1266.

Western Cape Government. (2014), Claiming compensation for occupational injuries or diseases. [Online]. Available at: www.westerncape.gov.za/service/claiming-compensation-occupational-injuries-or-diseases [accessed 5 May 2017].

Part V
Technologies for safety management

20 Design of flexible horizontal lifeline systems

Marcelo Fabiano Costella, Valéria Cristina da Motta and Letícia Nonnenmacher

Summary

Accidents involving falls from height in civil construction are still frequent and are the result of inadequate planning in the design of safety systems, such as horizontal lifelines. In this chapter, a protocol for the design of flexible horizontal lifelines (FHLLs) in construction work is presented. The protocol consists of data collection, specification calculations and detailed guidelines for the use of the system, developed according to an analysis of regulatory requirements and already implemented designs and interviews with system designers. The result is a protocol that standardises the different instructions for the design of FHLLs, including the degree of detail required to interpret national and international parameters. As such, it enabled the mapping of the diversity of rules, standards and practices required for the design of this system, filling an information gap and guiding the designers of FHLLs.

Requirements for individual protection equipment

The construction industry is one of the sectors with the most significant number of recorded work accidents because workers are exposed to risks during most of their activities. Among the variety of work carried out on a job site, those performed at heights have been pointed out as the main culprits of deadly incidents in different countries, including Brazil (Lana et al., 2014), the United States (Evanoff et al., 2016; Nguyen, Tran & Chandrawinata, 2016) and the United Kingdom (Health and Safety Executive, 2017). Data from Brazil (AEAT, 2017) reveal that, in 2017, a total of 1,796 accidents and 24 deaths were recorded as a result of falls from height on job sites.

The hierarchy of protection in workplace safety should be based on eliminating or replacing the risk of falls in an activity and, when this is not possible, collective protection equipment (CPE) should be adopted. If this is still insufficient to guarantee the safety of workers, then individual protection equipment (IPE) becomes necessary.

The horizontal lifeline system (HLS) is one of the individual protection measures that can be used during tasks performed at heights, such as maintenance, repair and demolition services, including the assembly of the safety systems themselves, such as guardrail systems and edge working platforms.

According to Wang, Hoe and Goh (2014), the conception of an HLS design must comply with regulatory requirements, such as the free-fall distance and impact loads. However, the Brazilian legislation related to the safety of workers, including NR 35 (Brazil, 2016) and NR 18 (Brazil, 2018), lacks calculation parameters related to the design of lifeline structures for civil construction, implying the use of an HLS designed without a technically consistent basis. As for the international standards that exist as a possibility to complement this information, the American standard ANSI Z359-6 (American National Standards Institute, 2009), which deals with calculation parameters for HLS, indicates the essential characteristics of the requirements.

Goh and Wang (2015) also have recognised the need for further studies and specific, detailed standards to improve the planning and design techniques in workplace safety and health. This lack of specific rules leads to the somewhat cursory specification of horizontal lifelines, causing non-compliances at the time of their implementation and potential risks of work accidents because of falls from height.

Local studies related to the specification of HLS systems present two kinds of problems: one relating to the gap between the design and the real working system, and the other relating to considering only local legislation.

In this chapter, the necessary information is brought together to fill this gap in the specification and design of active fall protection systems based on flexible horizontal lifelines (FHLLs). The aim is to provide support and guidance to professionals who want to design or adapt the system, providing further technical quality to future projects and promoting the subject for future research and improvements in the current standards.

Flexible horizontal lifeline (FHLL)

A horizontal lifeline is a horizontal line fixed at two extremities in a particular work area. It is stretched to serve as a connecting device for the equipment used by workers (Health and Safety Executive, 2004). The most frequently used typology is the FHLL, characterised by the use of steel lifelines attached to an anchoring system composed of anchor posts, which are installed at strategic locations to enable secure access to the areas of activity. There are two different models regarding the method of operation: fall restraint systems and fall arrest systems (Branchtein, 2013; Wang, Hoe & Goh, 2014).

The fall restraint system works by limiting the movement of the worker to within a certain distance, not allowing him to gain access to the risk zone and enter into a free fall. This system is usually applied to roof services (Branchtein, Souza & Simon, 2015). The fall arrest system, on the other hand, is made up of interconnected components and sub-systems that operate in such a way as to stop the fall of a worker and avoid his collision with floors, structures and other obstacles. This system is usually used during the construction phase of the structure of buildings (International Organization for Standardization, 2000; British Standards, 2008).

Lanyards and harnesses

In addition to the anchoring structure and the steel lifeline, an FHLL system has the body harness, the lanyard and the energy absorber as principal components (Goh & Wang, 2015).

Fixed at the intermediate anchoring of the personal fall protection system, the lanyard, which consists of a rope or cable of flexible material with latching devices at both ends, must tie the worker to the steel cable. The lanyard can be coupled to an energy absorber, which helps dissipate the kinetic forces generated during the fall, and its use is compulsory in fall arrest systems (International Organization for Standardization, 2000; British Standards, 2005).

As for the harnesses, these are defined as support devices that connect the worker to the anchoring system by linking to the lanyard. Harnesses are composed of buckles, straps and other elements that can support the weight of the user and the other existing loads (Workplace Safety and Health Council, 2012). Full body or simple belt harnesses can be used depending on the purpose and objectives of the protection system: fall restraint or fall arrest.

Research method

The performance requirements of the international standards covering work at heights and fall protection systems were reviewed as a way to develop a protocol for the specification of flexible horizontal lifelines, with the parameters of the analysis relating only to the FHLL system using steel lifelines and its components (lanyards, energy absorbers and harnesses). In terms of the operation typology, only fall arrest systems were observed, with the other typologies being excluded, such as the vertical lifeline, rigid horizontal lifeline and positioning systems.

The studied standards included:

- ANSI/ASSE Z359-1: The fall protection code – American National Standards Institute (2016)
- ANSI/ASSE Z359-6: Specifications and Design Requirements for Active Fall Protection Systems – American National Standards Institute (2009)
- BS 8437: Code of pratice for selection, use and maintenance of personal fall protection systems and equipment for use in the workplace – British Standard Institution (2005)
- BSEN 363: Personal Fall Protection Equipment – personal fall protection systems – British Standard Institution (2008)
- ISO 10333-2: Personal Fall-Arrest Systems – Part 2: Lanyards and energy absorbers – International Organization for Standardization (2000)
- ISO 16024: Flexible horizontal lifeline systems – International Organization for Standardization (2005)
- OSHA 1926.502: Fall protection systems criteria and practices – Ocupational Safety and Health Administration (1996)
- NR 35: Trabalho em altura (Work in height) – Brazil (2016).

Beyond the normative references, most detailed design information was developed according to the analyses of already implemented FHLL designs and through interviews with designers. These observations included the identification of points of failures, vulnerabilities and practical information to be considered in the protocol, including requirements that had not been considered yet.

The calculation and specification of FHLL systems were based on the mechanics of materials principles and the working stress design also adopted by Branchtein, Souza and Simon (2015) for steel lifelines and their components.

In this way, a direct and straightforward composition of the required steps and information for the specification of fall arrest FHLL systems was obtained with the following protocol:

1. Organise and develop the design that will receive the lifeline system
2. Specify the typology of the adopted lifeline in the design
3. Pre-determine the model of the flexible horizontal lifeline system
4. Determine the location of the anchoring system in the floor plans
5. Gather the required information for the specification of the steel lifeline
6. Calculate the specifications of the steel lifeline
    6a. Cable clamps
7. Gather the required information for the specifications of the anchoring system
8. Determine the forces acting on the anchoring system
9. Calculate the specifications of the anchoring system
    9a. Exerted bending stress
    9b. Exerted stress
    9c. Permissible stress
    9d. Buckling
10. Determine the fall factor
11. Calculate the free-fall zone
12. Create a detailed design
13. Write guidelines for using the system

Protocol for design of flexible horizontal lifelines

The protocol is presented next with the inclusion of all required support for the complete design and specification of a flexible horizontal lifeline fall arrest system.

1. Organise and develop the design that will receive the lifeline system

The plans of each floor must be separated to observe their differences, and the floor plans must be cleared of furniture and interior walls, keeping in mind that the location of stairs and open clearing spaces must be preserved. Observe the cut-away views repairing possible changes in the height of the floors in order to determine the dimensions of the system correctly.

Flexible horizontal lifeline systems 291

2. Specify the typology of the adopted lifeline in the design

The design must specify and declare the typology of the employed lifeline, defining whether it will be a fall restraint or fall arrest system.

The protocol of this chapter was developed for a fall arrest system.

3. Pre-determine the model of the flexible horizontal lifeline system

A model of the anchoring system to be designed, and its frames and support models for the anchor posts must be pre-determined. For example, one must establish whether to use a cross-arm brace or tripod. Figure 20.1 shows the model that will be specified, which is made of anchor posts, a support frame for the anchor posts, an anchor point for the steel line, a base frame and steel lifeline.

4. Determine the location of the anchoring system in the floor plans

The system must be placed in the floor plan, observing the maximum distance between the anchor posts, going around the entire floor. Represent all distances in the system, especially those referring to the edges of the building, observing the length of the lanyard and representing the area that the worker will reach when connected to the system in the design, ensuring sufficient distance to perform the work.

5. Gather the required information for the specification of the steel lifeline

Gathering the following characteristics is essential for the specification of the steel lifeline: total number of workers, total body weight (maximum 100 kg, according to NBR 15835 of the Associação Brasileira de Normas Técnicas, 2011;

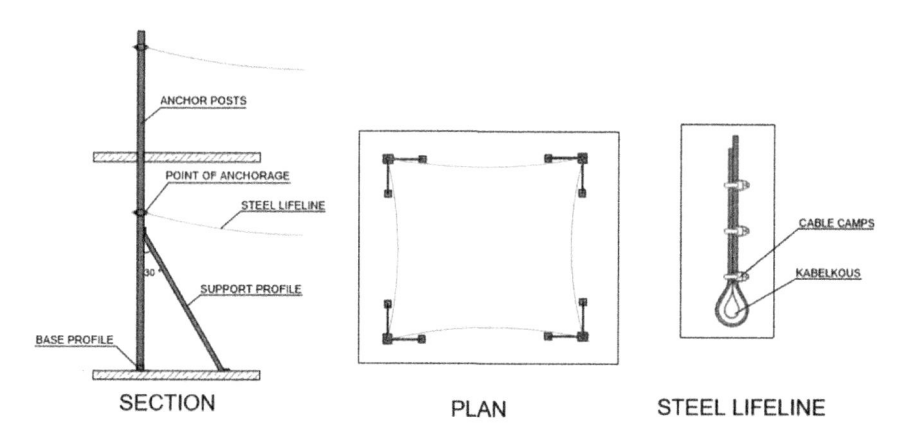

Figure 20.1 Anchoring system.

assembly sag; maximum space between the anchor posts; and total dynamic load (600 Kgf, according to the Brazilian NR 35, 2016). In addition to tensile strength, lifeline factor (F), lifeline diameter and elasticity module (E) can be obtained in the technical manual of the adopted steel lifeline. The usual recommendation is the use of a galvanised steel lifeline of 6 × 19 or 6 × 25.

6. Calculate the specifications of the steel lifeline

The specification of an FHLL can be done without considering the energy absorber. Figure 20.2 shows the nomenclature adopted in the formulas and calculations for the specification, where L is the gap between the anchor posts of the lifeline, and the other abbreviations are shown within their respective steps.

The first step is the calculation of the assembly sag (F1) with Equation 20.1. According to ANSI Z359.6 ASSE (American National Standards Institute, 2009), the bending must be at least 2% of the gap. The larger the assembly bending, the lower the stress will be.

$$F1 : 0.02L \tag{20.1}$$

The second step is to calculate the actual length of the lifeline (L1) with the assembly bending, using Equation 20.2, and the third step is to calculate the bending (F2) considering the actual length of the lifeline (L1) through Equation 20.3:

$$L1 : L\left(1 + \frac{2}{3}\left(\frac{f1}{L/2}\right)^2\right) \tag{20.2}$$

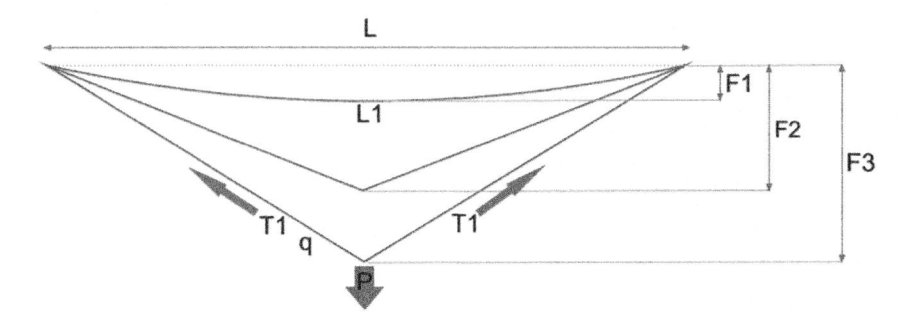

Figure 20.2 Horizontal lifeline diagram without energy absorber.

(Adapted from Branchtein, M. C., de Souza, G. L. & Simon, W. R., 2015, Sistema de Proteção Ativa Contra Quedas com Linha de Vida Horizontal Flexível, Em: V, in: A. Filgueiras (Ed.), *Saúde e segurança do trabalho na Construção Civil brasileira*, Procuradoria Regional do Trabalho da 20° região Sergipe, Sergipe: Ministério Público do Trabalho, pp. 159–176.)

$$F2: \sqrt{\left(\frac{L1}{2}\right)^2 - \left(\frac{L}{2}\right)^2} \qquad (20.3)$$

The fourth step is to calculate the stretch of the lifeline (L) with Equation 20.4, so the bending can be calculated (F3), which is the maximum bending when the dynamic load is at its maximum. However, to calculate the stretch of the lifeline, the tractive force (T) that acts upon it must be known, which depends on the dynamic load on the body (P) and the angle formed by the cable when subjected to the dynamic load. However, this same angle depends on the bending (F3). As a result, it necessary mathematically to adopt any value (T) and perform an iterative calculation:

$$\Delta L : \frac{T\ L1}{E\ Ac} \qquad (20.4)$$

where Ac is the area of only the metal part of the lifeline, determined by Equation 20.5, with the use of factor (F), which is provided in the technical manuals of each steel lifeline or in the NBR 2408 standard (Associação Brasileira de Normas Técnicas, 2009), and where dc is the diameter of the lifeline:

$$Ac : F\,dc^2 \qquad (20.5)$$

The next step is the calculation of the bending (F3) with Equation 20.6.

$$F3: \sqrt{\left(\frac{L1+\Delta L}{2}\right)^2 - \left(\frac{L}{2}\right)^2} \qquad (20.6)$$

The standard determining the vertical dynamic load that acts perpendicular to the lifeline can be found in NR 35 (Brazil, 2016), which stipulates 6 KN as the maximum value to be applied on the worker at the time of the fall arrest, which can also be called braking force or impact force. For situations in which more than one worker is connected to the system, ANSI Z359.6 ASSE (American National Standards Institute, 2009) explains that the value of 6 KN per person must be used for up to two workers and that an additional load of 1 KN per worker should be used when this limit is exceeded. For example, for three workers: (2 p × 6 KN) + 1 kn = 13 KN.

The last step is to determine the force on the steel lifeline, previously based on any tractive force (T), but now deduced from Equation 20.7 with the similar triangles method, through which one finds (T1):

$$T1: \frac{P(L1+\Delta L)}{4f3} \qquad (20.7)$$

$T1$ is compared with the previously adopted T. If the value of the two forces is not equal, an interpolation is performed and the value found is once again substituted in the place of T at the beginning of the calculations until $T1$ is equal to T. When the forces are equal, this will be the force considered in the design for the specification of the anchoring system.

According to NR 18 (Brazil, 2018), the safety lifelines must have a safety factor of at least five. The breaking stress of the lifeline that is listed in its technical manual, therefore, should be at least five times greater than the tractive force (T) on the lifeline.

Cable clamps

According to NBR 11900-4 (Associação Brasileira de Normas Técnicas, 2016), the attachments and patches between the lifelines must use cable clamps. The number varies according to the diameter of the lifeline, with the existing parameters listed in a table of Annexure I of the same standard. The distance between the clamps is approximately six times the diameter of the lifeline. The remaining part of the cable must have a minimum length equal to the width of the base of the clamp.

7. Gather the required information for the specification of the anchoring system

The model of all the components in the lifeline must be determined: steel lifeline, clamps for the lifeline, frames, bolts and nuts used for the structure of the lifeline.

The following information is needed for the specification of the frames: model and material of the frame, dimensions (length, width, thickness), section area, the largest gap between one structure and another, weight, elastic modulus, the yield stress of the steel, inertia, resistance modulus and safety coefficient. The angle that the supporting frame makes with the anchor post also needs to be known.

For the specification of the bolts, the following information is needed: where they are fixed, their quantity, type of bolt, dimensions, section area, supported shear or traction load, supported shear or traction stress, and safety coefficient.

8. Determine the forces acting on the anchoring system

Determine the forces exerted on all frames that make up the system, decomposing the traction force on the lifeline that goes through the anchoring system through the sum of forces. The forces for one of the anchoring structures of the system should be determined individually, as in Figure 20.3, which will be repeated for the entire floor, forming a comprehensive FHLL system in conjunction with the steel lifeline.

For the specification of all components of the system, such as the frames and bolts, the stress exerted on the part must be lower than the permissible stress of the employed material (exerted stress < permissible stress). If this criterion is

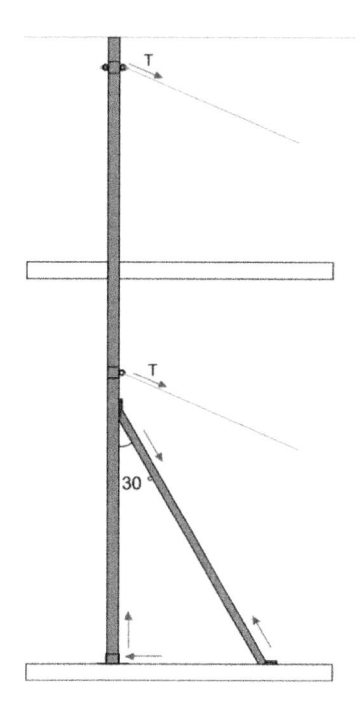

Figure 20.3 Forces calculated for the system.

not met, the part must once again be specified to obtain a greater area of steel, or another material must be adopted. According to OSHA 1926.502(d)(15)(i) (Occupational Safety and Health Administration, 1996), the lifeline design should consider a safety factor of at least 2.

9. Calculate the specifications of the anchoring system

Exerted bending stress

The calculation of the exerted bending stress according to Equation 20.8 should be used for the specification of the main anchor frame instead of the exerted tension because, generally, the support of the structure is not placed at the same height as the lifeline connection. This distance between the force caused by the steel cable and the force that the support exerts to resist this effort generates a bending moment.

$$\sigma \max : \frac{M}{W} \qquad (20.8)$$

where σx is the exerted bending stress, M is the bending moment and W is the resistance modulus.

Exerted stress

The exerted stress can be calculated through Equation 20.9:

$$\sigma : \frac{P}{A} \tag{20.9}$$

where σ is the exerted stress, P is the axial force and A is the cross-section area.

Permissible stress

The permissible stress can be calculated through Equation 20.10:

$$Permissible\ stress : \frac{\sigma e}{CS} \text{ or } \frac{\sigma r}{CS} \tag{20.10}$$

where σe is the yield stress of the steel, Σr is the breaking stress of the steel and CS is the safety coefficient.

For ductile materials like steel, the yield stress should be used for the calculation.

Buckling

Buckling must be calculated for parts where the cross-section area is small compared to the length (spare parts), which is the case for the main anchoring structure and the anchoring support structure. According to Pinheiro (2005), for a correctly specified system, the force that the part should support should not be greater than the critical buckling force called P_{cr} (Equation 20.11):

$$P_{cr} : \frac{\pi^2 x E x I}{Lfl^2} \tag{20.11}$$

where E is the elasticity modulus of the steel, I is the lowest moment of inertia of the bar and Lfl is the buckling length of the bar.

The buckling length of the bar is calculated with Equation 20.12:

$$Lfl : L x K \tag{20.12}$$

where L is the length of the bar and K is the buckling parameter according to the support condition of the frame.

10. Determine the fall factor

The fall factor is determined by the ratio between the distance the worker travels during the fall and the length of the lanyard. Ideally, its value should be less or equal to 1 and never exceed 2. According to the NR 35 standard (Brazil, 2016),

Flexible horizontal lifeline systems 297

when the fall factor is less than 1 or when the lanyard has a length of less than 0.9 m, the use of an energy absorber is not mandatory.

11. Calculate the free-fall zone

The calculation of the free-fall zone serves to determine the minimum installation height of the horizontal lifeline, ensuring that the worker does not collide with the lower floor of the work site. The diagram in Figure 20.4 shows how this distance should be determined in cases where the worker uses a lanyard or retractable fall arrester.

In Figure 20.4, a is the lanyard; b is the fully open energy absorber; c is the distance from the point of connection of the harness to the person's foot (1.5 m); d is 1 meter of safety margin as determined by the standards; A1 is distance between the ring attached to the lifeline to the carabiner of the retractable fall arrester in the fully retracted position; b1 is the retractable cable outside the housing in working position; and B1 is the retractable cable in the final position.

Based on Equation 20.13, the free-fall zone with the use of the lanyard is determined; if the worker does not make use of the energy absorber, the length should be disregarded:

$$ZLQ : F3 + a + b + c + d \qquad (20.13)$$

For those cases in which a retractable fall arrester is used, one should check if the worker will reach the floor using Equation 20.14:

$$Hp : F3 - F1 + B1 - b1 + d \qquad (20.14)$$

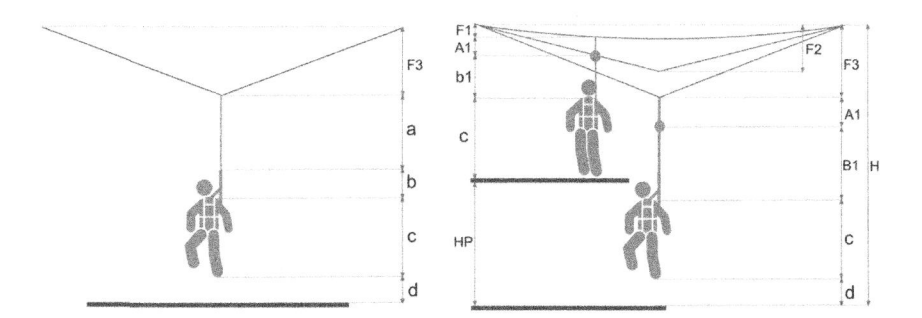

Figure 20.4 Free-fall zone (ZLQ) diagram.

(Adapted from Branchtein, M. C., de Souza, G. L. & Simon, W. R., 2015, Sistema de Proteção Ativa Contra Quedas com Linha de Vida Horizontal Flexível, Em: V, in: A. Filgueiras (Ed.), *Saúde e segurança do trabalho na Construção Civil brasileira*, Procuradoria Regional do Trabalho da 20° região Sergipe, Sergipe: Ministério Público do Trabalho, pp. 159–176.)

Subsequently, the minimum height for the installation of the retractable arrester should be verified according to Equation 20.15:

$$H tot : F3 + A1 + B1 + c + d \qquad (20.15)$$

12. Create a detailed design

All frames, steel lifelines and bolts used should be illustrated in the design, specifying the dimensions (width, length and diameters) and materials used as well as the mountings, number of screws and models to be used. This information should be attached to the location of the project and written in the complete specifications and calculation log. Cut-away views of the model system should also be provided, representing the installation floor and its use with the respective dimensions.

13. Write guidelines for using the system

A qualified person with extensive knowledge of the subject, capable of assessing and specifying the product, is required to develop the design (Health and Safety Executive, 2004). Also, the qualifications of the engineer who designed the system will be required, including a statement of whether the components can be used again in other locations.

The guidelines of the safety equipment should be included in the design: adopted harness, length and model of the lanyard, information plate stating the number of people that can be connected simultaneously to the lifeline in the same section, and warning about mandatory anchoring during activities involving fall risks. In addition, calculation logs, documented procedures for emergency rescue situations, and instructions for the assembly and disassembly should be established, following the technical manuals of the employed equipment.

Implications for practice and research

Owing to the high number of accidents caused by falls from height related to the ineffectiveness of fall protection systems, there is a need for further studies on designs and calculation of flexible horizontal lifelines that examine the interpretation of normative requirements and details applied in practice at the construction site, such as those presented in this chapter.

Although there is a considerable diversity of standards assisting the professional in the planning of FHLL systems, an analysis of the already implemented designs revealed that designers do not use these as a model. As such, the lack of information in the Brazilian standards regarding the specification of the system and the lack of studies addressing the theme tend to limit the design of FHLL systems to complying with only the minimum requirements, or forces professionals to adopt different methods and parameters, with different results, which is not conducive to safety.

Besides, the current legislation does not include the integration of the various parts involved in the conception of the design, which results in a design that is difficult or impossible to execute on the construction site. These adaptations in the assembly generate doubts about the safety and effectiveness of the system.

Thus, this protocol fills a gap caused by the absence or dispersion of information as well as the lack of standardisation of regulatory information and practices, resulting in a document that can guide professionals who want to design or optimise FHLL systems.

References

American National Standards Institute. (2016), ANSI/ASSE Z359-1: The fall protection code, American National Standard, United States of America.

American National Standards Institute. (2009), ANSI/ASSE Z359-6: Specifications and design requirements for active fall protection systems, American National Standard, United States of America.

Anuário Estatístico de Acidentes do Trabalho. (2017), EAT: Ministério da Fazenda, Brasília.

Associação Brasileira de Normas Técnicas. (2009), NBR 2408: *Cabos de Aço para Uso Geral – requisitos mínimos*, ABNT, Rio de Janeiro.

Associação Brasileira de Normas Técnicas. (2016), NBR 11900-4: *Terminal para Cabo de Aço – Parte 4: grampos leve e pesado*, ABNT, Rio de Janeiro.

Associação Brasileira de Normas Técnicas. (2011), NBR 15835: *Equipamento de Proteção Individual Contra Queda de Altura – cinturão de segurança tipo abdominal e talabarte de segurança para posicionamento e restrição*, ABNT, Rio de Janeiro.

Branchtein, M. C. (2013), Lifeline design: Calculation of the tensions. In: *Proceedings of International Society for Fall Protection Symposium*, 27–28 June, Las Vegas, Nevada.

Branchtein, M. C., de Souza, G. L. & Simon, W. R. (2015), Sistema de Proteção Ativa Contra Quedas com Linha de Vida Horizontal Flexível, Em: V. In: A. Filgueiras (Ed.), *Saúde e segurança do trabalho na Construção Civil brasileira, Procuradoria Regional do Trabalho da 20° região Sergipe*. Sergipe: Ministério Público do Trabalho, pp. 159–176.

Brasil. (2018), NR 18: *Condições e Meio Ambiente de Trabalho na Indústria da Construção*, Brasília.

Brasil. (2016), NR 35: *Trabalho em altura*, Brasília.

British Standard Institution. (2005), BS 8437: Code of pratice for selection, use and maintenance of personal fall protection systems and equipment for use in the workplace, BSI, England.

British Standard Institution. (2008), BSEN 363: Personal fall protection equipment – Personal fall protection systems, BSI, England.

Lana, L. D., de Quadros, J. D. N., Weise, A. D., Reis, R. P., Rosa, L. C. & Buligon, S. M. (2014), Avaliação dos Riscos do Trabalho em Altura na Construção Civil, *Revista Produção Online*, Vol. 14, pp. 344–363.

Evanoff, B., Dale, A. M., Zeringue, A., Fuchs, M., Gaal, J., Lipscomb, H. J. & Kaskutas, V. (2016), Results of a fall prevention educational intervention for residential construction, *Safety Science*, Vol. 89, pp. 301–307.

Goh, Y. M. & Wang, Q. (2015), Investigating the adequacy of horizontal lifeline system design through case studies from Singapore, *Journal of Construction Engineering and Management*, Vol. 141, pp. 1–8.

Health and Safety Executive. (2004), *A Review of Criteria Concerning Design, Selection, Installation, Use, Maintenance and Training Aspects of Temporarily-Installed Horizontal Lifelines*. Sudbury, UK: HSE Books.

Health and Safety Executive. (2017), Health and safety in construction in Great Britain, 2017. [Online]. Available at: www.hse.gov.uk/statistics/fatals.htm [accessed 20 February 2017].

ISO – International Organization for Standardization. (2000), ISO 10333-2: Personal fall-arrest systems – Part 2: Lanyards and energy absorbers.

ISO – International Organization for Standardization. (2005), ISO 16024: Flexible horizontal lifeline systems.

Ocupational Safety and Health Administration. (1996), 1926.502 – Fall protection systems criteria and practices. [Online]. Available at: www.osha.gov/pls/oshaweb/owadisp.show:document?p_table=STANDARDS&p_id=10758 [accessed 25 September 2017].

Pinheiro, A. C. F. B. (2005), *Estruturas metálicas: Cálculos, detalhes, exercício e projetos*. São Paulo: Blucher.

Nguyen, L. D., Tran, D. Q. & Chandrawinata, M. P. (2016), Predicting safety risk of working at heights using Bayesian Networks, *Journal of Construction Engineering and Management*, Vol. 142, pp. 1–11.

Wang, Q., Hoe, Y. P. & Goh, Y. M. (2014), Evaluating the inadequacies of horizontal lifeline design: Case studies in Singapore. CIB W099, 2–3 June, Lund, Sweden.

Workplace Safety and Health Council. (2012), *Workplace Safety and Health Guidelines: Personal Protective Equipment for Work at Height*. Singapore: WSH Council Books.

21 Development of RFID automation system to improve safety on construction sites in Brazil

*Victor Hugo Mazon de Oliveira and
Sheyla Mara Baptista Serra*

Summary

Data from the International Labour Organization (ILO) indicate that Brazil is among the four countries with the highest number of work-related accidents in the world. One of the alternatives to improve this situation is the use of information and communication technologies (ICT). This chapter presents the development of an automated monitoring system of collective protection equipment (CPE) and workers in risk zones through radio frequency identification (RFID) technology. Specific interface software, which manages the functions of the tool by storing, organising the database, and generating results through spreadsheets, graphs and plant images, was developed to enable this research. The testing routine combined laboratory experiments and validation of the proposal in a real construction environment. The system proved to be efficient in (1) identifying and locating safety CPE by indicating its position in the plant image and showing its characteristics; and (2) identifying the approach of workers to areas of risk of falling. Both proposals adopt RFID technology for a specific purpose.

The need for new techniques to manage safety in construction

The International Labour Organization (ILO, 2017) organised work accident reports with records submitted by more than 200 countries. In this ranking, Brazil occupies the fourth place for deaths caused by accidents at work, behind only the United States, Thailand and China. Another aspect is that the construction industry has a disproportionately high rate of recorded accidents compared to other industries.

According to the Ministry of Social Security (BRASIL, 2017), between 2006 and 2010 the construction sector in Brazil expanded with a consequent increase in the number of employees. During the same period, the number of work accidents increased from 29,054 to 54,664, representing an increase of 88.14%. These data reiterate the need to develop techniques, tools and methodologies capable of assisting in the management of safety conditions and mitigation of the risks of accidents.

The diversity of operations carried out at construction sites increases the risks to which workers are exposed, making it necessary to take preventive measures. The regulatory norms and technical standards guide the various measures to be taken for the prevention or reduction of industrial accidents.

Mehrjerdi (2008) indicated that the construction industry has shown an evolutionary interest in applying tools using new techniques of management including the use of information and communication technology (ICT).

For Demiralp, Guven and Ergen (2012), there is a clear deficit in the traditional managerial methodologies of construction. These authors argue that the use of automated data collection technologies such as laser scanners, wireless sensors and radio frequency identification (RFID) can help to efficiently track components in construction supply chains. According to Demiralp and colleagues (2012), the problems encountered in the current, manual methods of material tracking include late deliveries, missing components and incorrect installations.

RFID is a method of automatically identifying, retrieving and storing data remotely through radio signals using devices called RFID tags.

According to Domdouzis, Kumar and Anumba (2007), RFID technology is innovative, with many applications in various fields. Currently, there are attempts to harmonise the use of RFID frequencies among different countries to facilitate, for example, its use in international trade and aeronautics. According to Domdouzis and colleagues (2007), RFID technology is becoming ubiquitous, and with technological advances, there is a consequent cheapening of equipment. In their opinion, the proliferation of RFID usage suggests that most industrial sectors will experience the effects of this technology.

Giretti and colleagues (2009) reported a study dedicated to safety and health management on construction sites using new communication technologies powered by RFID. The deployed system would emit hazard warning signals when workers approached a region with a high risk of an accident. Thus, the initial function of the RFID system was to track the position of workers involved in construction activities in real time, leading to the development of software able to control access to non-authorised, hazardous areas.

The use of RFID to manage materials and components brings significant gains in efficiency and productivity for management activities on a construction site, including keeping the inventory up to date and accurate. The deployment of RFID enables the integration and communication of the information flow among project teams, materials acquisition and monitoring, the warehouse, the local office and the construction site, among others (Ren, Anumba & Tah, 2011).

Demiralp and colleagues (2012) presented a case study in the supply chain of precast concrete panels manufactured at a production plant. The RFID system was designed to include transport, the handling of the elements and assembly activities on a construction site. In the company's existing practice, after a panel is produced, workers mark it with tags that are used to identify the components automatically throughout the supply chain. RFID technology has been integrated with a Global Positioning System (GPS) system to locate the components.

There is indeed large scope for the applicability of RFID systems in building construction and its utilisation can help to address the occurrence of many accidents at work.

However, technology itself is only a tool, and needs orientations and procedures to be developed, disseminated and inserted according to its application in the construction process. The investigation of a new way to develop mechanisms using RFID is an opportunity to meet the demand for innovation required by the construction industry, especially in emerging countries that need to use more automated processes to improve construction management.

Despite the significant advantages, the use of ICT, and specifically RFID, on construction sites in Brazil is still very low. RFID technology is an example of innovation in construction, which should be studied further. The number of construction accidents in Brazil is considerable and the adoption of RFID technology can contribute to making the work environment safer and more productive, but its effective application is very low.

In this chapter, the development of a monitoring and control system for site workers and CPE using RFID technology is presented. The system consists of RFID components, specific software and the items to be managed. Software was developed and named Construction Control by RFID, which incorporates two basic functions: (1) inventory control of collective protection equipment, and (2) access control of workers to the areas of fall hazards. The software operates according to a menu that makes it possible to choose a function to use at any time. In addition, the software stores, compiles and organises independent databases providing the results of the equipment readings in the form of reports, charts and spreadsheets.

Perspectives and concepts

RFID systems store information in electronic data transport devices. The power supply and the data exchange between the transport device and the reader are performed using magnetic or electromagnetic fields instead of touch.

RFID technology essentially consists of four units or components: antenna, radio waves transmitter, RFID tag and storage software. An antenna emits radio signals continuously at a given frequency. When a tag configured to detect the transmission frequency used by the antenna detects these signals, it is activated and communicates with the reader through frequency transmittance modulation. The antenna stabilises the communication between the tag and the transmitter. The reader obtains and analyses the data and sends it to the system to which it is connected in order to recognise the communication protocols in the software that stores the information (Finkenzeller, 2010; Domdouzis et al., 2007; Lieshout et al., 2007).

Examples of applications of RFID in the construction industry demonstrate the variety of potential uses for technology in isolation, or in conjunction with other technologies. Several publications regarding the management of materials and components, logistics and supply chain are available. Wang and colleagues (2017) investigated the data-driven mechanisms and benefits of using RFID in knowledge-based, pre-cast construction supply chains with a barcode. The results of the construction of 100 pre-cast wall panels in a two-echelon, pre-cast, construction supply chain revealed that the knowledge-based RFID system could generate a 62.0% savings in operational costs, which is 29.0% higher than that of a barcode-based system. Lanko and colleagues (2018) considered the application feasibility of blockchain in logistics of construction materials using RFID technology. An example of the introduction to the manufacturing and delivery process of ready-mixed concrete is given. By combining RFID and GPS technology, it will be possible to track the delivery of ready-mix concrete in real time, which will make it possible to predict the time of delivery of the products to the customer accurately and optimise the construction process as much as possible, minimising losses from delays in delivery.

RFID also has potential uses in the area of security. Lee and colleagues (2012) have developed a real-time location system of workers based on RFID identification for safety management. Yang and colleagues (2012) present a requirement analysis and technical design of a real-time identity information tracking solution for pro-active accident prevention aimed at improving safety performance at construction sites.

Costin and colleagues (2015) presented a proposal for the integration of safety management through RFID technology, called building information modelling (BIM) for tracking workers in real time using visualisation in a 3D model. The system was deployed on a 900,000 square foot hospital project consisting of three major buildings, 125 contractors and 1200 workers. Access to the building was controlled by RFID, which ensured visibility and tracking of equipment, materials and personnel in the construction site. RFID readers were installed at the entry to the premises and the exit turnstiles, and the construction site was divided into control areas.

Therefore, great potential for the application of RFID technology in construction is described in the literature. In this sense, the creation of models, methods and strategies in the development of systems dedicated to the health and safety of workers using RFID technology must be studied. According to Wu and colleagues (2013), 66% of accidents observed during a case study could have been avoided in a pro-active and preventive way by using remote monitoring technologies, such as RFID for example.

Method

The research strategy adopted for this study was design science research (DSR). According to Hevner and colleagues (2004), DSR is a strategy to implement information systems within an organisation for the purpose of improving the effectiveness and efficiency of that organisation. The objective of DSR is to produce an artefact or even a limitation. In this sense, DSR is regarded as a rigorous process for problem-solving in the design of artefacts, evaluation of what was designed and communication of the results obtained.

The objective of this chapter is to present the development of a system for monitoring and controlling site workers and collective protection equipment (CPE) using RFID technology.

The installation of CPE is defined according to the characteristics of the activities being carried out at the time of construction. There are several solutions available in the market. However, CPE is not always installed and used efficiently, or properly controlled during use.

It is understood that the development of existing CPE management and control methods will help to maintain and preserve the safety of the sites.

One of the main reasons for the implementation of DSR in this study was the potential of this method to overcome the gap between the existing and widely applied knowledge of RFID technology, and its specific application in monitoring and managing people and equipment in hazardous areas in construction sites.

According to Van Aken (2004), the adoption of this research strategy is oriented towards the solution of problems, while, at the same time, producing knowledge that can serve as a reference for the development of theories. Figure 21.1 shows the conceptual framework for understanding, implementing and evaluating the information systems proposed by Hevner and colleagues (2004) adapted to the objectives considered for this study.

The research covered the development of a monitoring and control system that integrates RFID components, specific software for integration and two different items to be managed: the CPE and workers in building sites.

Table 21.1 shows the RFID-specific components used in the research. A passive tag model B-20070/C-20055, resistant to interference from metallic structures and bad weather in the operation of its data transmission, was used to identify the safety equipment (row 1 image). Passive tag, helmet stickers were

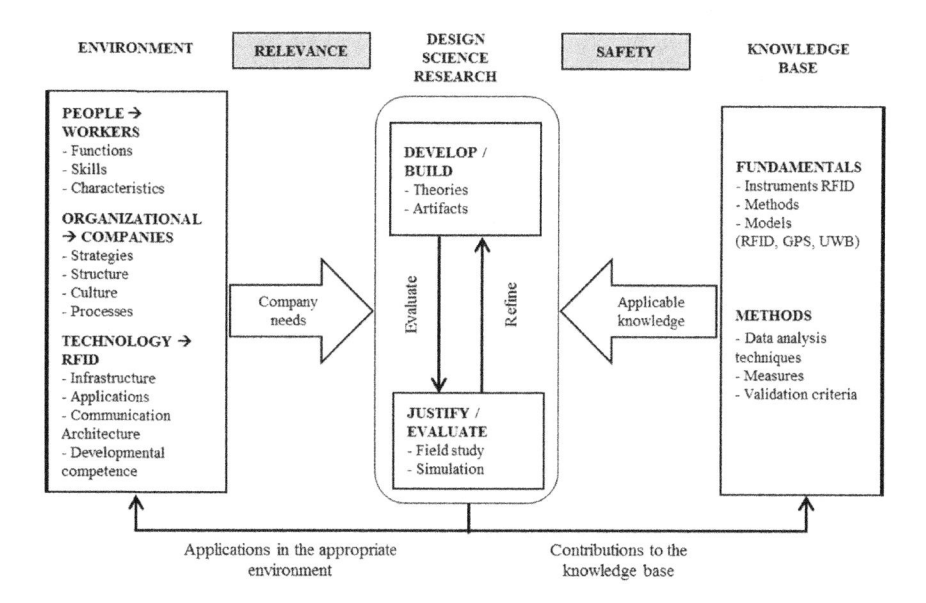

Figure 21.1 Investigative structure using the DSR for the application of RFID technology in this study.

(Adapted from Hevner, A. R., March, S. T., Park, J. & Ram, S., 2004, Design science in information systems research, MIS Quarterly, Vol. 28, pp. 75–105.)

Table 21.1 RFID equipment used in the tests

Type	Origin	Model	Description	Image
(a) Tag passive RFID	IF RFID	B - 20070/C-20055	Frequency: 915 MHz Memory: 128 bit CPE Dimensions: 140 × 31 mm	
(b) Tag passive RFID	Smart Parking	Adhesive	Frequency: 800–960 MHz Description Global CPE Class 1 Dimensions: 9.3 × 1.9 cm	
(c) Fixed RFID reader	Acura Global	UHF 900 BT	Frequency: 800–960 MHz Weight: 170 g Dimensions: 14.8 × 5.1 × 3.0 cm Interface: Bluetooth 2.0 EDR / USB 1.1 / UART 3.3V / 3- threads	
(d) Tablet	Tecnopia	T10Q10	Screen 10 Memory: 8 GB	

used to identify the location of employees (row 2 image). This passive tag does not have its own power source; its functions are activated when the tag enters the field of the reader, triggering communication through electromagnetic waves (Maghiros, Rotter & Lieshout, 2007). An RFID UHF 900 BT mobile reader was used to recognise the CPE tags (row 3 image). This is a lightweight, high mobility device that operates in the frequency range between 800 and 960 MHz. For reading the workers' locations, a fixed RFID reader, installed at a point on the floor of the building, was used. Access to the internal database of the developed software can be obtained via Bluetooth or a data communication cable. A tablet was also used to monitor specific software on construction sites (row 4 image).

The system validation tests were divided into two steps. The first was conducted in a laboratory environment where the purpose was to develop, evaluate and validate the procedures and functioning of the data collection and processing systems (Oliveira & Serra, 2016). The second step involved tests on a real construction site, reproducing all the established test routines carried out in the laboratory (Oliveira & Serra, 2019).

Finally, specific software was developed. The programme operation was simulated to test and validate the procedures and functioning of the management tool developed. The contributions to this study included the results of the tests, an evaluation of the functioning of the developed system using RFID technology and the development of new software. The evaluation of the artefact developed is presented next.

Results and discussion

System development

The development of a monitoring system through control software was intended to consolidate a non-intrusive remote monitoring tool with passive tags using RFID technology. For this, two distinct working functions were designed: (1) definition and identification of areas where workers are at risk of falling, such as areas near the peripheries or openings in the slabs; and (2) the management of safety equipment, such as the holders of the safety trays and the peripheral guardrails, previously registered in its database on the construction sites.

Operationally, the system's functionality consists of carrying out continuous iterations between the data previously registered and those obtained from the readings of the RFID tags. The product of all the interactions enables the system to generate audio alerts, create a continuous database, cross-check and compare data with the archive, generate numerical results, generate graphics and identify objects stored in the database. In this way, the system is able to locate the position of each element according to the demand of the operator. Moreover, the characteristics and data of each element can be compiled in reports and images to facilitate understanding and tool use.

The software is part of the system that combines RFID components to manage CPE and workers on building sites. The simplified representation of this software design is illustrated by the block diagram shown in Figure 21.2.

The interface developed for the software made operator communication accessible and easy to understand. The aim was to adapt and facilitate the software to the needs and characteristics of the construction managers. Visual programming made the functions available in the form of icons arranged according to the logical sequence of use. The software was developed for Windows and Android platforms, thus making it compatible with hosting on computers that use Windows, and tablets or smartphones that use Android operating systems.

CPE control

The first step in implementing the system was to register the safety equipment used in the study. The study was restricted to the 13th floor with services of

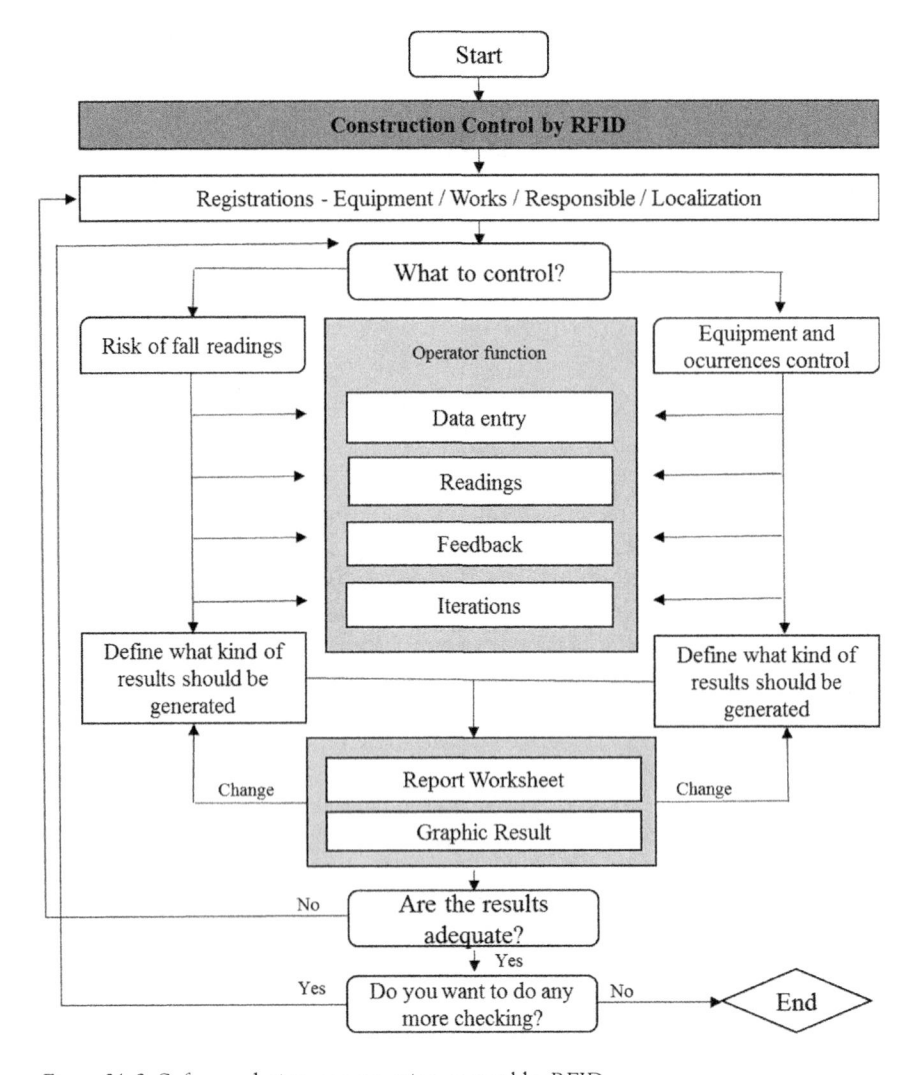

Figure 21.2 Software design: construction control by RFID.

disassembling the formwork and cleaning the floor. Each item of CPE and the helmets of the four workers received a passive RFID tag as described.

The system was provided with a database of information about the equipment and the place of operation. The database was formed, for example, by the register of equipment and details of the building design.

In order to register, the software provides a set of cadastre icons that open windows for the insertion of all the data necessary for this operation. The numeric and text information must be entered or selected in the options menu of each register box. The registration of the information follows a predetermined order that organises and makes the information entered available for the next step of the cadastre.

The images of the floor plans to be managed are inserted into the database via the file upload. The format of the image files should be bitmap (.bmp) so that the software recognises it and allows the upload. These files must be stored on the computer where the software will be used.

The second step was to establish connections among the devices. This process was divided into two stages. The first connection occurs for the RFID reader to establish the tag ID. After recognising the tag, the reader communicates via Bluetooth with a tablet or smartphone feeding the software database installed in it. The second connection transmits information obtained during the readings by the tablet or smartphone to the main computer. This connection happens when the devices are paired via Bluetooth using the same network and IP address. The synchronisation of the equipment updates the matrix database thus producing new interactions.

During configuration, the user can also define which specific CPE data are required through data filtering. The results can thus be organised in an orderly manner by the type of equipment, the place of application, the work or the person responsible (operator).

Results are presented on numeric worksheets obtained by accessing the icons for 'Risk of Falling Readings' and 'Equipment and Occurrences Control'. These two icons access the screens for 'Readings' or 'Occurrences' where the operator can set any filters for the final report. After defining the Report Grouping, Date, Order and Query, the operator will just click on the 'Report' button and the software will automatically generate a spreadsheet showing the data according to the settings established by the operator. This operational sequence is shown in Figure 21.3.

This tool provides an overall analytical view of all occurrences performed by the programme. The reports are editable and can be generated as extensions .xls, .txt or .pdf, allowing changes or improvements of the layout.

The same operating principle applies in obtaining the graphical results. For this, the access must be done through the icon for 'Work' (see Figure 21.4). The next screen allows the operator to define which option to access. Once the option is set, the operator clicks on the icon for 'Map'. In the same screen, a secondary window will appear showing the design of a building that should contain the number of floors pre-defined in the configurations for that work. The floors containing monitored equipment were painted yellow. The operator should click on the desired floor and another window will show that floor plan. This map will contain symbols that define each type of equipment (triangles, circles, squares, etc.) as pre-defined in the settings (Oliveira & Serra, 2019). These symbols will be located in the plan exactly where the CPE is on the 13th floor. To see the overall equipment information in the plan, the operator simply clicks on the specific symbol and a second window appears on the side of the plan displaying the equipment data.

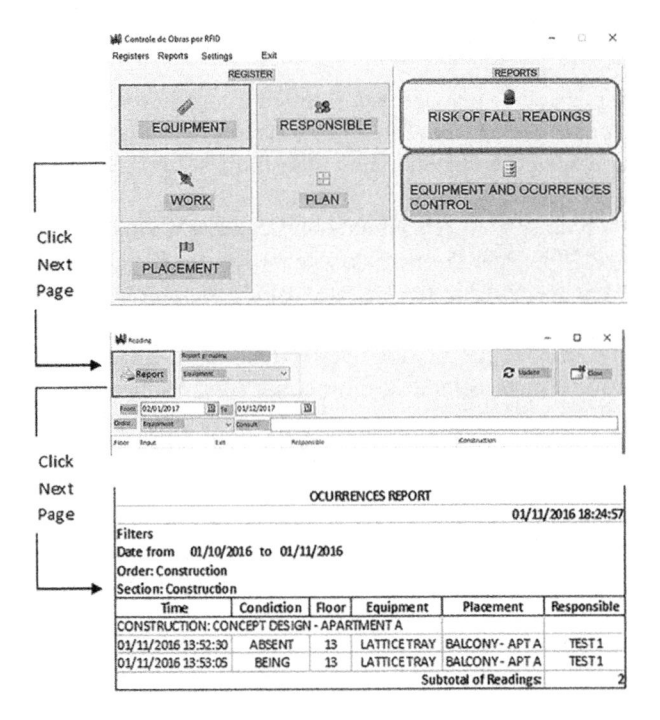

Figure 21.3 Operational sequence for the attainment of results on the CPE as reports.

Safety of the workers

From this phase on, the aim of the data collection was to test the efficiency of the RFID equipment in identifying the workers. The laboratory test demonstrated that the RFID readings are triggered from a triangular geometric formation within a maximum reading distance of 2.90 m and a minimum of 1.90 m, as can be verified in Oliveira and Serra (2016). Knowing the limiting distances, the risk of fall range was established as a 1 m zone between the distances of 2.90 and 1.90 m set to 'Warning' as, in this area, the reader will recognise a person's entry and issue a beep as an alert about the risk of falling. While in this range, the employee will be at least 1.90 m away from the fall limit. The 'Imminent Risk' represents the area of greatest risk to the worker. Therefore, the worker should withdraw or take preventive measures to avoid an imminent risk of a fall.

The solution adopted for the work site was to build a wooden box with equal dimensions of 30 cm, with a front door opening fixed to the structure through a profiled metal framework of 38 × 19 mm to resist the incidence of wind, rain and potential impact that could occur during operations. Internally, electric power was installed to enable the equipment to operate permanently over the days when the tests were conducted. To ensure the safety and integrity of the

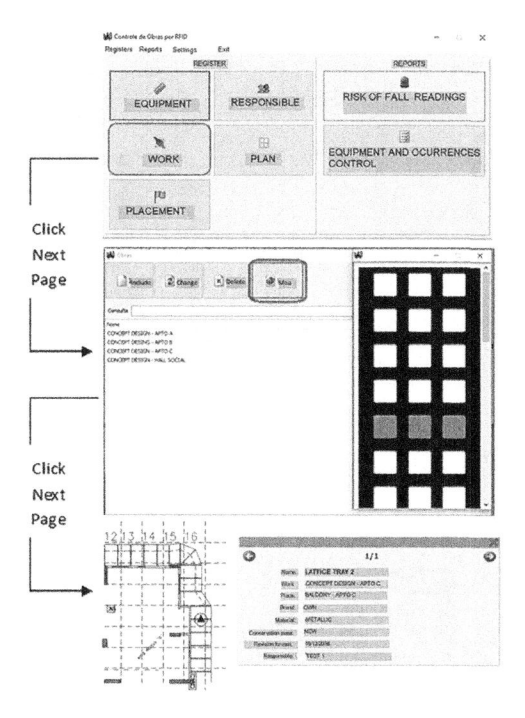

Figure 21.4 Operational sequence for graphical results.

tablet responsible for receiving and storing the readings, the box remained locked throughout the tests. The UHF 900BT reader was attached to the outside of the box door with the antenna aligned at a height of 1.85 m. The reader was positioned in alignment with the floor periphery so that its antenna was on the imaginary line of fall formed by the edge of the slab.

The total length of the field trials was 10 days, during 7 of which effective measurements were performed. A period of 2 hours per day was established for valid readings. Over 7 days, 232 measurement readings were performed, being the number of daily readings: day 1 = 30, day 2 = 11, day 3 = 7, day 4 = 8, day 5 = 8, day 6 = 49, day 7 = 119. The variability in the number of readings occurred because of the work activities in the area where the equipment was installed since the employees remained working on their usual activities during the testing period. Figure 21.5 shows the comparative RFID reading per employee and date.

An important result in this phase of the study was the identification of reading distances. From 232 measurements, 46 identifications were made within the area known as 'imminent risk' at less than 1.90 m from the falling edge, which represented 19.83% of the total. However, the shortest distance obtained was 1.65 m, thus keeping the employee at a safe distance from the falling edge.

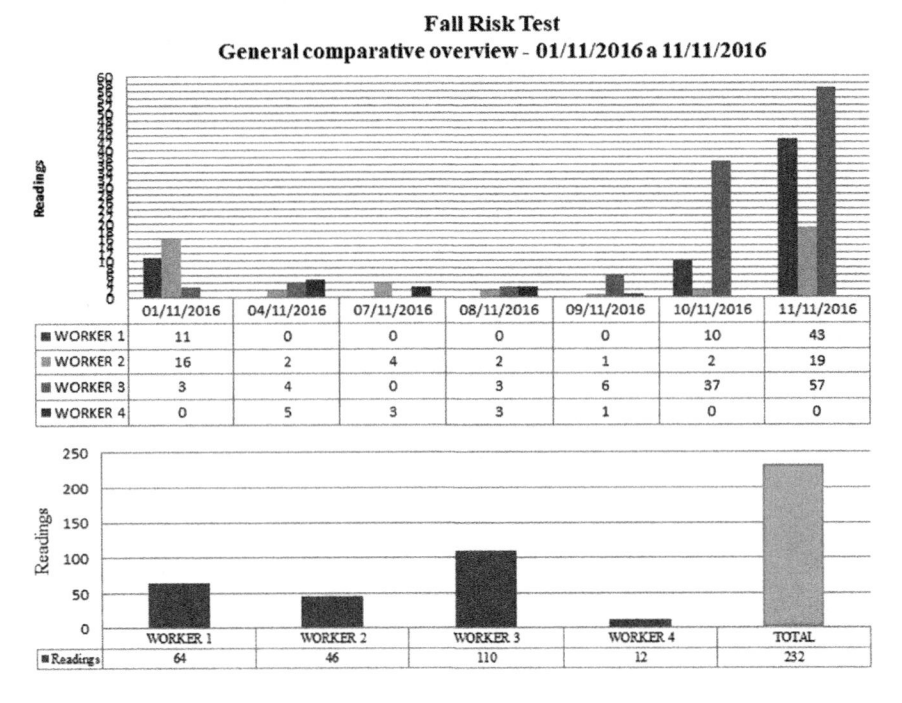

Fall Risk Test
General comparative overview - 01/11/2016 a 11/11/2016

	01/11/2016	04/11/2016	07/11/2016	08/11/2016	09/11/2016	10/11/2016	11/11/2016
WORKER 1	11	0	0	0	0	10	43
WORKER 2	16	2	4	2	1	2	19
WORKER 3	3	4	0	3	6	37	57
WORKER 4	0	5	3	3	1	0	0

	WORKER 1	WORKER 2	WORKER 3	WORKER 4	TOTAL
Readings	64	46	110	12	232

Figure 21.5 Comparative chart by employee group and measurement date.

The effectiveness of the readings confirmed that all the workers' approaches were identified by the RFID reader and, in all them, the audible warning was activated instantaneously. On the other hand, it was observed that the wind had a significant influence on the operation of the RFID system. The greater the wind incidence, the more difficult it was for the equipment to read longer distances and the more difficult it was to hear the equipment's sound alarm.

Implications for practice and research

This method could be adopted in other developing countries and contribute to reducing the risk of work accidents in construction. The aim of presenting the viability and accessibility of a system using ICT instead of the usual methods was to contribute to the evolution of new managerial tools incorporated into construction sites.

Another contribution was the effectiveness of the audible alert when workers access hazardous locations in the working area. In this way, the workers were more alert to the potential dangers.

One of the challenges to the deployment of ICT on construction sites is to make tools accessible to workers. This barrier was overcome by the simplicity of the operating system, accessed via manageable equipment. This reduced the steps in the safety management process with a consequent reduction of time, cost and errors without creating difficulties for a work environment of great complexity such as a construction site.

The contributions produced in the results actually presented a new alternative. However, when the exploratory research expanded from a single floor to an entire area of a construction site, it was found that the amplitude of the reading area of the fixed RFID equipment was limited. Therefore, in order to monitor an entire building for the risk of falling, it would be necessary to use many RFID readers, considering an average distance of 3.30 m between them. This is a solution that would generate high cost when compared to other technologies combined. This is a limiting situation mainly in countries in the process of economic and technological development where cost may be decisive.

Knowledge of technology is another factor. Although RFID technology is widely diffused in the automated industries, it is still used little in construction. In countries such as Brazil, the knowledge and technology might not be available to enable the incorporation of this technology in areas, sectors or even industries that do not usually use it, such as construction. This was the case during this study. During the process of designing a system, it might be necessary to import equipment because of the non-availability of the necessary model in the national industry.

Gaps in knowledge should encourage the development of new research to improve the solutions proposed in this study. Further research is required for more potent RFID readers and improvements to the information system proposed in this study. In addition, the design of the software can be upgraded to versions that would support the use of RFID systems on construction sites.

Future studies should extend the application of the system to the management of equipment other than just for safety. The diversity of equipment and software with high commercial value alone should encourage the continuation of the research.

Conclusions

In this chapter, the results of research into the application of RFID technology has been presented with the specific purpose of constituting an information system focused on the maintenance of collective protection equipment and improvement of risky activities and workers' safety on construction sites.

The tests that determined the functionality of the system and the development of the specific software were conducted simultaneously. The environment that hosted the trials on the construction site featured a series of interferences, such as elements of concrete, profiled metal and wood panels. Even in the face of these obstacles, common to the daily routine of a construction site, the RFID component and software set succeeded in performing the readings, with the consequent identification and location of the equipment and workers monitored. The use of tags with specific attributes, protected against the action and interference of metallic elements and weather were fundamental in achieving the results obtained.

The results have produced a system with functions that completely modify the way to administer equipment and workers. Innovations such as real-time monitoring, remote monitoring without the necessity of submitting to the existing safety risks when moving around the construction site, reduction of the risk of accidents through control of the equipment maintenance period, creation of a databank and the use of a track record are benefits and contributions of this research.

Simplicity in the software operation was a purpose pursued in the course of its development. The expectation is to stimulate the managers to use it. In this sense, the design of the interface using an Android operating system made a great contribution by making it possible to use the software on smartphones or tablets, making it accessible to the workforce, as these tools are commonly used as part of their routine, which humanises the adoption of this technology.

The possibility of issuing reports was an important function of the software developed. Through data filtering it was possible to reduce and/or group the amount of information contained in the reports, thus eliminating redundancies.

The results of the readings and the database, organised as reports and spreadsheets with filters for the choice of information to be transmitted, have enriched the presentation of information. Moreover, the graphical results in the form of plans with the identification of the equipment therein, have simplified and invigorated the way operators and managers have access to the information generated by the system.

Contributions to the continuity of new research are extensive. An example for a future and immediate application of the system would be to extend monitoring to other types of equipment and tools, such as drills, circular saws, wheelbarrows, etc., usually present on construction sites.

Preventive measures should be taken to reduce the risk of accidents on construction sites. Changes in strategic thinking and management systems are needed also to drive the industry to use new technologies, methods, and procedures that can provide an integrated, safe, and highly productive environment.

Acknowledgements

The authors would like to thank the research project Technologies for Sustainable Social Interest Housing Construction Sites (CANTECHIS), supported by the Studies and Projects Financing Agency (FINEP), and Coordination of Superior Level Staff Improvement (CAPES) in Brazil.

References

Brasil. (2017), Ministério da Previdência Social. Anuário Estatistico de Acidentes de Trabalho. [Online]. Available at: http://sa.previdencia.gov.br/site/2018/09/AEAT-2017.pdf

Costin, A. M., Teizer, J. & Schoner, B. (2015), RFID and BIM-enabled worker location tracking to support real-time building protocol and data visualization, *Journal of Information Technology in Construction (ITcon)*, Vol. 20, pp. 495–517. [Online]. Available at: www.itcon.org/2015/29.

Demiralp, G., Guven, G. & Ergen, E. (2012), Analyzing the benefits of RFID technology for cost sharing in construction supply chains: A case study on prefabricated precast components, *Automation in Construction*, Vol. 24(7), pp. 120–129.

Domdouzis, K., Kumar, B. & Anumba, C. (2007), Radio-frequency identification (RFID) applications: A brief introduction, *Advanced Engineering Informatics*, Vol. 21(4), pp. 350–355.

Finkenzeller, K. (2010), *RFID Handbook: Fundamentals and Applications in Contactless Smart Cards, Radio Frequency Identification and Near-Field Communication*. London: Wiley, p. 462.

Giretti, A., Carbonari, A., Naticchia, B. & Grassi, M. (2009), Design and first development of an automated real-time safety management system for construction sites, *Journal of Civil Engineering and Management*, Vol. 15(4), pp. 325–336.

Hevner, A. R., March, S. T., Park, J. & Ram, S. (2004), Design science in information systems research, *MIS Quarterly*, Vol. 28, pp. 75–105.

International Labour Organization. (2017), World statistic: The enormous burden of poor working conditions. [Online]. Available at: www.ilo.org/moscow/areas-of-work/occupational-safety-and-health/WCMS_249278/lang--en/index.htm.

Lanko, A., Vatin, N. & Kaklauskas, A. (2018), Application of RFID combined with blockchain technology in logistics of construction materials, *MATEC Web of Conferences*, Vol. 170 (Article Number 03032). [Online]. Available at: https://doi.org/10.1051/matecconf/201817003032.

Lee, H. S., Lee, K. P., Park, M. Baek, Y. & Lee, S. (2012), RFID-based real-time locating system for construction safety management, *Journal of Computing in Civil Engineering*, Vol. 26(3), pp. 366–377.

Lieshout, M. C., Grossi, L., Spinelli, G., Helmus, S., Kool, L., Pennings, L., Stap, R., Veugen, T., Waaij, B. V. & Borean, C. (2007), RFID technologies: Emerging issues, challenges and policy options. Institute for Prospective Technological Studies. Joint Research Centre. European Commission.

Maghiros, J., Rotter, P. & Lieshout, M. (2007), RFID technologies: Emerging issues, challenges and policy options. Institute for Prospective Technological Studies. Joint Research Centre. European Commission. [Online]. Available at: www.dhi.ac.uk/san/waysofbeing/data/data-crone-vanlieshout-2007.pdf.

Mehrjerdi, Y. Z. (2008), RFID-enabled systems: A brief review, *Assembly Automation*, Vol. 28(3), pp. 235–245.

Oliveira, V. H. M. & Serra, S. M. B. (2016), Desenvolvimento de sistemas de controle remoto em tempo real para canteiros de obras. In: *Proceedings of the VII Encuentro Latinoamericano de Economía y Gestión de la Construcción (ELAGEC)*, 16–17 November 2016, Bogotá, Colômbia, p. 16.

Oliveira, V.H. M. & Serra, S. M. B. (2019), Control of collective security equipment by RFID in the construction site. In: P. M. F. M. Arezes (Ed.), *9th International Conference on Applied Human Factors and Ergonomics (AHFE 2018)*, Advances in Intelligent Systems and Computing (AISC), Vol. 791, pp. 116–127. [Online]. Available at https://doi.org/10.1007/978-3-319-94589-7_12.

Ren, Z., Anumba, C. J. & Tah, J. (2011), RFID-facilitated construction materials management (RFID-CMM) – A case study of water-supply project, *Advanced Engineering Informatics*, Vol. 25(2), pp. 198–207.

Van Aken, J. E. (2004), Management research based on the paradigm of the design sciences: The quest for field-tested and grounded technological rules, *Journal of Management Studies*, Vol. 41(2), pp. 219–246.

Wang, Z., Hu, H. & Zhou, W. (2017), RFID enabled knowledge-based precast construction supply chain, *Computer-Aided Civil and Infrastructure Engineering*, Vol. 32(6), pp. 499–514.

Wu, W., Yang, H., Li, Q. & Chew, D. (2013), An integrated information management model for proactive prevention of struck-by-falling-object accidents on construction sites, *Automation in Construction*, Vol. 34, pp. 67–74.

Yang, H., Chew, D. A. S., Wu, W., Zhou, Z. & Li, Q. (2012), Design and implementation of an identification system in construction site safety for proactive accident prevention, *Accident Analysis and Prevention*, Vol. 48, pp. 193–203.

Index

Page numbers in **bold** denote tables, in *italic* denote figures

Abdelhamid, T. S. 220
Abdul, R. A. H. 104
Abubakar, U. 15, 16, 25, 205, 207, 210
accident: catastrophic 181; causation 6, 86–87, 89, 91, **92**, 98–99, 104, 109, **109**, 220, *221*; construction 5, 87, 98–99, 104, 181, 275, 303; cost of (COA) 99, 152–153, 273, **278**, 279; fatal 50, 54, 59, 85, 103, 110, 197, 224, 270, 272, 276, **278**, 279; industrial 271, 301; major 181; motor vehicle 9, 276, **277–278**, 279, **280**, 281; occupational 1, 9, 49, 118, 181, 219, *219*, **220**, **278**; preventing 155, 170, 180, 230, 304; reporting **51–52**, 98, 103, 105, **109**, 301; self-reported 85, 89, 96; serious 104, 173, 228; site 7, 108–109, 128, 169, 173; work 287–288, 301, 312; workplace 118, 230, 260; *see also* accident injury rate
accident frequency rate (AFR) 29–30, 35–40, *37–38*, **37**
accident injury rate (AIR) 106, 108, **109**, 110, **111–112**
accident investigation and near miss reporting (AINMR) 213
Adeyemo, O. 60
Agumba, J. N. xiv, 9, 273; *co-author of Chapter 19*
AIDS/HIV 9, 237–242, *241*, **243**, 245, **246–247**, 248–253; pandemic 237–238, 240, 252–253; -related knowledge 9, 237–238, 240–242, **243**, 244–245, **246–247**, 248–252
Ajayi, M. O. 131
Ajayi, S. O. 134
Aje, I. 129–131, 135

Akanbi, M. 133
Akanmu, A. 135
Akinkunmi, G. A. 142–143, **145–147**
Aksorn, T. 105–107, 110
alcohol use/alcohol consumption 237–240, 242–245, **246**, **247**, 248–252
Alinaitwe, H. xiv, 104–106, 109; *co-author of Chapter 7*
Amended Compensation for Occupational Injuries and Diseases Act (COIDA) (1997) 274–275
analytical hierarchy process (AHP) 214–215
analysis of moment structures (AMOS) 90, **92–93**, 96, 244–245
Annan, J. 16, 142, 148
Antonsen, S. 209
Arndt, V. 272
Asah-Kissiedu, M. xiv, 6; *co-author of Chapter 8*

Bangladesh 8, 218, **225**
Barth, S. P. 274
Bartle, I. 18, 25
Behm, M. 130, 133
benevolence 8, 179, 183, *184*, 185, 187–188
Blazsin, H. 187
Bobek, A. 259, 263
Booth, C. xiv, 117, 120; *co-author of Chapter 8*
Boschman, S. J. 271
Bowen, P. A. xv, 9, 238, 241; *co-author of Chapter 17*
Braun, V. 19
Brazil 10, 287, 288, 289, 293, 294, 301, 303, 313, 314

318 *Index*

Brenner, H. 270–271, 279
British construction industry 18
Building and Other Construction Workers (BOCW) Act (1995) 59–63, 64, 65–68, **69–72**, 70–73, **74**, 77–80
built environment practitioners (BEPs) 152–154, 156; councils (BEPCs) 154, 161, 164–166
Bulgaria 258–260, 262
Bureau of Indian Standards 63, **70–72**
Bust, P. D. 261, 263–264

Carey, M. P. 242
Castillo-Montoya, M. 19
Castro, D. 16–18, 32–33, 40
Centre for the Improvement of Working Conditions & Environment (CIWC&E) 56
Chan, A. P. C. xv, 6, 8, 89, 128, 261; *co-author of Chapters 6 and 14*
Chau, N. 271, 274
Cheung, C. M. xv, 8; *co-author of Chapter 15*
Chinese Pharmacopoeia Commission 186
Chinese: contractors 8, 191; foremen 198–199, **199**; international contractors (CIC) 8, 191–195, **194**, 197–202; workers 191, 199–202, **201**
Chinwuba, M. S. 170–171, 173
Chong, H.-Y. xv, 224; *co-author of Chapter 13*
Ciccarelli, M. xv; *co-author of Chapter 13*
Claessen, H. 272
coefficient of variation (CV) 107, **108**
collective protection equipment (CPE) 10, 287, 301, 303–309, **306**, *310*, 313
command-and-control 17, 18, 31
communication 3, 7, 8, 49, 87, 91, 95, 98, 119, **147**, 149, 169–174, 176, 177, 191, 193, 199, 200, **201**, 202, 218, 220, 227–229, 250, 252, 258, 261, 263, 265–267, 301–304, 306, 307
Compensation for Occupational Injuries and Diseases Act (1993) (COID) 30, 32, **157**, **163**
Compensation for Occupational Injuries and Diseases Act, Amended (1997) (COIDA) 274–275
Conference of Occupational Safety and Health, 18th (COSH) 230
confirmatory factor analysis (CFA) 90–91, **92–93**, 244–245, 248

construction design and management (CDM) 17; regulations 22, 147, 169
construction health and safety (CH&S) 15, 17–20, *21*, 22–23, **24**, 25–26, 121, 152, 156, 279
construction industry commission (CICO) 113
Construction Industry Development Agency (CIDA) 132–134, **133**, 136
Construction Industry Development Board (cidb) 30, 32, 35, 152–153, 165, 230
construction managers (CMs) 154, 307
construction project managers (CPMs) 153–154
contractors: large-sized (LCs) 19–20, 22–23; medium-sized (MCs) 19–20, 22–23, **24**, 25; micro-sized (MiCs) 19–20; small-sized (SCs) 19–20, 22, **24**; *see also* self-regulation, subcontractors
contractor safety management (CSM) 213
contractors that constantly self-regulate (CSRCs) 20, 22
contractors that do not constantly self-regulate (NCSRCs) 20, 22–23
contractors that do not self-regulate (non-SRCs) 20
Costella, M. F. xv, 9; *co-author of Chapter 20*
Costin, A. M. 304
countries: African 9, 142, 170, 177, 205–206, 210, 212; Asian 200; developed 1–4, 6, 9, 46, 87, 109, 116, 122, 141, 193, 258–260, 262–263, 266–267; developing (DCs) 1–9, 15–16, 19, 26, 29, 43–45, 47, 47, 49, 53, 55, 57, 60, 96, 99, 116–120, 123–124, 129, 132, 134, 136, 141–142, 149–150, 164, 171, 173–174, 176, 179, 192–193, 197, 252–253, 258–260, 262–263, 312; European 259–260; foreign 213; high-income 259–260; home 201, 259, 262–264; host 201, 258–259, 261–262, 266–267; low-income 9, 43, 258–259, 262–263, 267; middle-income 43; underdeveloped 43, 192, 263
Croatia 258–260, 262

Dainty, A. R. J. 263–264, 266
Davis, K. 132
Davis, N. 264
Deacon, C. H. 141, 145, 154
DeArmond, S. 87
De-chun, N. 87

Deming, W. E. 44; Cycle 44, 120
Demiralp, G. 302
Department of Occupational Safety and Health (DOSH) 219, 222–223, *222*, 227, 229–230
disability: non-permanent (NPD) 219, **220**, 274–275; permanent (PD) 9, 219, **220**, 270–276, 279, **280**, 281–282; retirement 271–272
disabling injury incidence rate (DIIR) 152
distributed occupational safety and health (DOSH) 15–16
Domdouzis, K. 302–303
drug usage (DU) 9, 237–242, **243**, 244–245, **246–247**, 248–253
Duddukuru, S. D. xvi, 5; *co-author of Chapter 5*

Eco-Management and Auditing Scheme (EMAS) 120
economies 1, 205, 211, 259–260; developed 108; developing 123, 129
Edwards, P. J. xvi, 9; *co-author of Chapter 17*
Ejohwomu, O. xvi; *co-author of Chapter 15*
emergency preparedness planning (EPP) 105, **108, 111**
Emuze, F. i, 2–3, 8; *co-author of Chapters 1 and 12 and co-editor*
energy isolation (EI) 213
Engineers (Registration etc.) Act 16
environmental management systems (EMS) 6, 116–117, 120–123
equivalent full-time workers (EFTW) 103, 106, 108, 113
Eriksson, P. 141
Evangelinos, K. xvi; *co-author of Chapter 2*
exploratory factor analysis (EFA) 90–91, **92–93**

Fagbenle, O. I. 144, **145–146**
Faimau, G. 240
Falkstedt, D. 272
fall arrest system 288–291
Famuyiwa, F. 18
Faremi, F. A. 133
Fatal Accidents Act (1855) 50
fatal injury rate (FIR) 106, 108, 113
Fatima, S. K. 133
fault tree analysis (FTA) 214–215
Federal Ministry of Labour and Employment Inspectorate Division (FMLEID) 16

Federated Employers' Mutual Assurance Company (FEM) 30, 270–273, 275–276, 282
Fellini, I. 259, 263
Finneran, A. 2, 15, 25, 39–40
flexible horizontal lifelines (FHLLs) 9, 287–292, 294 298–299

Gambatese, J. A. 130, 133, 141
Gao, R. xvi, 8, 192; *co-author of Chapter 14*
Geller, E. S. 221, 228
Geminiani, F. L. 30
Ghana 6, 16, 116, 118–120, 123, 142; construction industry 119, 123
Gibb, A. 2–3, 15, 25, 39–40, 181, 264; *author of the Foreword*
Gilbert, L. 238
Gillen, M. 134
Giretti, A. 302
Glendon, I. 209, 215
Goetsch, D. L. 30, 129
Goh, Y. M. 180, 288–289
gross domestic product (GDP) 2, 48–49, **48**, 103, 116
Guldenmund, F. W. 187, 206, 209–210, 215
Gunningham, N. 16–18, 25, 29–33, 40

Hadikusumo, B. H. W. ii, 5, 8, 59–60, 105–107, 110, 130, 221, 227–228; *co-author of Chapters 1, 4, 5, and 16 and co-editor*
Hale, A. R. 181–182, 207–208
Hämäläinen, P. 1, 205
Hamid, A. R. A. 30, 123, 220, 224
Hardy, G. 142–143, **147**
Hassan, H. 260
Haupt, T. C. 275, 279
hazard 44, 95, 98, 105, 110, 155, **158**, 161, 185, 186, 197, 198, 230, 266, 302
health 1–7, 15–17, 20, 29–33, 39, 43–50, **51**, 53–57, 60–63, 64, **70–72**, 104–106, 113, 116–119, 121–123, 129–134, **135**, 140–142, 152, 154, 155, 161, 169, 171, 174, 183, 185, 186, 196, 197, 205, 207, 209–213, 218, 219, 222, 223, **226**, 227, 229–231, 237, 238, 240, 250, 258–261, 263–267, 270–273, 276, 279, 281, 302, 304
Health and Safety at Work Act (HASWA) (1974) 17, 45
Health and Safety Executive (HSE) 1–3, 15, 18, 25, 46, 104, 109–110, 140,

320 Index

169–173, 208, 210, **211**, 261, 263, 271, 279, 287–288, 298
heat stress 182–184, 186
Hevner, A. R. 304–305
Hinze, J. 43, 198, 200, 273
Hofstede, G. 194
Holt, G. D. 131, 135
Hong, S. Y. 240
horizontal lifeline system (HLS) 287–288
Hudson, P. 207
Hui-Nee, A. 223
human error 30, 207

Idoro, G. I. 16, 18, 25, 105, 118, 130, 213
Idubor, E. E. 60
illiteracy 4, 8, 49, 170, 172–174, 176
immigration 259
India 5, 53, 59–63, *64*, 65, 77–80, 193; construction industry 59–61, 63, 65, 68, 77–78
individual protection equipment (IPE) 287
information and communication technologies (ICT) 301–303, 312
injury: fatal 2, 18, 103, 106, **109**, 140, 270–271, 281; non-fatal 103, 106, 271; occupational 1–2, 46, 48, 53–54, 113, 271, 279, 281–282; serious 104, 224, **225**, 273; workplace 271; *see also* accident injury rate, disabling injury incidence rate, fatal injury rate, non-fatal injury rate
integrated management system (IMS) 120–123
International Labour Organization (ILO) 43–48, 47, 49–50, 53–55, 171–172, 211, 275, 301
International Labour Standards 47, 48–50, 53
International Standards Organization (ISO) 32, 120

Japan Industrial Safety & Health Association (JISHA) 56
Järvholm, B. 272
Jaselskis, E. J. 193
Jia, A. Y. xvi, 8, 181, 183, 186; *co-author of Chapter 13*
job hazard analysis (JHA) 95, 98, 105, **108**, 109–110, **112**

Kabwama, S. N. 251
Kader, R. 251
Kalichman, S. C. 239, 242, 250
Kalumba, D. xxvi; *co-author of Chapter 7*

Kartam, N. A. 104
Keng, T. C. 228
Kirby, T. 239, 251
Kjellén, U. 192–193
Korea Occupational Safety & Health Agency (KOSHA) 56
Korosec, R. P. 31

Labib, A. 213
Labour Administration Convention (1978) 55
Labour Inspection (Agriculture) Convention (1969) 48, 55
labour-only procurement (LoP) 7, 140, 142–150, **145–147**
Labour Statistics Convention (1985) 44, 46, 55
Labour Welfare and Human Resource Department 49
Lagos State 18, 22, **24**
Lagos State Building Control Agency (LSBCA) *21*, 22
Lagos State Safety Commission (LSSC) *21*, 22
Lanko, A. 304
learning from failures (LFF) 205–206, 210, 212–215, *214*
Lee, H. S. 304
lifeline system 9, 87, 289, 290
Letamo, G. 240
Leveson, N. G. 180–181
Levinson, A. 32–33, 39–40
Li, B. xvii; *co-author of Chapter 13*
Lin, J. 104, 132–133
Littau, P. 132
Loosemore, M. xvii, 193, 266; *co-author of Chapter 13*
Lopez-Fresno, P. 121
Love, P. E. D. 3, 128, 179

McKay, S. 261
McPhail, C. 239, 241
Mahamadu, A.-M. xvii, 141; *co-author of Chapter 8*
Malaysia 2, 8, 56, 218–219, *219*, 220, 222–224, *222*, 229–230; construction industry 8, 218, 220–224, 227, 229–230
management commitment and employees' involvement (MC&EI) 91, 95–96, 98
management systems (MSs) 6, 32–33, 40, 85–86, 116–124, 155, 160, **160**, 170, 179, 186, 193, 207, 314
Manu, P. i, 3–4, 122; *co-author of Chapters 1 and 8 and co-editor*

Massey, D. 263
Master Builders Association (MBA) 39, 156
Master Builders South Africa (MBSA) 154–155
maturity models (MMs) 118, 122
Meardi. G. 259–260
Mearns, K. 194, 197, 209
Medical Examination of Young Persons (Industry/Non-industry Occupations) Convention (1946) 55
Mehrjerdi, Y. Z. 302
mental health 134, 258–261, 265–267
migrant workers 9, 258–267
migration 9, 237, 263, 266–267
Ministry of Overseas Pakistanis and Human Resource Development (OPHRD) 47–48, *47*, 50, 53–54
Mitra, B. 273
Mitra, S. 274
Mohamed, S. 105, 194
Mohammad, M. Z. xvii, 8, 221, 223, 228, 230; *co-author of Chapter 16*
Mohler, D. G. 133
Morojele, N. K. 238–239
Mosanawe, J. O. 118
Motta, V. C. da xvii; *co-author of Chapter 20*
multiple criteria decision making (MCDM) 214–215
Murie, F. 260
Musonda, I. xvii, 9, 105; *co-author of Chapter 19*
Mwakali, J. A. 103–106

National Industrial Court of Nigeria (NICN) 212, **212**, 214
National Institute of Labour Administration Training (NILAT) 56
National Institute of Occupational Safety and Health (NIOSH) 56, 223, 230
National Tripartite Council for Occupational Safety & Health 50, 55
national workforce size per inspector ratio (NWSPIR) 211
negligence 30, 220, 224
Nepal 8, 218, **225**
New Zealand 18, 30, 142–142
Nigeria 5, 7–8, 15–20, 22–23, **24**, 25–26, 39, 118, 128–130, 132–134, 141–144, 173–174, 205, 207, 210–213, **211–212**, 215; construction industry 5, 8, 15, 20, 25, 39, 129–131, 136, 142, 213–214, 216

Nigerian Ministry of Labour and Productivity (NMLP) 210–211
Nigerian Nuclear Regulatory Authority (NNRA) 16
non-fatal injury rate (NIR) 106, 108, 113
Nonnenmacher, L. xviii; *co-author of Chapter 20*
Nuclear Safety and Radiation Protection Act (1995) 16

Occupational Cancer Convention (1974) 55
Occupational Hygiene Safety and Associated Professionals (OHSAP) 39
occupational safety and health (OSH) 1–7, 9, 16, 43–50, **44**, **52**, 53–57, 60, 62–63, 65, 67, **69**, 79, 89, 104, 197, 218, 229–230
Occupational Safety and Health Act (1970) 213
Occupational Safety and Health Act (1994) 222–224, 227
Occupational Safety and Health Act (2006) 106
Occupational Safety and Health Administration (OSHA) 32, 78, **211**, 273, 289, 295
Occupational Safety and Health Convention (1981) 43–44, 55
Occupational Safety and Health Council of Hong Kong (OSHC) 90
Occupational Safety and Health Master Plan for Malaysia 2015 (OSH-MP15) 229
Occupational Safety and Health Master Plan for Malaysia 2016–2020 (OSH-MP20) 229
Occupational Safety and Health Regulatory Commission (OSHRC) 212, *212*
Occupational Safety and Health Regulatory Commission (OSHRC) 212, *212*
Ogunde, A. O. 142–143, **147**, 148
Ogunsanmi, O. E. 142–143, **145–147**
Ogunsemi, D. 130–131
Okorie, V. N. xviii, 8, 170, 172–174; *co-author of Chapter 12*
Okoye, P. U. 129
Okwel, M. xviii, 6; *co-author of Chapter 7*
Olaru, M. 121
Olatunji, O. A. xviii, 7, 129, 132, 134; *co-author of Chapters 3, 9 and 10*
Oliveira, V. H. M. de xviii, 10, 307, 309–310; *co-author of Chapter 21*

Omeife, C. A. 16, 18
Oregon Occupational Safety and Health Division (Oregon OSHA) 106
Oswald, D. xviii–xix, 9, 259, 261; *author of Chapter 18*

Pakistan 5–6, 43–44, 47–50, 47, **48**, 51, 53–57, 85–89, 86, 98, **225**; construction industry 5–6, 85, 87–88, 91, **92–93**, 95, 98–99
Pakistan Bureau of Statistics 44, 47–48, 47
Pakistan Engineering Council (PEC) 89, 98
Pakistan Environmental Protection Act (PEPA) (1997) **52**
Pakistan Factories Act (1934) 44, 47, 49, **51**, 54, 56
Pakistan Factories Act (1961) 53–54
Pakistan Plant Quarantine Act (1976) 50
Pakistan Public Works Department Contractor's Labour Regulation 50
Peltzer, K. 239, 251
personal protection programmes (PPP) 105, **108**, 109–110, **111**
personal protective equipment (PPE) 8, 22, 32–33, 91, 95, 105, 141, 170, 185–186, 196–197, 213, 223–224, **225**, 227–228, 230
Personal Protective Equipment at Work Regulations (1992) 213
Philippines 8, 193, **225**
Phoya, S. 169–171, 173–174
Pinheiro, A. C. F. B. 296
Plan-Do-Check-Act (PDCA) Cycle 32, 44, 44, 46, 120–121; *see also* Deming
polyvinyl chloride (PVC) 133
Prevention of Major Industrial Accidents Convention (1993) 55
proactive behaviour-based safety (PBBS) 224
Professionals and Project Register 33, 35
Promotional Framework for Occupational Safety and Health Convention (2006) 44, 55

Qayoom, A. xviii, 5; *co-author of Chapter 4*
quality management systems (QMS) 120–121
quantity surveyors (QSs) 20, 38, 153–156, **157–159**, 160–161, **160**, **162–163**

radio frequency identification (RFID) 302–313, **306**, 308; technology 10, 301–305, 305, 307, 313
Reader, T. 87
Rebelo, M. F. 117, 120–123
Rees, J. V. 17–18, 31–33, 40
reliability block diagrams (RBDs) 214–215
Reporting of Injuries, Diseases and Dangerous Occurrences Regulations (RIDDOR) 46
risk 15, **24**, 25, 39, 40, 44, 87, 104, 129, 133, 140, 144, 145, **147**, 148–150, 155, 161, 170, 171, 177, 180, 186, 194, 211, 215, 228, 238–240, 251–253, 260, 262–267, 271–272, 281, 287, 288, 302, 307, 309–314
risky sexual behaviour (RS) 9, 237–242, 241, **243**, 244–245, **246–247**, 248–252
Robroek, S. J. 270–271
Robson, L. S. 32
Romania 258–260, 262
Rowlinson, S. xix, 3, 130, 181, 187; *co-author of Chapter 13*
Rural Workers' Organizations Convention (1975) 55

safety: auditing 105, **108**, 110, **112**, **133**, **135**; behaviour 8, 96, 98, 170, 177, 180, 194, 215, 218–224, 222, **225–226**, 227–231; belts and nets 63, **70–72**; budget 85, 91, 98, 227–228; climate (SC) 6, 85, 87–91, **92–93**, 94–95, 95–96, 98–99, 194, 209–210; communication 7–8, 87, 98, 119, 169–174, 176–177, 193, 199–200, **201**, 202; compliance 6, 85, 87, 89, 96, **133**, 133–134, **135**, 227; consideration 130, 132; construction 8, 15, 60–62, **75**, 76–80, 104, 117, 130, 134, 169–171, 173–174, 177, 183, 187, 192–193, 197–198, 200–202, 218, 221; culture 7–8, 25, 29–30, 113, 119, 128–130, 134, 169–171, 176, 176–177, 198, 205–210, 213, 215–216, 218–224, 222, **225**, 227–231; education 98; enforcement 91, 98–99; enhanced 6, 8, 95, 98, 213; integrated 6, 118, 122; law 5, 47, 47, 50, 60–61, 63, 68, 76–77, 79, 212; legislation 4–5, 30, 43–44, 46–47, 53, 55, 60, 78; management 5, 7–9, 85–86, 119, 132, 169–171, 173, 176,

Index   323

185, 187, 191–193, 195–196, 198–199, 201, 207, 209, 211, 223, **226**, 227, 230, 263, 304, 313; manager 39, 61, 65, 67, 68, **69**, 196, 198–201; officer **69**, 172, 199, 223, **225**, 228; organization 63, **70–72**; outcomes 128, 130, 136, 193, 205, 208, 216; participation 6, 85, 87, 89, 96, 227; performance (SP) 6, 8, 29, 32, 85, 87, 91, **93**, 98, 130, 170–172, 174, 177, 192–193, 197–198, 200, 205–207, 210–211, 213, 218, 221, 223, 229, 231, 281, 304; personnel 67, 76, 78, 199, 218, 223, 227, 230; records 60, 79, 105, **108**, **112**, 129, **133**, **135**, 136, 205, 253; rules/regulations 5, 30, 59–61, 63, 65–68, 68, 70–73, **70–72**, **74–75**, 76–80, 85, 91, 95–96, 98–99, 105, **108**, 110, **111**, 133, 183, 186–188, 199, 209, **211**, 213–214; systems 61, 63, 68, 77, 87, 133, **133**, **135**, 136, 208, 214, 264, 287; training 56, 60, 78–79, 85–87, 91, 95, 98, 172, 174, 198, **199**, 221, 223–224, **225–226**, 227, 261; workplace 7–8, 169–171, 174, 287; *see also* contractor safety management, safety, health and environmental management capability maturity model, site safety supervisor, workplace safety and health, workplace safety failure modes effects analysis
safety consciousness and responsibility (SC&R) 91, 98
safety enforcement and promotion (SE&P) 91, 95, 98–99
Safety and Health in Agriculture Convention (2001) 55–56
Safety and Health in Construction Convention (1988) 44, 49, 55–56
safety and health management systems (SHMS) 116–123
safety, health and environmental (SHE) 6, 116–119, 121–124
safety, health and environmental management capability maturity model (SHEM-CMM) 116, 118, 122–124
safety rules and safe work practices (SR&WP) 91, 95, 98
Saunders, M. 19, 33
Saurin, T. A. ii; *co-author of Chapter 1 and co-editor*
Scharrer, A. 30–33, 40
Schein, E. H. 209, 215

Schubert, U. 193
Scott-Sheldon, L. A. J. 240, 251
Serra, S. M. B. xix, 10, 307, 309–310; *co-author of Chapter 21*
self-interest 8, 179, 183, *184*, 185, 187
self-regulation: client-led (CLSR) 20, *21*, 22–23; community-led *21*, 25; contractor 5, 29; crusader-led 23, 25; enforced/mandatory 18, *21*, 22, 25, 31, 33, 39; full 18, 31; industry (ISR) 18–20, *21*, 22–23, 32; level of 29–30, *34*, 35–40, *36*, **37**, *38*; partial 18, 31, 33; pure *17*, 18, 20, *21*, 22, 25, 31, 33; voluntary 29, 31, 33
sensemaking 179–183, *184*, 186–188
Seth, P. 239
severity rate (SR) 153
Shisana, O. 237–239, 250–251
Simon, A. 120–121
site safety supervisor (SSS) 223, **225–226**, 227
Smallwood, J. J. xix, 2, 7, 60, 141, 144–145, 154–155, 172, 239; *author of Chapter 11*
South Africa 5–7, 9, 29–33, 37, 39–40, 142, 144, 152–154, 164–165, 237–242, 252–253, 271–272, 274–276, 279, 281–282; construction industry 9, 29–30, 32, 35, 39, 152–154, 240, 252, 270, 272–273, 276, 281
South African Business Coalition on HIV/AIDS (SABCOHA) 238
South African Construction Regulations 32, 153–154, **157**, 161, **162**, 164
South African Federation of Civil Engineering Contractors (SAFCEC) 39, 156
South African Institute of Occupational Safety and Health (SAIOSH) 39
Spangenberg, S. 196
stakeholder 3, 39, 88, 90–91, 98–99, 121, 129–130, 132, 140–142, 146, 148, 152–154, 164, 179, 181, 187–188, 215, 281; construction 32, 95–96, 99, 279; industry 4–5, 117, 279; multi- 6, 120, 152–153, 164, 181; project 129, 134, 141, 153
Starren, A. 194, 197
Statistical Package for the Social Science (SPSS) 90, **92–93**, 107, 244
Stern, F. 133
stress 290, 292, 294–296

subcontractors 3, 23, 85, 88–91, 95, 98, **108**, **111**, 131, 143, **145–146**, 146, **158**, **163**, 193, 197–198, 266
suggestion scheme 7–8, 169–171, 173–174, 176–177
Suraji, A. 200
systems thinking 180–181
Szubert, Z. 272

Tadesse, S. 141–142
Takala, J. 1–2
tertiary education 155, 160, 164, 165
trade unions 5, 43, 45, 53, 55, 57

Uganda 6, 103–104, 108, 113, 142; construction industry 103, 105, 113
Uganda Bureau of Statistics (UBOS) 103
Umeokafor, N. xix, 4–5, 7, 15–16, 19, 25–26, 39–40, 141–144, 148, 205–206; *co-author of Chapters 2, 3, 9 and 10*
UNAIDS 237–238
uncertainty avoidance (UA) 194, 206, 210
United Kingdom (UK) 2, 9, 17, 45–46, 104, 108–110, 142, 169, 208, 210, **211**, 258–267, 287; *see also* British construction industry
unsafe acts (UA) 56, 104–107, 110, **111–112**, 113, 218, 220, *221*, 230
unsafe conditions (UC) 56, 85–87, 104–108, 110, **111–112**, 113, 220, *221*, 230

Van Aken, J. E. 305
Van Heerden, M. S. 240
Vazquez, R. F. 172, 174
Vecchio-Sadus, A. M. 169–171, 173

Waehrer, M. G. 271
Walls, C. B. 18

Wanberg, J. 3, 128
Wang, Q. 288
Wang, Z. 303
Weber, K. 182
Weick, K. E. 179–182, 186
well-being 3, 9, 117, 134, 169, 171, 173, *184*, 218, 263
Wells, J. 141
West Pakistan Hazardous Occupation Rule (1963) 50
Wills, J. 145–148
Windapo, A. O. xix, 5, 16–18, 30–31, 35–36, 39; *co-author of Chapters 2, 3, 9 and 10*
work at height (WAH) 213, 289
workforce 1, 6–7, 169, 171–172, 174, 186, 192, 209, 211, 252, 258, 262, 267, 314; general 9, 272; national 194, **211**, 261; safe 132; skilled 271; transient 170; unskilled 170, 174
workplace safety and health (WSH) 61–62, 229, 288
workplace safety failure modes effects analysis (WPSFMEA) 214–215
Wouters, E. 238
Wu, W. 304

Xu, M. xix; *co-author of Chapter 13*

Yang, H. 304
Yunusa-Kaltungo, A. xix–xx; *co-author of Chapter 15*

Zahoor, H. xx, 6, 49, 86, 89, 91; *co-author of Chapter 6*
Zhang, R. P. xx; *co-author of Chapter 15*
Zohrabi, M. 19

Printed in Great Britain
by Amazon

38501112R20196